Interaction between Atmospheric and Oceanic Dynamics at Mesoscale and Small Scale

Interaction between Atmospheric and Oceanic Dynamics at Mesoscale and Small Scale

Guest Editors

Weizeng Shao
Jian Shi

MDPI

Basel • Beijing • Wuhan • Barcelona • Belgrade • Novi Sad • Cluj • Manchester

Guest Editors

Weizeng Shao
College of Oceanography and Ecological Science
Shanghai Ocean University
Shanghai
China

Jian Shi
College of Meteorology and Oceanography
National University of Defense Technology
Changsha
China

Editorial Office
MDPI AG
Grosspeteranlage 5
4052 Basel, Switzerland

This is a reprint of the Special Issue, published open access by the journal *Journal of Marine Science and Engineering* (ISSN 2077-1312), freely accessible at: https://www.mdpi.com/journal/jmse/special_issues/IZ84W10J6U.

For citation purposes, cite each article independently as indicated on the article page online and as indicated below:

Lastname, A.A.; Lastname, B.B. Article Title. *Journal Name* **Year**, *Volume Number*, Page Range.

ISBN 978-3-7258-5779-1 (Hbk)
ISBN 978-3-7258-5780-7 (PDF)
https://doi.org/10.3390/books978-3-7258-5780-7

Contents

Journal of
Marine Science and Engineering

MDPI

Editorial

Interaction Between Atmospheric and Oceanic Dynamics at Mesoscale and Small Scale

Weizeng Shao [1,*] and Jian Shi [2,*]

1 College of Oceanography and Ecological Science, Shanghai Ocean University, Shanghai 201306, China
2 College of Meteorology and Oceanography, National University of Defense Technology, Changsha 410073, China
* Correspondence: wzshao@shou.edu.cn (W.S.); shijian@nudt.edu.cn (J.S.)

1. Introduction

The exchange of momentum and heat between ocean dynamic processes at meso- and small scales is an interesting issue for the oceanography community [1]. Against the background of climate change, air–sea interaction plays an important role in economic activities [2]. Moreover, with the development of artificial intelligence (AI), the promotion of AI-based oceanography could deepen the understanding of multiscale air–sea interactions [3,4]. With this background, the newly launched Special Issue "The Interaction between Atmospheric and Oceanic Dynamics at Mesoscale and Small Scale" is founded. This Special Issue present relevant research on the interaction between atmospheric and oceanic dynamics at meso- and small scales utilizing traditional techniques, i.e., remote sensing [5] and numerical models [6–8]. Moreover, it is worthwhile to research these dynamics using AI.

2. An Overview of Published Articles

This Special Issue focuses on the interaction between atmospheric and oceanic dynamics in complex conditions, containing 11 published articles. The main contributions are delineated below.

Contribution 1 investigated how the semi-enclosed Sea of Japan (SJ) influenced Typhoon Songda's (TY-0418, 4–8 September 2004) intensity changes using high-resolution three-dimensional Weather Research and Forecasting (3D-WRF)/UM-KMA models and Geostationary Operational Environmental Satellite (GOES)-IR imagery. The approach identified key mechanisms, i.e., terrain-induced friction, enhanced latent heat release over warm currents, moisture convergence, and orographically modulated vorticity, explaining the typhoon's structural evolution, and rapid intensification within the confined SJ basin.

To understand the seasonal variability and long-term trends of Sea Surface Temperature (SST) fronts in Zhoushan and adjacent seas (1982–2021) and their driving mechanisms, Contribution 2 employed frontal intensity (SST horizontal gradient) and frequency analysis using a defined threshold (0.03 °C/km). This quantified frontal activity and linked it to river discharge, wind, upwelling, and currents. The method effectively revealed distinct seasonal patterns (strong winter bands, summer hotspots), key drivers (e.g., Changjiang Diluted Water), and significant winter intensification trends over decades.

To understand the mesoscale eddy regulation of Subsurface Chlorophyll Maximum Depth (SCMD) seasonality in the Kuroshio–Oyashio confluence, composite averaging and normalization of satellite altimetry and reanalysis data isolated SCMD responses to cyclonic and anticyclonic eddies. Contribution 3's superiority lies in effectively quantifying distinct eddy-induced SCMD monopole patterns (shallow/deep centers), revealing asymmetric seasonal correlations and demonstrating eddy-driven westward Chl-a transport over differing distances.

Addressing severe local scour at monopile foundations in the central Bohai Sea, an integrated numerical–experimental approach was proposed by Contribution 4, analyzing two decades of tidal currents, sediment transport, and scour. Results quantified minimal background erosion (<0.02 m/year), revealed 80% of scour occurs initially, and demonstrated that collar protection reduces scour depth by 50%, providing a crucial basis for design and mitigation in similar regions.

Spatiotemporal variations in NW Pacific tropical cyclones (frequency, genesis, Accumulated Cyclone Energy (ACE)) and links to warm pool/monsoon trough were resolved using wavelet analysis, correlation, and Mann–Kendall tests on typhoon, SST, and wind data. This method proposed by Contribution 5 specifically detects trends, mutation points (e.g., 1996 frequency shift), and periodicities in cyclone behavior. Its superiority lies in revealing divergent trends (rising frequency vs. declining ACE), quantifying how warm pool thermal states inversely modulate monsoon trough intensity and genesis locations, and identifying synchronized multiscale climate oscillations driving cyclone activity.

The characterization of complex wind–wave dependence for deep-sea floating structure design employed elliptical, Archimedean, and vine copula models with 20-year hindcast data. Modeling trivariate distributions quantified tail dependencies, demonstrating C-vine copula's optimal performance for asymmetry. The approach in Contribution 6 reduced engineering costs by 20–30% via contour-based extreme value identification and provided robust environmental criteria aligned with actual sea conditions.

In Contribution 7, siltation challenges at Binhai Port's entrance were addressed by integrating field data with a 2D tidal sediment model. Quantifying hydro-sedimentary dynamics and typhoon impacts (e.g., Lekima/Muifa) revealed critical mechanisms: tidal asymmetry-driven transport and lagged sedimentary responses. Predicting extreme-event siltation (e.g., 1.0 m during Muifa) provides actionable insights for port management and sustainable mitigation strategies.

Accurate wind stress parameterization in air–sea coupled models, critical under typhoons, was addressed by systematically evaluating five momentum flux partitioning schemes in Contribution 8. Analyzing how drag coefficient formulations alter feedback and surge predictions showed that wave-state-aware schemes outperform wind-speed-dependent approaches in surge accuracy. Higher drag coefficients weaken winds while amplifying currents/surges, providing guidelines to improve forecasting fidelity and reliability.

Regulatory mechanisms of Wyrtki Jets (WJs) under Indo-Pacific climate (ENSO/IOD) interactions were analyzed by partitioning events into distinct phases. Contribution 9 isolated phase-locked responses (e.g., spring-positive anomalies during El Niño decay; fall-negative anomalies during dipole events), revealing opposing seasonal patterns governed by ENSO stages and amplified fall anomalies under IOD, providing insights for predicting Indian Ocean heat/matter redistribution.

In Contribution 10, accurate extreme wind–wave estimates for Bohai Sea engineering were generated using a chained 30-year (1993–2022) WRF-SWAN simulation. Applying the Gumbel distribution to 100-year return periods revealed a ring-shaped spatial pattern (offshore peaks: 37 m/s winds, 6 m waves) and steep nearshore gradients (20–25 m/s winds, 2–3 m waves), providing robust design criteria unattainable via observations.

Drivers of anomalous sea-level rise in the Bohai Sea were quantified using a coupled WRF-FVCOM-SWAVE model, validated against satellite/tide-gauge data. In Contribution 11, sensitivity experiments identified tides as the primary driver, with wind/elevation forcing and Stokes transport amplifying extremes (e.g., 1.1 m surges). This prioritizes essential processes (marginalizing boundary currents) for coastal hazard prediction.

3. Conclusions

The 11 contributions in this Special Issue systematically quantified the complex mechanism of air–sea interaction through multiscale coupling models, long-term data mining, and innovative statistical methods, providing scientific basis for typhoon prediction, coastal engineering protection, port management, and climate response strategies. These significantly improve the risk prevention and control capabilities of extreme environmental events. Major conclusions reveal how semi-enclosed basins like the Sea of Japan modulate typhoon intensification via terrain friction, warm-core eddies, and moisture convergence (Contribution 1). Research quantifies the long-term intensification of winter SST fronts in East China Sea (linked to river discharge/upwelling) and demonstrates mesoscale eddies' asymmetric seasonal control on chlorophyll distribution in the Kuroshio–Oyashio confluence (Contributions 2 and 3). For engineering applications, studies establish that collar protection reduces monopile scour by 50% in the Bohai Sea, while vine copula models optimize extreme wind–wave design criteria, cutting costs by 20–30% (Contributions 4 and 6). Climate analyses identify divergent NW Pacific tropical cyclone trends (rising frequency vs. declining intensity) governed by warm pool–monsoon trough interactions and reveal the phase-locked ENSO/IOD regulation of Wyrtki Jet anomalies. Methodologically, wave state-dependent wind stress schemes improve storm surge forecasts, and chained model simulations enable robust extreme event estimation (e.g., 100-year Bohai Sea winds: 37 m/s) (Contributions 5, 8, 9, and 10). Port siltation dynamics are attributed to tidal asymmetry and typhoon lag effects, while Bohai sea-level rise is primarily tide-driven, amplified by wind and Stokes transport (Contributions 7 and 11). Collectively, these studies provide critical insights for predicting marine hazards and optimizing coastal resilience.

Acknowledgments: We thank the contributors and anonymous reviewers of the Special Issue "Interaction between Atmospheric and Oceanic Dynamics at Mesoscale and Small Scale".

Conflicts of Interest: The authors declare no conflicts of interest.

List of Contributions:

1. Choi, S.-M.; Choi, H. Typhoon Intensity Change in the Vicinity of the Semi-Enclosed Sea of Japan. *J. Mar. Sci. Eng.* **2024**, *12*, 1638. https://doi.org/10.3390/jmse12091638.
2. Chen, H.; Ji, Q.; Wu, Q.; Peng, T.; Wang, Y.; Meng, Z. Seasonal Variability and Underlying Dynamical Processes of Sea Surface Temperature Fronts in Zhoushan and Its Adjacent Seas. *J. Mar. Sci. Eng.* **2024**, *12*, 2335. https://doi.org/10.3390/jmse12122335.
3. Chuang, Z.; Zhang, C.; Fan, J.; Yang, H. Response of Subsurface Chlorophyll Maximum Depth to Evolution of Mesoscale Eddies in Kuroshio–Oyashio Confluence Region. *J. Mar. Sci. Eng.* **2025**, *13*, 24. https://doi.org/10.3390/jmse13010024.
4. Hu, X.; Wang, Z.; Ma, X. Tidal Current with Sediment Transport Analysis and Wind Turbine Foundation Pile Scour Trend Studies on the Central Bohai Sea. *J. Mar. Sci. Eng.* **2025**, *13*, 180. https://doi.org/10.3390/jmse13010180.
5. Guo, J.; Wang, S.; He, X.; Song, J.; Fu, Y.; Cai, Y. Interannual Characteristics of Tropical Cyclones in Northwestern Pacific Region in Context of Warm Pool and Monsoon Troughs. *J. Mar. Sci. Eng.* **2025**, *13*, 334. https://doi.org/10.3390/jmse13020334.
6. Wu, Y.; Feng, Y.; Zhao, Y.; Yu, S. Joint Probability Distribution of Wind–Wave Actions Based on Vine Copula Function. *J. Mar. Sci. Eng.* **2025**, *13*, 396. https://doi.org/10.3390/jmse13030396.
7. Deng, X.; Wang, Z.; Ma, X. Impact of Silted Coastal Port Engineering Construction on Marine Dynamic Environment: A Case Study of Binhai Port. *J. Mar. Sci. Eng.* **2025**, *13*, 494. https://doi.org/10.3390/jmse13030494.

8. Cai, L.; Wang, B.; Wang, W.; Feng, X. The Impact of Air–Sea Flux Parameterization Methods on Simulating Storm Surges and Ocean Surface Currents. *J. Mar. Sci. Eng.* **2025**, *13*, 541. https://doi.org/10.3390/jmse13030541.
9. Feng, Q.; Zhou, J.; Han, G.; Xie, J. Variation of Wyrtki Jets Influenced by Indo-Pacific Ocean–Atmosphere Interactions. *J. Mar. Sci. Eng.* **2025**, *13*, 691. https://doi.org/10.3390/jmse13040691.
10. Zhang, H.; Wang, Z.; Ma, X. Wind and Wave Climatic Characteristics and Extreme Parameters in the Bohai Sea. *J. Mar. Sci. Eng.* **2025**, *13*, 826. https://doi.org/10.3390/jmse13050826.
11. Pan, S.; Liu, L.; Hu, Y.; Zhang, J.; Jia, Y.; Shao, W. Analysis of Abnormal Sea Level Rise in Offshore Waters of Bohai Sea in 2024. *J. Mar. Sci. Eng.* **2025**, *13*, 1134. https://doi.org/10.3390/jmse13061134.

References

1. Galperin, B.; Sukoriansky, S.; Qiu, B. Seasonal oceanic variability on meso- and submesoscales: A turbulence perspective. *Ocean Dyn.* **2021**, *71*, 475–479. [CrossRef]
2. Cronin, M.F.; Swart, S.; Marandino, C.A.; Anderson, C.; Browne, P.; Chen, S.; Joubert, W.R.; Schuster, U.; Venkatesan, R.; Addey, C.I.; et al. Developing an observing air–sea interactions strategy (OASIS) for the global ocean. *ICES J. Mar. Sci.* **2023**, *80*, 367–373. [CrossRef]
3. Cao, X.H.; Liu, C.; Zhang, S.Q.; Gao, F. A Method for Predicting High-Resolution 3D Variations in Temperature and Salinity Fields Using Multi-Source Ocean Data. *J. Mar. Sci. Eng.* **2024**, *12*, 1396. [CrossRef]
4. Chen, G.; Huang, B.X.; Chen, X.Y.; Ge, L.Y.; Radenkovic, M.; Ma, Y. Deep blue AI: A new bridge from data to knowledge for the ocean science. *Deep-Sea Res. Part I-Oceanogr. Res. Pap.* **2022**, *190*, 103886. [CrossRef]
5. Shao, W.Z.; Zhou, Y.H.; Hu, Y.Y.; Li, Y.; Zhou, Y.S.; Zhang, Q.J. Range current retrieval fromsentinel-1 SAR ocean product based on deep learning. *Remote Sens. Lett.* **2024**, *15*, 145–156. [CrossRef]
6. Marrow, R.; Le Traon, P.Y. Recent advances in observing mesoscale ocean dynamics with satellite altimetry. *Adv. Space Res.* **2012**, *50*, 1062–1076. [CrossRef]
7. Pan, S.; Liu, L.; Hu, Y.; Zhang, J.; Jia, Y.; Shao, W. Analysis of Abnormal Sea Level Rise in Offshore Waters of Bohai Sea in 2024. *J. Mar. Sci. Eng.* **2025**, *13*, 1134. [CrossRef]
8. Hu, Y.Y.; Shao, W.Z.; Jiang, X.W.; Lin, G.X. Performance of the different parameterization of depth-induced breaking during a typhoon. *J. Sea Res.* **2025**, *206*, 102602. [CrossRef]

Journal of
*Marine Science
and Engineering*

MDPI

Article

Analysis of Abnormal Sea Level Rise in Offshore Waters of Bohai Sea in 2024

Song Pan [1,2], Lu Liu [1], Yuyi Hu [1], Jie Zhang [3,*], Yongjun Jia [4] and Weizeng Shao [1,*]

1 College of Oceanography and Ecological Science, Shanghai Ocean University, Shanghai 201306, China; ps2006201@163.com (S.P.); m240300908@st.shou.edu.cn (L.L.); yy-hu@shou.edu.cn (Y.H.)
2 National Marine Data and Information Service, Ministry of Natural Resources, Tianjin 300171, China
3 East China Sea Forecasting and Disaster Reduction Center, Ministry of Natural Resources of the People's Republic of China, Shanghai 201306, China
4 National Satellite Ocean Application Service, Ministry of Natural Resources of the People's Republic of China, Beijing 100081, China; jiayongjun@mail.nsoas.gov.cn
* Correspondence: zhangjie0198@163.com (J.Z.); wzshao@shou.edu.cn (W.S.); Tel.: +86-21-61900326 (W.S.)

Abstract: The primary contribution of this study lies in analyzing the dynamic drivers during two anomalous sea level rise events in the Bohai Sea through coupled numeric modeling using the Weather Research and Forecasting (WRF) model and the Finite-Volume Community Ocean Model (FVCOM) integrated with the Simulating Waves Nearshore (SWAN) module (hereafter referred to as FVCOM-SWAVE). WRF-derived wind speeds (0.05° grid resolution) were validated against Haiyang-2 (HY-2) scatterometer observations, yielding a root mean square error (RMSE) of 1.88 m/s and a correlation coefficient (Cor) of 0.85. Similarly, comparisons of significant wave height (SWH) simulated by FVCOM-SWAVE (0.05° triangular mesh) with HY-2 altimeter data showed an RMSE of 0.67 m and a Cor of 0.84. Four FVCOM sensitivity experiments were conducted to assess drivers of sea level rise, validated against tide gauge observations. The results identified tides as the primary driver of sea level rise, with wind stress and elevation forcing (e.g., storm surge) amplifying variability, while currents exhibited negligible influence. During the two events, i.e., 20–21 October and 25–26 August 2024, elevation forcing contributed to localized sea level rises of 0.6 m in the northern and southern Bohai Sea and 1.1 m in the southern Bohai Sea. A 1 m surge in the northern region correlated with intense Yellow Sea winds (20 m/s) and waves (5 m SWH), which drove water masses into the Bohai Sea. Stokes transport (wave-driven circulation) significantly amplified water levels during the 21 October and 26 August peak, underscoring critical wave–tide interactions. This study highlights the necessity of incorporating tides, wind, elevation forcing, and wave effects into coastal hydrodynamic models to improve predictions of extreme sea level rise events. In contrast, the role of imposed boundary current can be marginalized in such scenarios.

Keywords: sea level rise; Bohai Sea; couple numeric model

1. Introduction

Since the early 20th century, global sea level has exhibited a gradual rise under the influence of climate change [1–3]. This accelerated rise poses a significant threat to coastal regions, amplifying risks of storm surge [4], flooding [5], and large-scale socio-economic damage [6]. Sea level monitoring primarily relies on two approaches: remote sensing and in situ observations. Satellite-based technologies, such as radar altimeters [7] and GPS-derived measurements [8], enable rapid, large-scale assessments of sea surface height. Meanwhile, traditional methods—including shipborne surveys, moored buoys, and tide gauge stations—remain critical for providing high-resolution, localized data. Advances in oceanographic modeling and computational power have further enhanced our capacity to study sea level dynamics. Numerical models are now widely employed for hindcasting historical trends and predicting future scenarios, with sensitivity experiments proving particularly effective for isolating drivers of extreme sea states [9,10].

During the Seasat mission in 1978 [11], the capabilities of various sensors—including a microwave scatterometer [12], radar altimeter [13], infrared radiometer [14], and synthetic aperture radar (SAR) [15]—were tested. Today, satellites such as MetOp, ERS, Haiyang-2 (HY-2), and Sentinel carry these sensors, providing continuous global ocean observations for atmospheric and oceanographic research. Scatterometers and microwave radiometers [16] are primarily used for sea surface wind monitoring and operate at radar incidence angles greater than 20°. In contrast, radar altimeters have a near-nadir beam angle (<2°) [17], directly measuring sea surface height from reflected signals. Additionally, significant wave height (SWH) and sea level anomaly (SLA) are derived from altimeter data [18]. These remotely sensed products are widely applied in mesoscale and large-scale oceanography, including studies of oceanic eddies [19,20], tropical cyclones [21], and long-term climate analysis [22]. However, the spatial resolution of these products remains relatively coarse—typically 12.5 km for scatterometers, 10 km along altimeter footprints, and 18 km for the Surface Wave Investigation and Monitoring (SWIM) instrument aboard the Chinese–French Oceanography SATellite (CFOSAT) [23]. As a result, they are insufficient for coastal water monitoring. While SAR (e.g., Gaofen-3 (GF-3) [24,25] and Sentinel-1 (S-1) [26]) offers finer resolution and wide swath coverage, its long-term data acquisition over specific regions remains challenging.

Numerical models are powerful tools for hindcasting and predicting atmospheric and oceanic dynamics by solving theoretical equations. Several well-known numerical models are used for atmospheric simulations, including the Weather Research and Forecasting (WRF) model [27], the PSU/NCAR MM5 model [28], the Rapid Refresh (RAP) model [29], and the Global and Regional Assimilation and Prediction System (GRAPES) [30]. Leveraging these models, agencies such as the European Centre for Medium-Range Weather Forecasts (ECMWF) and the National Centers for Environmental Prediction (NCEP) Global Forecast System (GFS) operationally produce global atmospheric reanalysis data [31]. For ocean wave modeling, the predominant numerical models are WAVEWATCH-III (WW3) [32] and Simulating Waves Nearshore (SWAN) [33]. In ocean circulation simulations, commonly used models include the Hybrid Coordinate Ocean Model (HYCOM) [34], the Finite-Volume Community Ocean Model (FVCOM) [35], and the Regional Ocean Modeling System (ROMS) [36]. Notably, FVCOM and SWAN utilize unstructured triangular meshes, which provide better conformity to coastlines. Despite their advantages, numerical forecasting models encounter several technical challenges, such as limited spatial and temporal resolution, uncertainties in initial conditions, and computational constraints—particularly under extreme weather conditions [37].

Numerical models serve as indispensable tools for analyzing ocean dynamics, providing critical insights into underlying mechanisms. Through parameter adjustments and boundary condition modifications, these models enable the simulation of various ocean dynamic processes and their environmental impacts. In this study, the WRF model is employed to hindcast wind fields, which serve as forcing fields for FVCOM and FVCOM-SWAVE. We specifically analyze two seawater backflow events in the offshore waters of the Bohai Sea caused by abnormal sea level rise, i.e., 20–21 October and 25–26 August 2024. The paper is organized as follows: Section 2 describes the datasets, including forcing fields and open boundary conditions for the coupled numerical models, along with validation sources. This section also details the model configurations. Sections 3 and 4 present the results and discussion, respectively. Section 5 provides concluding remarks.

2. Materials and Methods

In this section, we briefly describe the basic principles of the numerical models WRF and FVCOM-SWAVE. We then present the forcing fields and boundary conditions used in the coupled numerical model (FVCOM-SWAVE). Additionally, we outline the sources for validating the hindcasting results, including HY-2 operational products and tide gauge station data. Finally, we introduce the setup of sensitivity experiments in FVCOM-SWAVE.

2.1. Setups of the WRF

The WRF model, a collaborative atmospheric modeling framework jointly developed by the National Center for Atmospheric Research (NCAR) and the National Oceanic and Atmospheric Administration (NOAA), integrates NCAR climate science expertise with NOAA operational forecasting infrastructure. In this study, WRF version 4.2 was configured to simulate wind fields at a horizontal spatial resolution of 5 km with 34 vertical layers. Simulations were conducted for two distinct periods: (1) 1 August 2024 (00:00 UTC) to 31 August 2024 (00:00 UTC) (30 days) and (2) 1 October 2024 (00:00 UTC) to 30 November 2024 (00:00 UTC) (61 days). Key physical parameterizations included the WSM6 microphysics scheme, the RRTMG radiation scheme [38], the Yonsei PBL scheme [39], the MM5 Monin–Obukhov surface layer scheme [40], and the Noah LSM for land–atmosphere interactions [41]. Initial and lateral boundary conditions for the WRF simulations were derived from the NCEP GFS analysis dataset, which provides data at a 0.25° spatial resolution and 3-hourly temporal intervals. These inputs included vertical profiles of wind speed, air temperature, humidity, and pressure, as well as land surface variables such as terrain height and land-use classification.

2.2. Setups of FVCOM-SWAVE

FVCOM utilizes an unstructured triangular grid system [42] that provides enhanced geometric flexibility for resolving complex coastal geometries and shallow-water bathymetry. The model's finite-volume computational framework ensures the accurate representation of hydrodynamic processes across highly irregular topographies while supporting both Cartesian and spherical coordinate systems. For horizontal discretization, FVCOM employs triangular grid elements integrated with the Mellor-Yamada 2.5 (MY-2.5) turbulence closure scheme [43] for subgrid-scale parametrization. Vertical resolution is achieved through generalized terrain-following coordinates (e.g., sigma layers) in conjunction with the Smagorinsky turbulent mixing scheme. FVCOM incorporates a dynamic wet–dry grid cell treatment that accurately simulates tidal flat inundation dynamics. The model solves the fully coupled three-dimensional primitive equations, including momentum balance, mass continuity, and thermodynamic conservation for temperature, salinity, and density fields. This comprehensive approach enables the robust simulation of coastal and oceanographic processes across multiple spatiotemporal scales.

The FVCOM-SWAVE model is a spectral wave modeling system developed by integrating the widely used SWAN framework [44] with the unstructured-grid, finite-volume computational architecture of FVCOM. This synthesis enables fully coupled simulations of hydrodynamic and spectral wave processes in complex coastal and oceanic domains. By retaining the robust spectral wave-solving algorithms of SWAN while incorporating the adaptive spatial discretization of FVCOM, the model achieves enhanced accuracy in resolving irregular geometries, bathymetric gradients, and dynamic wave–current interactions critical for applications such as storm surge propagation and nearshore wave dynamics. The governing equations, adapted from the spectral formulation of SWAN, resolve wave energy evolution, nonlinear wave–wave interactions, and bidirectional wave–current feedback within a three-dimensional hydrodynamic framework. Central to the FVCOM-SWAVE system is the wave action density conservation equation:

$$\frac{\partial N}{\partial t} + \nabla \cdot \left[\left(\vec{C_g} + \vec{V} \right) N \right] + \frac{\partial C_\sigma N}{\partial \sigma} + \frac{\partial C_\theta N}{\partial \theta} = \frac{S_{tot}}{\sigma} \quad (1)$$

where N denotes the wave energy density spectrum; t represents the computational time; C_g is the wave group velocity; V is the ambient water velocity; σ denotes the relative wave frequency; θ is the wave direction; C_σ and C_θ represent the wave speed in relative frequency and wave direction, respectively; and S_{tot} represents the source–sink terms in SWAN. The FVCOM-SWAVE model differentiates itself from SWAN through its numerical schemes. Specifically, it employs the Flux-Corrected Transport (FCT) algorithm in frequency space to minimize numerical diffusion, adopts the implicit Crank–Nicolson method in directional space to enhance stability, and provides explicit and implicit second-order upwind finite-volume schemes in geographic space to handle advection on unstructured grids. In this study, the coupled FVCOM-SWAVE model operates with a temporal resolution of 1 h outputs and an unstructured triangular grid at 0.05° (~5 km) resolution (Figure 1). Leveraging its capability to adapt to complex coastal geometries, the model effectively captures small-scale nearshore phenomena, including waves and currents, while maintaining high computational accuracy in wave dynamics simulations. The simulation period aligns with that of the WRF model, as previously described.

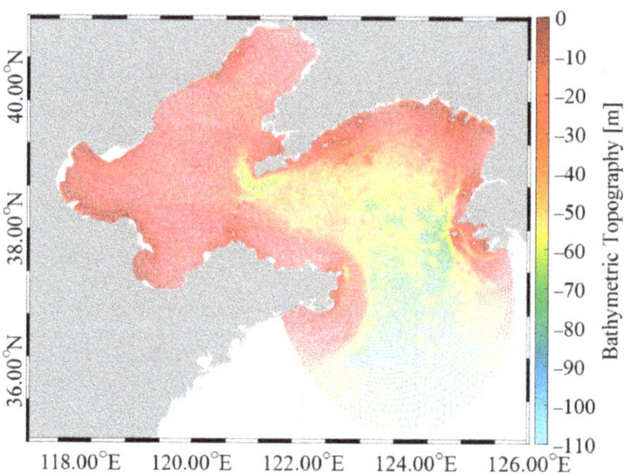

Figure 1. The unstructured triangular grid covers the study area, with the contours representing water depth from the General Bathymetry Chart of the Oceans (GEBCO).

2.3. Forcing Fields and Boundary Conditions in FVCOM-SWAVE

Wind forcing plays a critical role in hindcasting ocean dynamics. However, the spatial resolution of global meteorological datasets such as ECMWF and GFS (0.25° grid) is inadequate for resolving fine-scale processes in coastal waters. The WRF model, designed for meteorological applications across spatial scales ranging from tens of meters to thousands of kilometers, can generate simulations driven by either observational/analysis data or idealized atmospheric conditions. In this study, WRF was configured to simulate wind fields at an enhanced horizontal resolution of 0.05°. To ensure consistency in forcing field data and avoid interpolation errors, the spatial resolution (5 km) of the WRF model was matched to that of FVCOM-SWAVE. These outputs were coupled with the FVCOM-SWAVE system, which employs an unstructured grid architecture to simulate sea surface dynamics. As demonstrated in previous studies [45], wave simulations are sensitive to sea level variability and current interactions. Consequently, the FVCOM-SWAVE simulations were forced using WRF-derived wind fields, along with CMEMS datasets providing sea surface current velocity, salinity, temperature, and sea level at 0.08° spatial resolution and 3-hourly temporal intervals. These inputs were applied as boundary conditions and dynamic forcing fields. Figure 2 illustrates the spatially distributed CMEMS-derived sea surface current speed, sea level, salinity, and temperature fields at 12:00 UTC on 20 October 2024.

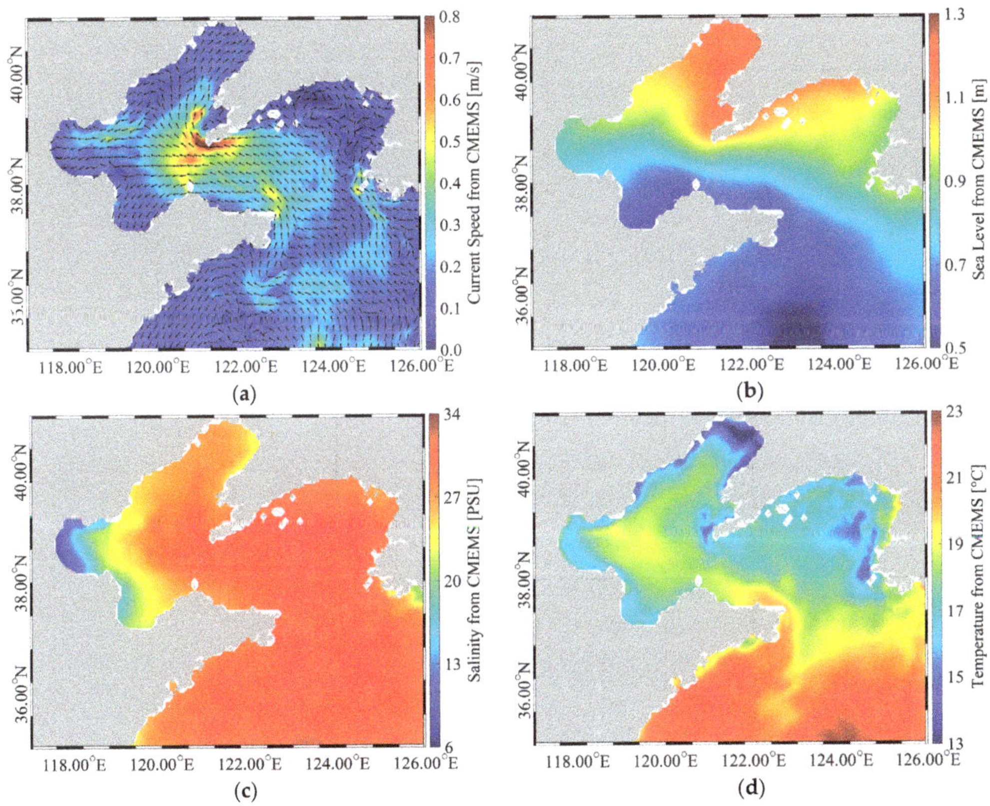

Figure 2. The CMEMS wind current vector maps on 20 October 2024 at 12:00 UTC: (**a**) sea surface current, (**b**) sea level, (**c**) salinity, and (**d**) sea surface temperature.

2.4. Validation Sources

The HY-2B satellite, China's first civilian mission equipped with a scatterometer and altimeter, was launched in 2018. In our prior study [7], the performance of HY-2B data products—categorized into four processing levels: Operational Geophysical Data Records (OGDRs), Interim Geophysical Data Records (IGDRs), Sensor Geophysical Data Records (SGDRs), and Geophysical Data Records (GDRs)—was systematically evaluated. Results demonstrated that the GDR products exhibited superior accuracy. Since 2021, HY-2B's successor satellites, HY-2C and HY-2D, have been successively launched and operationalized, forming a constellation to enhance oceanographic monitoring. To validate hindcasted wind and wave parameters, non-interpolated operational scatterometer and altimeter data from the HY-2(2B/2C/2D) satellites were obtained from the National Satellite Ocean Application Center (accessible via https://osdds.nsoas.org.cn, accessed on 3 June

2025). HY-2 altimeter and scatterometer products achieve high precision through rigorous processing, including precise orbit determination, instrument calibration, geophysical model integration, and ground validation. Additionally, hourly sea level observations from tide gauge stations along the Chinese coast, provided by the National Marine Information Center and subjected to quality control and analytical processing, were utilized to validate hindcasted sea levels. Figure 3a depicts the HY-2B scatterometer-derived wind speed field at 09:00 UTC on 19 October 2024, while Figure 3b illustrates the concurrent SWH distribution from the HY-2 altimeter. Red five-pointed stars denote the geographical locations of the tide gauge stations used in this study.

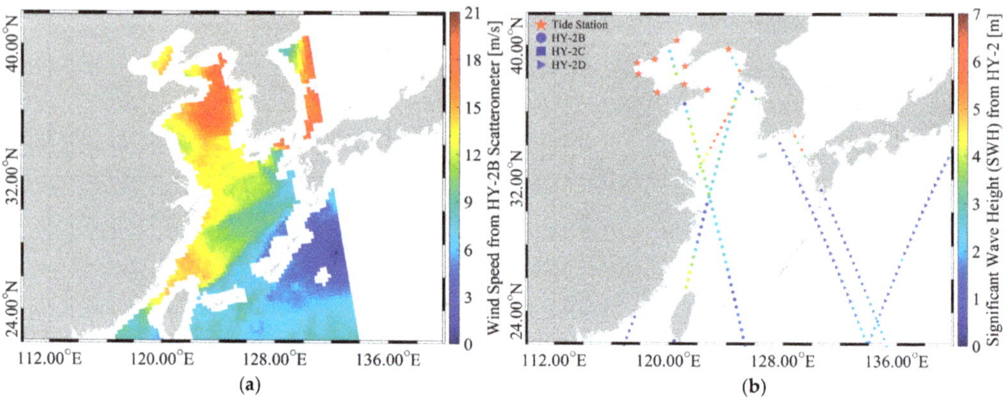

Figure 3. Satellite-derived oceanographic observations from 19 October 2024: (**a**) HY-2B scatterometer wind speed at 09:00 UTC and (**b**) HY-2 altimeter SWH, with red five-pointed stars indicating tide gauge station locations.

2.5. Stokes Transport Calculation

The Stokes drift speed, $u_{st}(z)$, describes how waves push water particles at different depths. Using a 1D wave spectrum model, it can be expressed as [46]:

$$u_{st}(z) = \frac{2}{g}\int_0^{\infty} \omega^3 S_\omega(\omega) e^{2kz} d\omega \tag{2}$$

where g denotes gravitational acceleration, ω represents angular frequency, k is the wavenumber, z is the vertical depth coordinate, and S_ω is the 1D wave spectrum. Earlier work [31] simplified this using SWH (H_s) and mean wave period (T_m), through the linear dispersion relation:

$$\vec{u_{st}}(z) = \vec{u_0} e^{\frac{8\pi^2 z}{g T_m^2}} \tag{3}$$

The Stokes drift induces a net Lagrangian transport, commonly referred to as Stokes transport. Integrating Stokes drift velocity over depth gives the Stokes transport vector (T_s) and water depth D in the vertical direction:

$$T_s = \pi \frac{H_s^2}{T_m} \cdot \vec{D} \tag{4}$$

2.6. Statistic Parameter Formula

The model–observation agreement was quantified using three statistical metrics:

$$\text{RMSE} = \sqrt{\frac{1}{N}\sum_{i=1}^{N}(p_i - q_i)^2} \tag{5}$$

$$\text{Cor} = \frac{\sum_{i=1}^{N}(p_i - \overline{p})(q_i - \overline{q})}{\sqrt{\sum_{i=1}^{N}(p_i - \overline{p})^2 \sum_{i=1}^{N}(q_i - \overline{q})^2}} \tag{6}$$

$$\text{Bias} = \frac{1}{N}\sum_{i=1}^{N}(p_i - q_i) \tag{7}$$

where p_i denotes the observed values (e.g., satellite measurements), q_i represents the simulated values (e.g., model outputs), N is the total number of collocated data points, and $\overline{p}$ and $\overline{q}$ are the arithmetic means of the observed and simulated values, respectively.

3. Results

This section presents validation results for the hindcasted variables. Subsequently, a sensitivity experiment is conducted to examine sea level variations during seawater backflow events. Finally, the relationships between wind speed, SWH, current velocity, and sea level are analyzed.

3.1. Validation

Figure 4a presents the WRF-simulated wind field on 19 October 2024 at 09:00 UTC, while Figure 4b displays the corresponding validation using HY-2B/2C/2D scatterometer wind speed measurements. Over 10,000 collocated data points were obtained within the study domain. The temporal mismatch between model outputs and satellite observations was maintained below 10 min to ensure comparability. Quantitative analysis reveals excellent model performance, with a root mean square error (RMSE) of 1.88 m/s and a correlation coefficient (Cor) of 0.85. These results demonstrate strong agreement between WRF simulations and HY-2 scatterometer wind speed observations. Similarly, Figure 5a shows that FVCOM-SWAVE simulated the SWH field on 26 October 2024 at 00:00 UTC, with colored triangles indicating concurrent HY-2 altimeter measurements. The validation yields an RMSE of 0.67 m and Cor of 0.84, indicating good consistency between model results and satellite observations (model outputs and satellite observations were maintained below 30 min). This confirms the model's capability to accurately reproduce wave conditions as observed by the HY-2 altimeter.

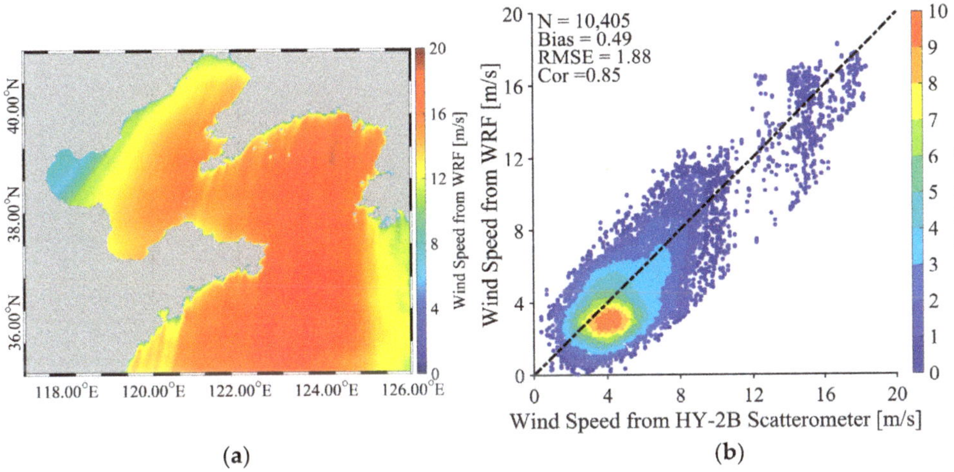

Figure 4. Wind field analysis for 19 October 2024 at 09:00 UTC: (**a**) WRF-simulated wind field and (**b**) validation against HY-2B/C/D scatterometer measurements.

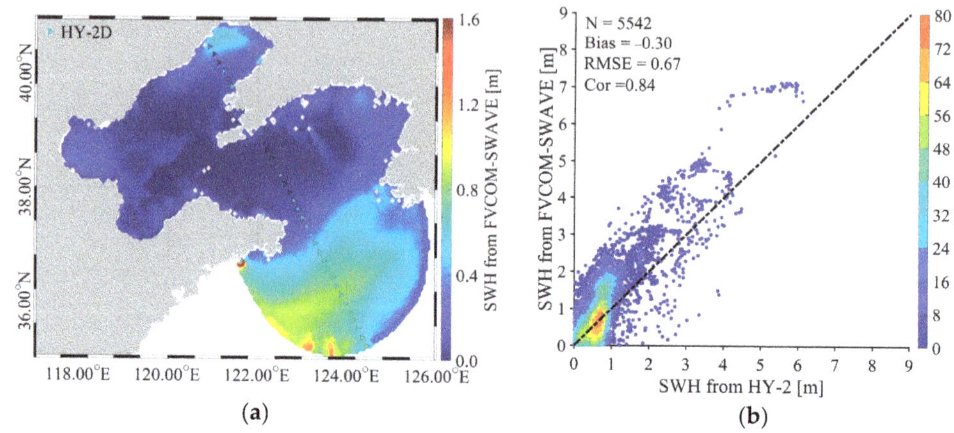

Figure 5. SWH validation on 26 October 2024 at 00:00 UTC: (**a**) modeled SWH field from FVCOM-SWAVE with triangular markers indicating coincident HY-2 altimeter measurements and (**b**) comparison between FVCOM-SWAVE simulations and HY-2 altimeter observations.

3.2. Sensitive Experiment

To examine the influence of different forcing conditions on model performance, we conducted four sensitivity experiments. All four forcing modes utilize temperature and salinity data from CMEMS as the initial conditions, with distinct configurations applied incrementally: Mode A employs tide forcing exclusively; Mode B integrates both wind and tide forcing; Mode C enhances this framework by combining wind and tide forcing while prescribing water level constraints at the open boundary; finally, Mode D extends the complexity of Mode C by incorporating additional velocity boundary conditions alongside water level specifications at the open boundary.

Figure 6 presents the time series validation between FVCOM-SWAVE-simulated sea levels and tide gauge observations. The results demonstrate that simulations incorporating wind, tide, and elevation forcing fields effectively reproduce observed high sea level conditions. Conversely, simulations with tidal forcing alone accurately capture low sea level variations, particularly at the tide station located at (117.79° E, 38.97° N) (Figure 6a). The comparison of different forcing configurations reveals that elevation and wind are key drivers of water level variability in the study area, while tidal forcing represents the fundamental mechanism governing sea level changes. Notably, currents appear to have negligible influence on sea level variations. Quantitative validation against tide gauge data (Figure 7) shows optimal model performance when considering only tidal effects, yielding an RMSE of 0.43 m and a Cor of 0.70. All four sensitivity experiments maintained RMSE values below 0.70 m, demonstrating consistent model performance across different forcing configurations. These results confirm the reliability of the model for future applications in the study region.

To investigate the influence of seawater backflow events in the offshore waters of the Bohai Sea, sea level simulations from FVCOM-SWAVE on 20 October 2024 and 26 August 2024 were analyzed under distinct forcing configurations. Figure 8 illustrates the spatial differences in sea level between three experimental cases and the baseline tide-only simulation on 20 October 2024: (a) wind and tide, (b) wind, tide, and elevation, and (c) wind, tide, elevation, and current. The results demonstrate a pronounced sea level rise in the northern and southern regions of the Bohai Sea. Notably, the imposed forcing field (elevation and current) exerted a significantly stronger influence than wind forcing, inducing sea level deviations of up to 0.6 m. Similarly, the sea level simulations from

FVCOM-SWAVE on 26 August 2024 are presented in Figure 9. The comparison result in Figure 9 shows the similar trend of sea level difference in Figure 8, with sea level deviations of up to 1.1 m. In contrast, the inclusion of the current forcing field exhibited negligible impact on sea level variability, consistent with the statistical findings presented earlier.

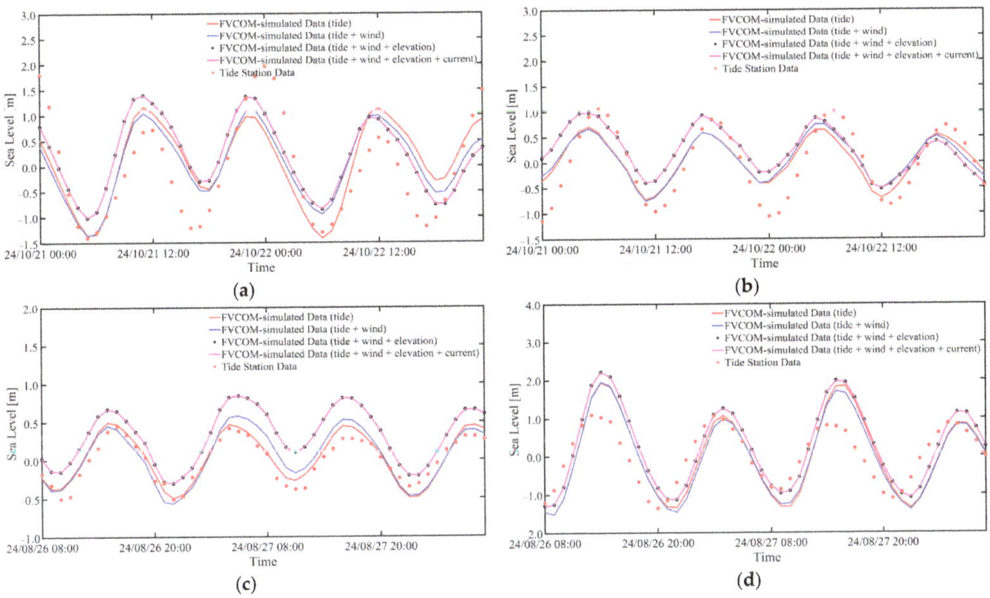

Figure 6. Time series comparison of simulated and observed sea levels: (**a**,**b**) FVCOM-SWAVE-simulated sea levels from 21 to 23 October 2024 at (**a**) station (117.79° E, 38.97° N) and (**b**) station (121.07° E, 37.72° N); (**c**,**d**) simulation from 08:00 UTC 26 August to 08:00 UTC 28 August 2024 at (**c**) station (121.07° E, 37.72° N) and (**d**) station (124.16° E, 39.82° N). Red markers denote tide gauge measurements.

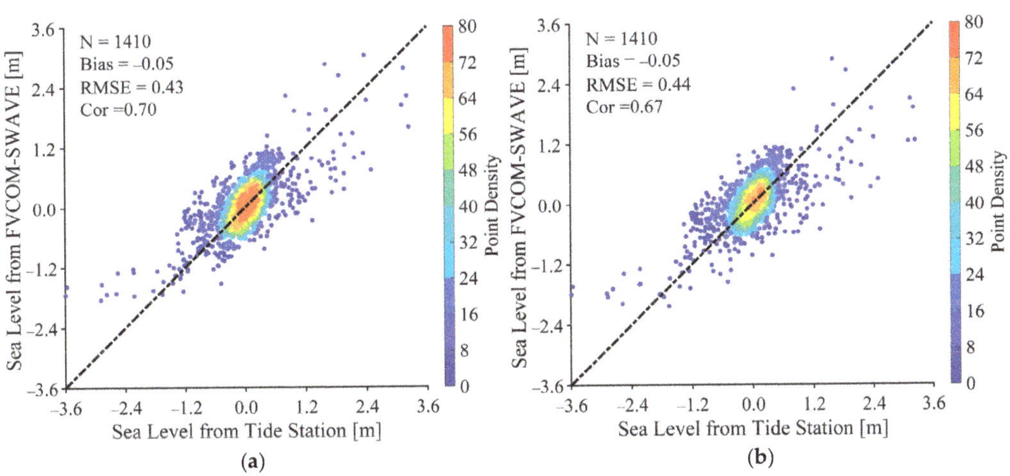

Figure 7. *Cont.*

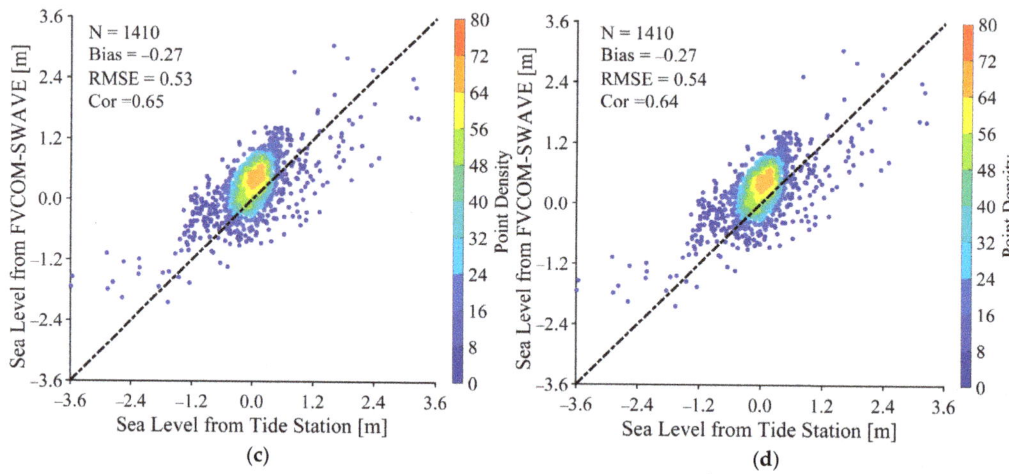

Figure 7. Model–observation comparison of sea level simulations under different forcing configurations: (**a**) tide, (**b**) wind + tide forcing, (**c**) wind + tide + elevation forcing, and (**d**) wind + tide + elevation + current forcing.

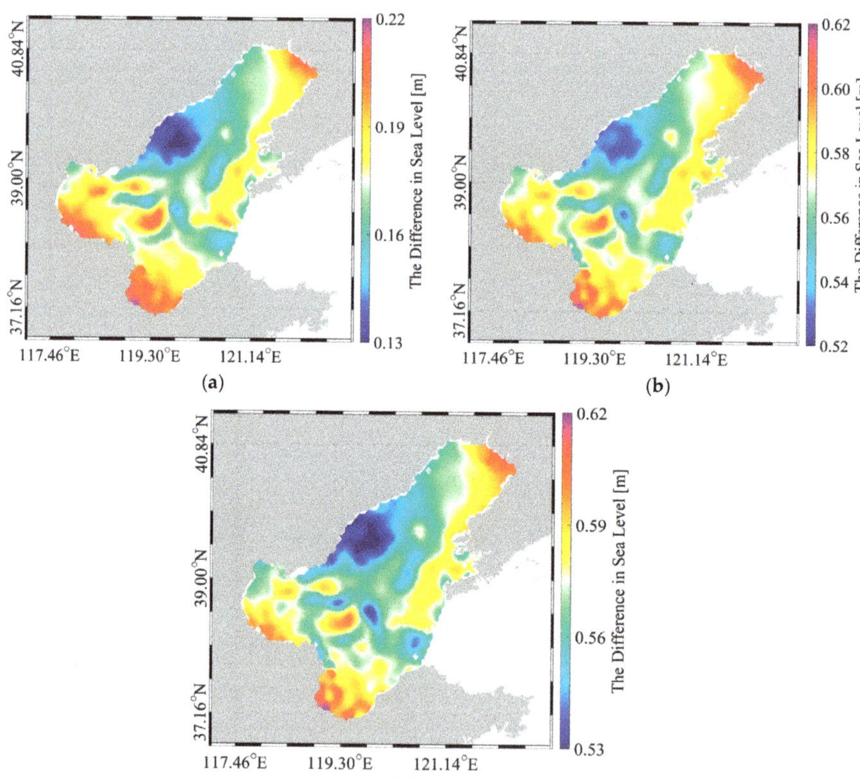

Figure 8. Sea level anomalies (relative to tide-only forcing) for 20 October 2024 under different forcing conditions: (**a**) wind + tide forcing, (**b**) wind + tide + elevation forcing, and (**c**) wind + tide + elevation + current forcing.

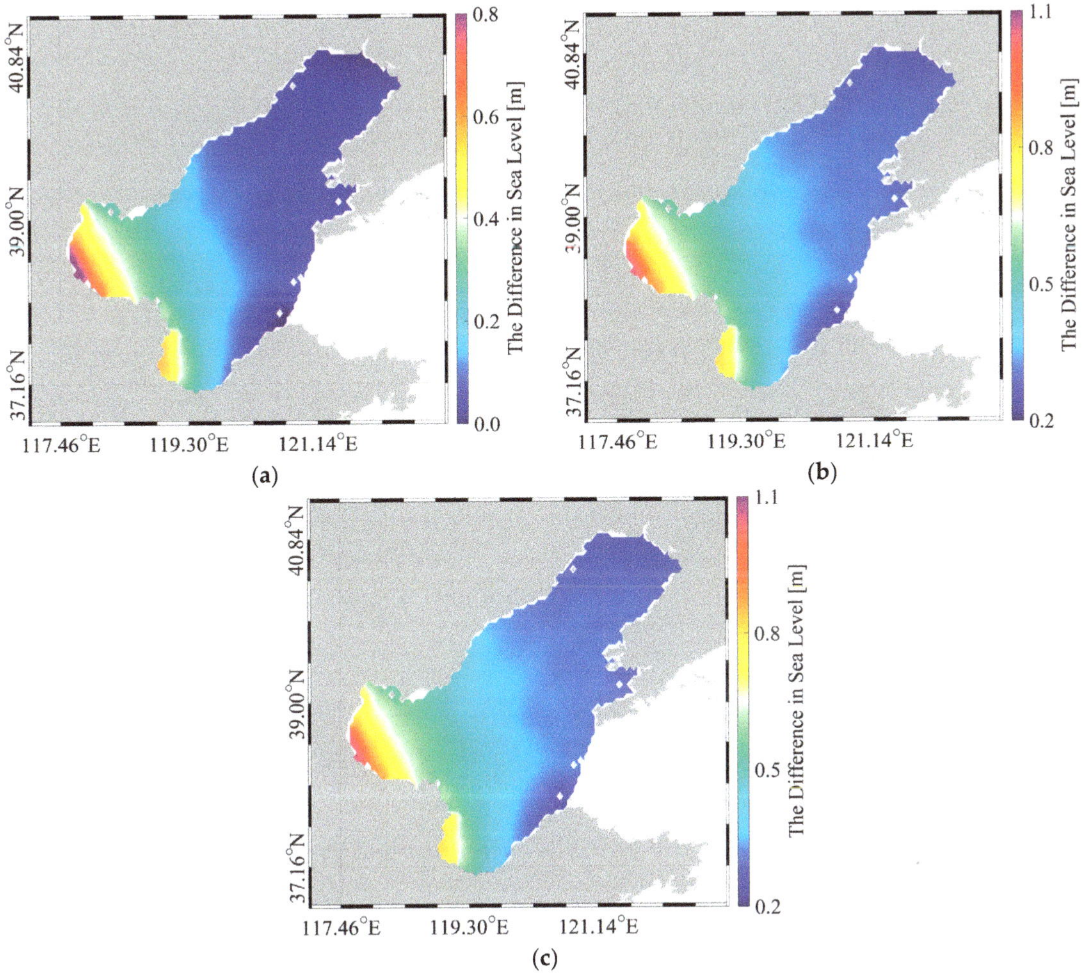

Figure 9. Sea level anomalies (relative to tide-only forcing) on 26 August 2024 under different forcing conditions: (**a**) wind + tide forcing, (**b**) wind + tide + elevation forcing, and (**c**) wind + tide + elevation + current forcing.

3.3. The Relationship Between the Wind Speed, SWH, Imposed Boundary Current, and Sea Level

To investigate the influence of sea surface parameters—wind speed, SWH, and current (mainly imposed boundary current) velocity—on sea level dynamics, we analyzed differences in simulated sea level (calculated by subtracting 20 October results from 21 October outputs), alongside concurrent wind speed simulated by WRF, SWH, and imposed boundary current velocity simulated by FVCOM-SWAVE fields for 20 October 2024. The sea level anomaly map (Figure 10a) reveals a pronounced rise of up to 1 m in the northern Bohai Sea, consistent with observed seawater backflow events in offshore regions. This anomaly correlates with elevated wind speeds (>20 m/s) and SWH (>5 m) in the adjacent Yellow Sea (Figure 10b,c), suggesting that wind- and wave-driven Stokes transport may have displaced water masses toward the Bohai Sea. In contrast, imposed boundary current velocities exhibited no significant correlation with sea level fluctuations (Figure 10d), underscoring the limited role of advective processes in this event. Time series analysis (Figure 11) further demonstrates that the pre-event period (20 October) was characterized by extreme wind and wave conditions, which subsided prior to the peak sea level rise on 21 October. Additionally, Figure 12 shows the time series analysis of the FVCOM-SWAVE simulation during the August 2024 results. The wind speed, imposed boundary current, sea level, and SWH in the Bohai area exhibit climatological periodic variations [47,48], and there is a certain correlation between sea level height and wind speed, SWH, and imposed boundary current. During periods of sea level rise (18–22 October and 26–27 August), these parameters simultaneously peak, and the consistent direction of wind speed and flow velocity drives water masses toward the Bohai Sea. These findings suggest that the combined effects of wind, waves, and currents contribute to anomalous sea level rise. This highlights a complex ocean–atmosphere interaction, where wind speed variations influence wave generation and sea level changes, subsequently altering current velocity and direction.

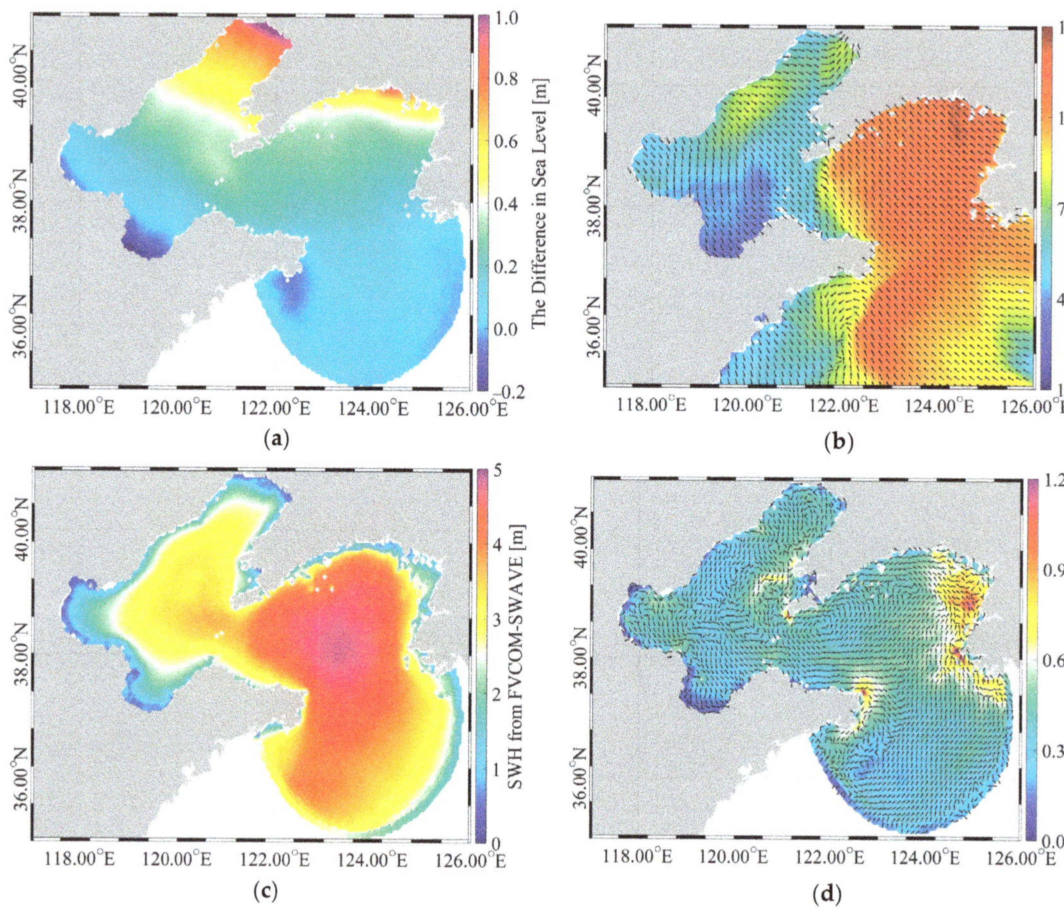

Figure 10. Comparison of marine conditions preceding sea level changes (20 October 2024): (**a**) sea level anomaly (21 October minus 20 October simulation results), (**b**) WRF-modeled wind speed, (**c**) FVCOM-SWAVE-simulated SWH, and (**d**) current (mainly imposed boundary current) velocity.

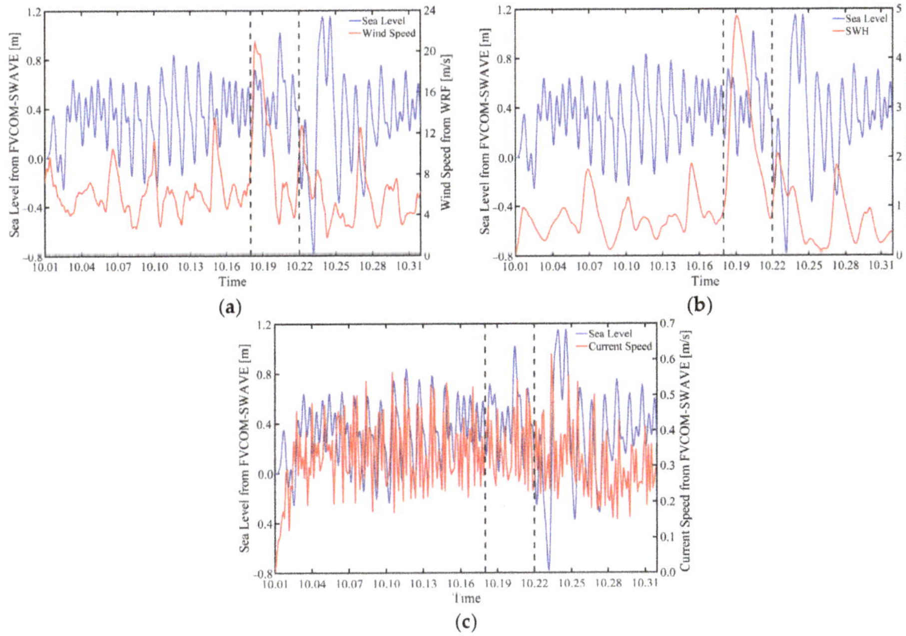

Figure 11. Time series of sea surface parameters in the Bohai Sea on 1–31 October 2024: (**a**) wind speed, (**b**) SWH, and (**c**) imposed boundary current speed. The dash lines represent the occurrence of sea level rapid rise.

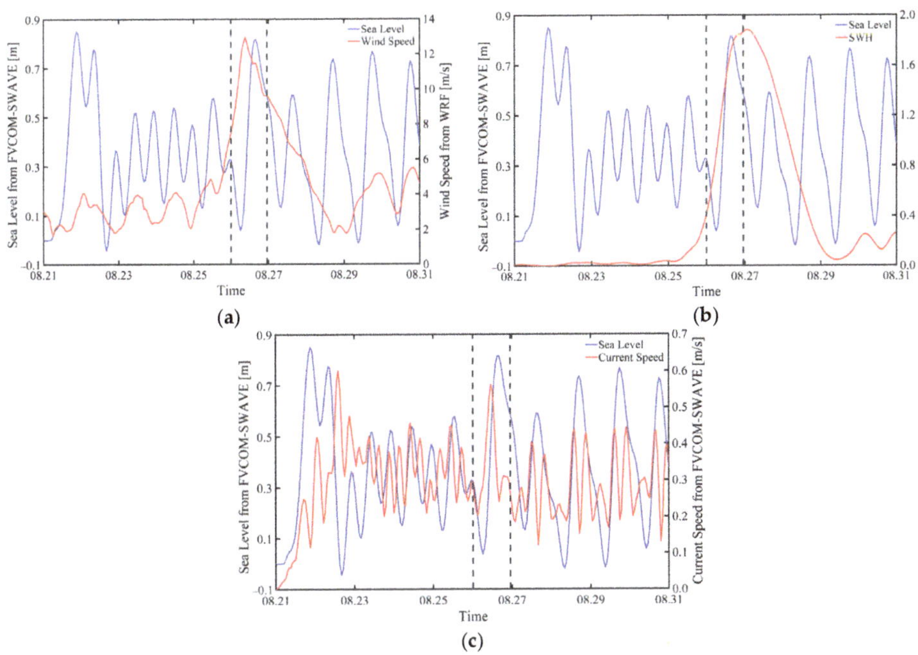

Figure 12. Time series of sea surface parameters in the Bohai Sea on 21–31 August 2024: (**a**) wind speed, (**b**) SWH, and (**c**) imposed boundary current speed. The dash lines represent the occurrence of sea level rapid rise.

4. Discussion

Stokes transport along the transect marked in Figure 11a (black line) was computed using wave parameters simulated by the FVCOM-SWAVE model. Figure 13a provides a schematic representation of cross-regional transport dynamics between the Bohai and Yellow Seas, where net inflow into the Bohai Sea is defined as positive. Figure 13b compares the time series of net transport under three distinct forcing configurations during October 2024: (1) tide-only, (2) combined wind, tide, current, and elevation boundary conditions, and (3) wave-induced Stokes transport. Similarly, Figure 13c shows the time series of net transport under three distinct forcing configurations in August 2024. The analysis demonstrates that tidal forcing dominates the net water flux into the Bohai Sea. However, Stokes transport contributions on 21 October 2024 and 26 August 2024 exhibit a pronounced secondary influence, amplifying sea level rise during this period. These results underscore the primacy of tidal dynamics in governing Bohai Sea hydrodynamics, with wave-induced Stokes transport providing secondary but non-negligible contributions to water level variability. The temporal decoupling between hydrodynamic forcing and sea level response highlights the cumulative contribution of wave-induced Stokes drift to coastal water accumulation.

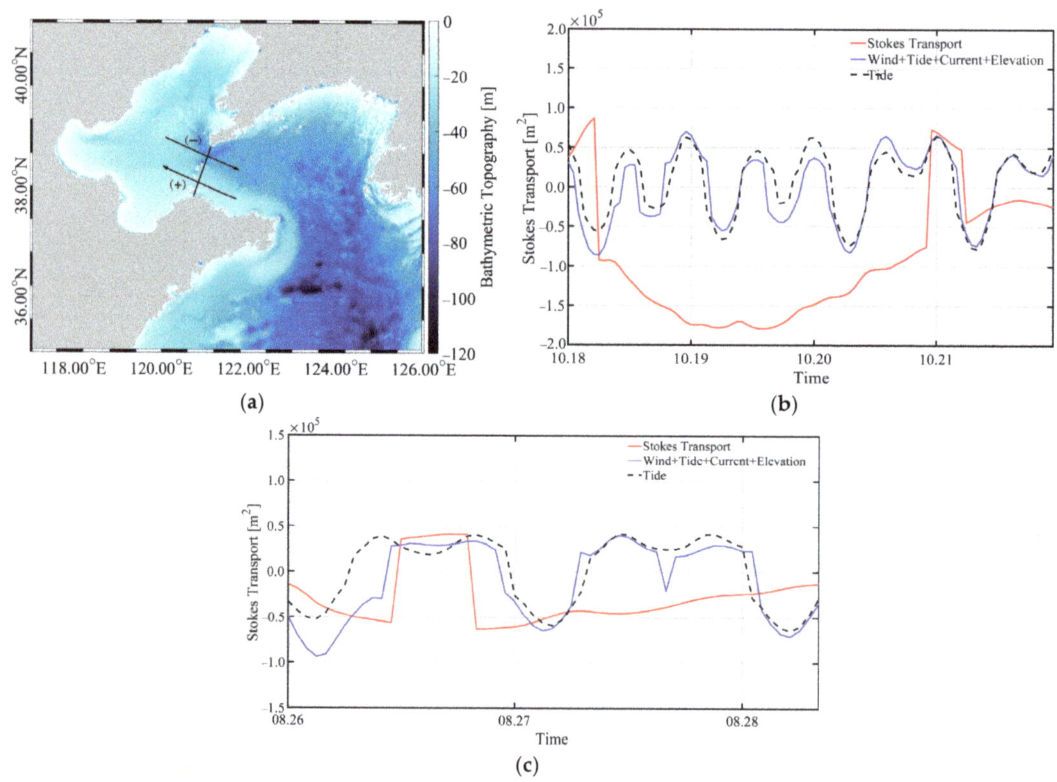

Figure 13. (**a**) Schematic diagram illustrating transport dynamics between the Bohai Sea and Yellow Sea regions, in which the arrows represent the movement of water mass and black line is the selected horizontal profile. (**b**) Time series of transport from 18 to 22 October under three forcing conditions: tide alone; combined effects of wind, tide, current, and elevation; and Stokes transport. (**c**) Time series of transport from 26 to 28 August under the same three forcing conditions.

5. Conclusions

In 2024, a pronounced sea level anomaly precipitated a seawater backflow event (on 20–21 October and 25–26 August) in the offshore region of the Bohai Sea, significantly impacting coastal hydrodynamics. The purpose of this study is to investigate the ocean dynamic during two anomalous sea level rise events in the Bohai Sea through the coupled application of two high-resolution numerical models: the WRF atmospheric model and FVCOM, integrated with its spectral wave component, FVCOM-SWAVE. Focusing on two events in 2024, on 20–21 October and on 25–26 August, the framework synergistically integrates atmospheric forcing, ocean circulation, and spectral wave dynamics to elucidate the mechanisms driving coastal hydrodynamic anomalies in the Bohai Sea, China.

The wind fields simulated by the WRF model were validated against near-real-time measurements from the HY-2 scatterometers, yielding an RMSE of 1.88 m/s and a Cor of 0.85. Similarly, HY-2 altimeter data validated the SWH simulations from FVCOM-SWAVE, with an RMSE of 0.67 m and a Cor of 0.84. Both models demonstrated robust performance when evaluated against HY-2 satellite observations. To assess the influence of dynamic drivers on sea level variability, four sensitivity experiments were conducted using the FVCOM-SWAVE model, validated against tide gauge station data. The results indicate that elevation and wind are critical parameters affecting water level rise, while current (mainly imposed boundary current) exhibited negligible impact on sea level dynamics. Tides were confirmed as the fundamental driver of sea level rise in the Bohai Sea, with experiments underscoring their dominant role in variability, supplemented by wind and elevation effects. Notably, simulations of the 20–21 October 2024 and the 25–26 August 2024 events identified elevation forcing as critical, driving localized sea level rises of up to 0.6 m in the northern and southern Bohai Sea and 1.1 m in the southern Bohai Sea. Relationship analysis demonstrated that a 1 m sea level rise in northern Bohai correlated with strong Yellow Sea winds (20 m/s) and SWH of 5 m, which facilitated water influx into the region. Time series analysis revealed these extreme conditions preceded the surge, while currents showed no correlation. Transport analysis confirmed tides as the dominant driver of sea level rise in the Bohai Sea. On 21 October and 26 August, Stokes transport significantly amplified water level increases, highlighting its critical role during extreme events. This study underscores the interplay between tidal forces and wave-driven Stokes transport in shaping coastal hydrodynamics.

A limitation of this study is the lack of climatic statistics on storm surges in the Bohai Sea. In future work, this coupled modeling system will be used to investigate the mechanisms underlying climate change.

Author Contributions: Conceptualization, S.P., Y.J. and W.S.; methodology, S.P. and Y.H.; software, L.L.; validation, S.P., Y.H. and L.L.; formal analysis, S.P. and W.S; investigation, W.S. and J.Z.; resources, Y.J. and W.S.; data curation, J.Z.; writing—original draft preparation, S.P. and L.L.; writing—review and editing, Y.H., J.Z. and W.S.; visualization, S.P. and L.L; supervision, Y.J.; project administration, J.Z.; funding acquisition, W.S. All authors have read and agreed to the published version of the manuscript.

Funding: This research was funded by the National Natural Science Foundation of China under contract, grant number 42376174, and the Natural Science Foundation of Shanghai under contract, grant number 23ZR1426900.

Data Availability Statement: The original contributions presented in this study are included in the article. Further inquiries can be directed to the corresponding author.

Acknowledgments: We appreciate the provision of original code of the Finite-Volume Community Ocean Model (FVCOM) with the Simulating Waves Nearshore (SWAN) module (FVCOM-SWAVE) by the Marine Ecosystem Dynamics Modeling Laboratory (MEDML). The code of Weather Research and Forecasting (WRF) is released by the National Oceanic and Atmospheric Administration. In addition, the National Satellite Ocean Application Service releases the Haiyang-2 (HY-2) product for scientific use. The Copernicus Marine Environment Monitoring Service (CMEMS) dataset is available at http://marine.copernicus.eu (accessed on 3 June 2025).

Conflicts of Interest: The authors declare no conflicts of interest.

References

1. Woodworth, P.L.; Blackman, D.L. Changes in extreme high waters at Liverpool since 1768. *Int. J. Clim.* **2002**, 22, 697–714. [CrossRef]
2. Wang, X.L.; Zwiers, F.W.; Swail, V.R. North Atlantic Ocean wave climate change scenarios for the twenty-first century. *J. Clim.* **2004**, 17, 2368–2383. [CrossRef]
3. Hallegatte, S.; Ranger, N.; Mestre, O.; Dumas, P.; Corfee-Morlot, J.; Herweijer, C.; Wood, R.B. Assessing climate change impacts, sea level rise and storm surge risk in port cities: A case study on Copenhagen. *Clim. Change* **2011**, 104, 113–137. [CrossRef]
4. Yin, J. Rapid decadal acceleration of sea level rise along the U.S. East and Gulf coasts during 2010-22 and its impact on hurricane-induced storm surge. *J. Clim.* **2023**, 36, 4511–4529. [CrossRef]
5. Kleinosky, L.R.; Yarnal, B.; Fisher, A. Vulnerability of hampton roads, virginia to storm-surge flooding and sea-level rise. *Nat. Hazards* **2007**, 40, 43–70. [CrossRef]
6. Joyce, J.; Chang, N.B.; Harji, R.; Ruppert, T.; Singhofen, P. Cascade impact of hurricane movement, storm tidal surge, sea level rise and precipitation variability on flood assessment in a coastal urban watershed. *Clim. Dyn.* **2018**, 51, 383–409. [CrossRef]
7. Yang, Y.C.; Qi, J.P.; Yang, Q.S.; Fan, C.Q.; Zhang, R.; Zhang, J. Reconstruction of wide swath significant wave height from quasi-synchronous observations of multisource satellite sensors. *Earth Space Sci.* **2024**, 11, e2023EA003162. [CrossRef]
8. Woeppelmann, G.; Miguez, B.M.; Bouin, M.N.; Altamimi, Z. Geocentric sea-level trend estimates from gps analyses at relevant tide gauges world-wide. *Glob. Planet. Change* **2007**, 57, 396–406. [CrossRef]
9. Sun, Z.F.; Shao, W.Z.; Yu, W.P.; Li, J. A study of wave-induced effects on sea surface temperature simulations during typhoon events. *J. Mar. Sci. Eng.* **2021**, 9, 622. [CrossRef]
10. Hu, Y.Y.; Shao, W.Z.; Shen, W.; Zuo, J.C.; Jiang, T.; Hu, S. Analysis of sea surface temperature cooling in typhoon events passing the Kuroshio Current. *J. Ocean Univ. China* **2024**, 23, 287–303. [CrossRef]
11. Malardé, J.; Mey, P.D.; Perigaud, C.; Minster, J.F. Observation of long equatorial waves in the Pacific Ocean by Seasat altimetry. *J. Phys. Oceanogr.* **2010**, 17, 2273–2279. [CrossRef]
12. Stoffelen, A.; Verspeek, J.A.; Vogelzang, J.; Verhoef, A. The CMOD7 geophysical model function for ASCAT and ERS wind retrievals. *IEEE J. Sel. Top. Appl. Earth Obs. Remote Sens.* **2017**, 10, 2123–2134. [CrossRef]
13. Wang, J.; Zhang, J.; Yang, J. The validation of HY-2 altimeter measurements of a significant wave height based on buoy data. *Acta Oceanol. Sin.* **2013**, 32, 87–90. [CrossRef]
14. Adler, R.F.; Markus, M.J.; Fenn, D.D.; Szejwach, G.; Shenk, W.E. Thunderstorm top structure observed by aircraft overflights with an infrared radiometer. *J. Appl. Meteorol.* **1983**, 22, 1408. [CrossRef]
15. Fan, S.R.; Kudryavtesv, V.; Zhnag, B.; Perrie, W. Radar scattering features under high wind conditions from spaceborne quad-polarization SAR observations. *IEEE Trans. Geosci. Remote Sens.* **2024**, 62, 4207309. [CrossRef]
16. Wang, Z.X.; Stoffelen, A.; Zou, J.H.; Lin, W.M.; Verhoef, A.; Zhang, Y.; He, Y.J.; Lin, M.S. Validation of new sea surface wind products from scatterometers onboard the HY-2B and Metop-C satellites. *IEEE Trans. Geosci. Remote Sens.* **2020**, 58, 4387–4397. [CrossRef]
17. Armitage, T.W.K.; Davidson, M.W.J. Using the interferometric capabilities of the ESA Cryosat-2 mission to improve the accuracy of sea ice freeboard retrievals. *IEEE Trans. Geosci. Remote Sens.* **2013**, 52, 529–536. [CrossRef]
18. Guo, Y.Y.; He, Y.J.; Li, M.K. Multi-scale wavelet analysis of TOPEX/Poseidon altimeter significant wave height in eastern China seas. *Chin. J. Oceanol. Limnol.* **2006**, 24, 81–86.
19. Yi, J.; Du, Y.; Zhou, C.; Liang, F.; Yuan, M. Automatic identification of oceanic multieddy structures from satellite altimeter datasets. *IEEE J. Sel. Top. Appl. Earth Obs. Remote Sens.* **2015**, 8, 1555–1563. [CrossRef]
20. Ma, X.D.; Zhang, L.; Xu, W.S.; Li, M.L.; Zhou, X.Y. A mesoscale eddy reconstruction method based on generative adversarial networks. *Front. Mar. Sci.* **2024**, 11, 1411779. [CrossRef]
21. Zhong, J.; Fei, J.; Huang, S.; Du, H.; Zhang, L. An improved QuikSCAT wind retrieval algorithm and eye locating for typhoon. *Acta Oceanol. Sin.* **2012**, 31, 41–50. [CrossRef]
22. Chen, G.; Chapron, B.; Ezraty, R.; Vandemark, D. A global view of swell and wind sea climate in the ocean by satellite altimeter and scatterometer. *J. Atmos. Ocean. Technol.* **2002**, 19, 1849–1859. [CrossRef]
23. Luo, Y.X.; Xu, Y.; Qin, H.; Jiang, H.Y. Wavelength cut-off error of spectral density from MTF3 of SWIM instrument onboard CFOSAT: An investigation from buoy data. *Remote Sens.* **2024**, 16, 3092. [CrossRef]

24. Hao, M.Y.; Hu, Y.Y.; Shao, W.Z.; Migliaccio, M.; Jiang, X.W.; Wang, J. Advance in sea surface wind and wave retrieval from synthetic aperture radar image: An overview. *J. Ocean Univ. China* **2025**, *24*, 1–19. [CrossRef]
25. Xie, X.T.; Zhang, J.L.; Zheng, Y.; Chen, K.H. Wind vector retrieval from Gaofen-3 SAR imagery using the method of iterative line fitting in image power spectrum domain. *All Earth* **2024**, *36*, 1–15. [CrossRef]
26. Mouche, A.; Chapron, B. Global C-band Envisat, RADARSAT-2 and Sentinel-1 SAR measurements in copolarization and cross-polarization. *J. Geophys. Res.* **2016**, *120*, 7195–7207. [CrossRef]
27. Shin, H.; Hong, S. Intercomparison of planetary boundary-layer parametrizations in the WRF model for a single day from cases-99. *Bound. Layer Meteor.* **2011**, *139*, 261–281. [CrossRef]
28. Reisner, J.; Rasmussen, R.; Bruintjes, R. Explicit forecasting of supercooled liquid water in winter storms using the MM5 mesoscale model. *Q. J. R. Meteorol. Soc.* **1998**, *124*, 1071–1107. [CrossRef]
29. Pinto, J.; Grim, J.; Steiner, M. Assessment of the High-Resolution Rapid Refresh Model's ability to predict mesoscale convective systems using object-based evaluation. *Weather Forecast.* **2015**, *30*, 892–913. [CrossRef]
30. Sun, M.H.; Duan, Y.H.; Zhu, J.R.; Wu, H.; Zhang, J.; Huang, W. Simulation of Typhoon Muifa using a mesoscale coupled atmosphere-ocean model. *Acta Oceanol. Sin.* **2014**, *33*, 123–133. [CrossRef]
31. Sun, Z.F.; Shao, W.Z.; Wang, W.L.; Zhou, W.; Yu, W.P.; Shen, W. Analysis of wave-induced stokes transport effects on sea surface temperature simulations in the western Pacific Ocean. *J. Mar. Sci. Eng.* **2021**, *9*, 834. [CrossRef]
32. The WAMDI Group. The WAM model-A third generation ocean wave prediction model. *J. Phys. Oceanogr.* **1998**, *18*, 1775–1810.
33. Holthuijsen, L. The continued development of the third-generation shallow water wave model 'SWAN'. *TU Delft Dep. Hydraul. Eng.* **2001**, *32*, 185–186.
34. Yao, F.C.; Johns, W.E. A HYCOM modeling study of the Persian Gulf: 1. Model configurations and surface circulation. *J. Geophys. Res.* **2010**, *115*, C11017. [CrossRef]
35. Chen, C.S.; Liu, H.; Beardsley, R.C. An unstructured, finite-volume, three-dimensional, primitive equation ocean model: Application to coastal ocean and estuaries. *J. Atmos. Ocean. Technol.* **2003**, *20*, 159–186. [CrossRef]
36. Rezende, L.F.; Silva, P.A.; Cirano, M.; Peliz, Á.; Dubert, J. Mean circulation, seasonal cycle, and eddy interactions in the eastern Brazilian margin, a nested ROMS model. *J. Coast. Res.* **2011**, *27*, 329–347.
37. Yao, R.; Shao, W.Z.; Zhang, Y.G.; Wei, M.; Hu, S.; Zuo, J.C. Feasibility of wave simulation in typhoon using WAVEWATCH-III forced by remote-sensed wind. *J. Mar. Sci. Eng.* **2023**, *11*, 2010. [CrossRef]
38. Bae, S.; Hong, S.; Lim, K. Coupling WRF double-moment 6-class microphysics schemes to RRTMG radiation scheme in weather research forecasting model. *Adv. Meteorol.* **2016**, *1*, 5070154. [CrossRef]
39. Hong, S.; Noh, Y.; Dudhia, J. A New vertical diffusion package with an explicit treatment of entrainment processes. *Mon. Weather Rev.* **2006**, *134*, 2318–2341. [CrossRef]
40. Salesky, S.; Chamecki, M. Random errors in turbulence measurements in the atmospheric surface layer: Implications for monin-obukhov similarity theory. *J. Atmos. Sci.* **2012**, *69*, 3700–3714. [CrossRef]
41. Chen, F.; Dudhia, J. Coupling an advanced land surface-hydrology model with the penn state-NCAR MM5 modeling system. Part I: Model implementation and sensitivity. *Mon. Weather Rev.* **2001**, *129*, 569–585. [CrossRef]
42. Shao, W.Z.; Chen, J.L.; Hu, S.; Yang, Y.Q.; Jiang, X.W.; Shen, W.; Li, H. Influence of sea surface waves on numerical modeling of an oil spill: Revisit of symphony wheel accident. *J. Sea Res.* **2024**, *201*, 102529. [CrossRef]
43. Mellor, G.L.; Yamada, T. Development of a turbulence closure model for geophysical fluid problems. *Rev. Geophys.* **1982**, *20*, 851–875. [CrossRef]
44. Qi, J.H.; Chen, C.S.; Beardsley, R.C.; Perrie, W.; Cowles, G.W.; Lai, Z.G. An unstructured-grid finite-volume surface wave model (FVCOM-SWAVE): Implementation, validations and applications. *Ocean Model.* **2009**, *28*, 153–166. [CrossRef]
45. Yao, R.; Shao, W.; Hao, M.; Zuo, J.; Hu, S. The respondence of wave on sea surface temperature in the context of global change. *Remote Sens.* **2023**, *15*, 1948. [CrossRef]
46. Breivik, Ø.; Bidlot, J.R.; Janssen, M. A stokes drift approximation based on the Phillips spectrum. *Ocean Model.* **2016**, *100*, 49–56. [CrossRef]
47. Wang, Y.P.; Liu, Y.L.; Mao, X.Y.; Chi, Y.T.; Jiang, W.S. Long-term variation of storm surge-associated waves in the Bohai Sea. *J. Oceanol. Limnol.* **2019**, *37*, 1868–1878. [CrossRef]
48. Zheng, C.W.; Pan, J.; Tan, Y.K.; Gao, Z.S.; Rui, Z.F.; Chen, C.H. The seasonal variations in the significant wave height and sea surface wind speed of the China's seas. *Acta Oceanol. Sin.* **2015**, *34*, 58–64. [CrossRef]

Journal of
Marine Science and Engineering

MDPI

Article

Wind and Wave Climatic Characteristics and Extreme Parameters in the Bohai Sea

Huayan Zhang [1], Zhifeng Wang [1] and Xin Ma [2,3,*]

[1] College of Engineering, Ocean University of China, Qingdao 266100, China; zhanghuayan@stu.ouc.edu.cn (H.Z.); wzf1984@ouc.edu.cn (Z.W.)
[2] Physical Oceanography Laboratory, Ocean University of China, Qingdao 266100, China
[3] Laboratory for Ocean Dynamics and Climate, Qingdao Marine Science and Technology Center, Qingdao 266100, China
* Correspondence: maxin@ouc.edu.cn

Abstract: The Weather Research and Forecasting (WRF) model is employed to conduct numerical simulations and simulated acquisition of a 30-year (1993–2022) wind field dataset for the Bohai Sea. The simulated WRF wind field is subsequently used to drive the Simulating Waves Nearshore (SWAN) model, producing a corresponding wave field dataset for the same period in the Bohai Sea. Using these datasets, we analyzed the extreme value distributions of wind speed and significant wave height in the study area. The results reveal that both the annual mean wind speed and significant wave height exhibit a ring-like spatial pattern. The highest values are concentrated in the southern Liaodong Bay to the central Bohai Sea region, with a gradual radial decrease toward the periphery. Specifically, values decline from the center outward, from southeast to northwest, and from offshore to nearshore regions. The Gumbel extreme value distribution is applied to estimate 100-year return period extremes, yielding maximum wind speeds of 37 m/s and significant wave heights of 6 m in offshore areas. In nearshore regions, the 100-year return period wind speeds range between 20–25 m/s, while significant wave heights vary from 2 to 3 m. This study provides important scientific basis and decision-making reference for the design of offshore extreme conditions.

Keywords: Bohai Sea; wind velocity; significant wave height; climatic characteristics

1. Introduction

Extensive research has been conducted on the extreme environment of wind and waves, particularly in the context of the global economy and the ocean as a strategic space. The significance of the ocean for international economic development has become increasingly evident [1]. However, with the continuous deepening and expansion of marine development in fisheries, energy, and other fields, the frequency and intensity of marine disasters are also increasing significantly [2–4], which poses a grave threat to the economic stability and ecological security of coastal regions. The statistical analysis of the "China Marine Disaster Bulletin" found that the number of deaths or disappearances caused by marine disasters in China from 2000 to 2015 was as high as 2599, of which the loss caused by wave disasters accounted for 73.7% [5]. Statistical data indicated that the direct economic loss incurred due to marine disasters in China in 2022 amounts to approximately CNY 24 billion [6]. Consequently, there is an imperative need to undertake rigorous research on the extreme environment of wind and waves.

In marine science, a comprehensive understanding of wind–wave characteristics, accurate prediction of extreme events, and assessment of climate change impacts on marine environments are crucial. For wind–wave studies in the Bohai Sea, the Weather Research and Forecasting (WRF) model has been widely employed to simulate wind waves driven by high-resolution wind fields. For example, Luo [7] combined WRF-generated high-resolution wind fields with the Simulating Waves Nearshore (SWAN) model to investigate seasonal spatial distributions of wind and wave characteristics in the Bohai Sea. Similarly, Liu [8] utilized the WRF model for long-term simulations to examine spatiotemporal characteristics of low-level atmospheric ducts in the South China Sea. Furthermore, significant progress has been achieved in optimizing the coupling between WRF and wind–wave models. Wu et al. [9] developed a comprehensive typhoon modeling system by constructing and implementing a coupled WRF–SWAN model, demonstrating its effectiveness in supporting typhoon forecasting. Du et al. [10] systematically examined the impacts of physical parameterization schemes in the WRF model on wind speed prediction accuracy and wind energy resource assessment, subsequently optimizing the model's performance for wind–wave prediction in the Bohai Sea. Chen [11] conducted a numerical hindcast of the most intense tropical cyclone recorded in the eastern South China Sea, quantifying extreme values of wind speed, wave height, current velocity, and water level while analyzing their characteristic patterns. Wang et al. [12] utilized the SWAN wave model to perform a numerical simulation of the wind field and waves in the Bay of Bengal and conducted an analysis of the temporal and spatial distribution characteristics and extreme parameters of wind and waves. Islek et al. [13] utilized the SWAN model to evaluate the long-term changes of wave characteristics in the Black Sea and analyzed the differences of wave characteristics in different regions. The study provided fundamental data for understanding the characteristics of wind waves in specific sea areas and demonstrated the effectiveness of the model in wind wave simulation.

Extreme wind speeds and wave conditions are critical parameters for coastal engineering design. In China's coastal regions, statistical models utilizing meteorological station data have been conventionally employed to determine design wind speed criteria [14]. The Gumbel distribution has proven particularly effective for characterizing extreme wind and wave conditions during severe weather events [15–17]. Through Gumbel distribution fitting, reliable estimates of extreme wind speeds and significant wave heights can be obtained, providing essential data for structural assessment and disaster prevention strategies. However, these methods face limitations in data-scarce regions or areas with poor observational infrastructure, as they fundamentally depend on long-term, high-quality wind speed measurements. Wei [18] demonstrated successful applications in storm wave prediction by forecasting significant wave height, mean wave period, and related parameters, highlighting the practical value of extreme value analysis. Furthermore, the investigation of climate change impacts on marine environments has emerged as a significant research focus in this domain. Lobeto et al. [19] systematically investigated future trends of extreme waves under various climate change scenarios, employing a wave climate simulation ensemble to project changes in extreme significant wave heights across global ocean surfaces. Yuksel [20] conducted comprehensive wind–wave simulations in the Marmara Sea by integrating wind field data with the SWAN model, specifically analyzing extreme wave modifications and associated topographic effects. These studies collectively underscore the critical role of extreme value analysis in assessing climate change impacts on marine environments, while particularly highlighting the methodological significance of Gumbel extreme value theory applications.

The research on wind and wave modeling under specific conditions in the Bohai Sea is limited. This study is not only limited to the application of a single model, but also uses the WRF model and wind and wave models (such as SWAN), which provides a more comprehensive perspective for the simulation of the marine environment and meteorological conditions in the Bohai Sea. The traditional extreme value distribution method was optimized, which improved the accuracy and reliability of wind speed prediction. Through the simulation and prediction of extreme ocean wave and wind speed events, the prediction ability of extreme events in a complex marine environment is improved, which provides strong support for offshore areas to cope with climate change and marine disasters.

2. Data Sources and Research Methods

2.1. Numerical Simulation of Wind Field

WRF is used to simulate the wind field in the study area as shown in Figure 1a. The water depth is white, and the water depth is blue. Tropical cyclone tracks are identified through systematic screening of tropical cyclone data obtained from the China Typhoon Network (http://typhoon.weather.com.cn/, accessed on 27 March 2025), as shown in Figure 1b. To ensure simulation accuracy while maintaining computational efficiency, a two-layer nested grid configuration is implemented in the model set-up [21]. This approach optimally balances numerical precision with computational resource requirements, ensuring both reliable results and practical feasibility. The numerical simulation employs a two-domain nested grid system with horizontal resolutions of 30 km (outer domain) and 10 km (inner domain), maintaining a 3:1 refinement ratio. To ensure numerical stability, the time integration step is set to 50 s for the inner domain and correspondingly 150 s for the outer domain, consistent with the spatial resolution ratio. The computational grids consist of 63 × 75 (outer domain) and 112 × 130 (inner domain) grid points, respectively. Output frequency is configured at 3 h intervals for the outer domain (d01) and 1 h intervals for the inner domain (d02). The model utilizes a Lambert conformal projection with 37 vertical layers extending to a top pressure of 5000 Pa. As shown in Figure 1, the simulation covers geographical ranges of 25° N–46° N, 112° E–134° E (d01) and 30° N–42° N, 116° E–128° E (d02). The physical parameterization schemes, validated through previous simulation analyses and literature review, are summarized in Table 1.

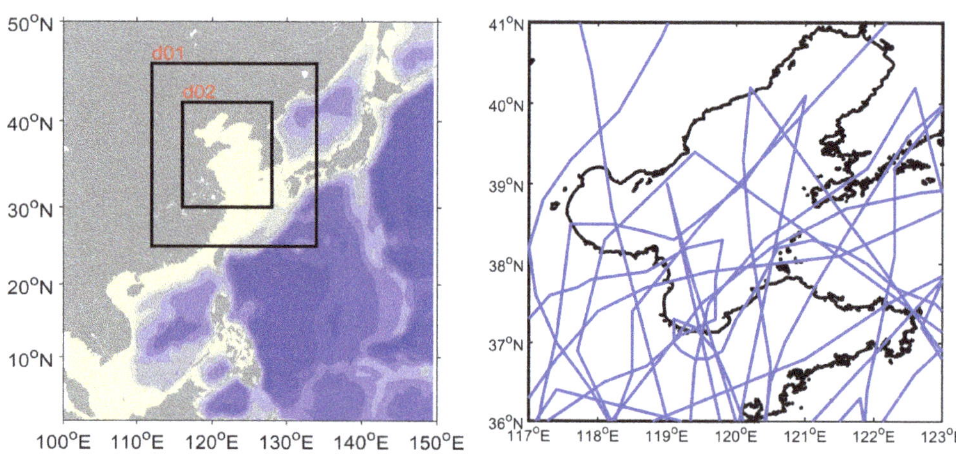

(a) (b)

Figure 1. (**a**) WRF simulation area map; (**b**) typhoon track map within a radius of 400 km at the site.

Table 1. WRF model parameter settings.

Name	Setting
Microphysical scheme	Lin et al. scheme [22]
Cumulus convection scheme	Kain–Fritsch (new eta) scheme
Longwave radiation scheme	rrtmg scheme
Shortwave radiation scheme	rrtmg scheme
Planetary boundary layer scheme	YSU scheme
Near-ground layer scheme	Revised MM5 Monin–Obukhov scheme
Land-surface process scheme	Unified Noah land-surface model scheme

2.2. Numerical Simulation of Wave Field

SWAN (Simulated Waves Nearshore) is the third generation of the nearshore shallow water wave numerical model [23–27]. It was first developed by the Delft University of Technology in the Netherlands and has grown after years of development and improvement. Due to the influence of the flow field, the density of the spectral energy is not conserved, but the action density, the variance density divided by the relative frequency conserved, is only changed with time and space. The equilibrium equation of wave action in a Cartesian rectangular coordinate system is expressed as follows [28]:

$$\frac{\partial N}{\partial t} + \frac{\partial C_x N}{\partial x} + \frac{\partial C_y N}{\partial y} + \frac{\partial C_\sigma N}{\partial \sigma} + \frac{\partial C_\theta N}{\partial \theta} = \frac{S}{\sigma} \tag{1}$$

where σ is the phase frequency, and θ is the phase frequency. C_x and C_y are propagation velocities in the x and y directions; C_σ and C_θ are the propagation velocities in the σ and θ directions. The source term S represents the impacts of the wind generation, dissipation, and non-linear wave–wave interactions. The source term S consists of the following parts [29]:

$$S = S_{in} + S_{nl3} + S_{nl4} + S_{ds,w} + S_{ds,b} + S_{ds,br} \tag{2}$$

where S_{in} represents wind energy input; S_{nl3} and S_{nl4} represent the energy projections of the non-linear interactive waves of three components and four components, respectively; $S_{ds,w}$ represents the white capping dissipation; $S_{ds,b}$ represents the bottom friction; and $S_{ds,br}$ represents the wave breaking due to shallow water.

The wind input term is expressed as follows [30]:

$$S_{in}(\sigma,\theta) = A + BE(\sigma,\theta) \tag{3}$$

where A represents linear growth and BE represents exponential growth [31].

$$S_{n/3}(\sigma,\theta) = S^{-}_{n/3}(\sigma,\theta) + S^{+}_{n/3}(\sigma,\theta) \tag{4}$$

$$S_{n/4}(\sigma,\theta) = S^{*}_{n/4}(\sigma,\theta) + \delta S^{**}_{n/4}(\sigma,\theta) \tag{5}$$

The white cap dissipation term represents the energy loss of waves in deep water, and the degree of energy dissipation is represented by wave steepness. The expression is as follows [31]:

$$S_{ds,w}(\sigma,\theta) = -\Gamma\overline{\sigma}\frac{k}{\overline{k}}E(\sigma,\theta) \tag{6}$$

where $\overline{k}$ and $\overline{\sigma}$ represent the average wave number and the average frequency of waves, respectively.

The formula for bottom friction is expressed as follows [31]:

$$S_{ds,b}(\sigma,\theta) = -C_{bottom}\frac{\sigma^2}{g^2\sinh^2(kd)}E(\sigma,\theta) \quad (7)$$

where C_{bottom} represents bottom friction coefficients.

Battjes and Janssen's formula is mainly used to describe the crushing effect of shallow water in the SWAN model [31]:

$$S_{ds,br(\sigma,\theta)} = \frac{D_{tot}}{E_{tot}}E(\sigma,\theta) = \frac{a_{BJ}Q_b\overline{\sigma}}{\beta^2\pi}(\sigma,\theta) \quad (8)$$

where D_{tot} represents energy dissipation rate per unit horizontal area. Q_b represents wave breaking factor. $\overline{\sigma}$ represents average speed.

In this study, a total of 30 years of wind field data from 1993 to 2022 were simulated by WRF, and the SWAN model is driven by the hindcast to calculate the waves in the Bohai Sea. In all cases, the SWAN model is used for non-stationary two-dimensional models to simulate wave propagation during 1993~2022. The calculation area is consistent with the WRF area, the resolution is 0.25° × 0.25°, the calculation time step is 1 h, and the output time step is 3 min. The physical process in SWAN mode is shown in Table 2. The water depth required for the calculation of the SWAN model uses the water depth data of the ETOPO1 global terrain database. ETOPO1 includes land terrain data and ocean water depth data (https://ngdc.noaa.gov/mgg/global/relief/ETOPO1/tiled/, accessed on 27 March 2025).

Table 2. Physical processes in the SWAN mode.

Physical Processes	SWAN Mode
Wind speed index growth part	Komen et al. [32]
White cap	Komen et al. [32]
Coefficient of friction	Hasselmann et al. [33]
Fragmentation caused by water depth	The ratio of maximum significant wave height to depth gamma = 0.73
Nonlinear wave–wave interactions	E1deberky

2.3. Verification of Wind and Wave Fields

The Bohai Sea experienced an extreme wave event from 10 to 12, November 2012, triggered by the interaction between a cold air mass and a cyclonic system. This rare meteorological–oceanic phenomenon is designated as the "11·10" wave event. In the historical data, the "11·10" wave is not common, and the event is successfully simulated. We can infer that the simulation results of the model under other similar complex conditions are also reliable, so this event is selected for research. The location of the observation station is depicted in Figure 2. The observation station information is shown in Table 3. Utilizing the hourly wind and wave data from 1993 to 2022, derived from WRF simulations and SWAN numerical modeling, the simulated results are validated against available observational data. A comparative analysis between the simulated and measured data is conducted [34], with the results shown in Figure 3 and Table 4. The findings indicate that the simulation demonstrates a high degree of accuracy, with the modeled data exhibiting strong agreement with the observational records. This validation confirms the reliability of the simulated dataset, ensuring its suitability for subsequent extreme value analysis.

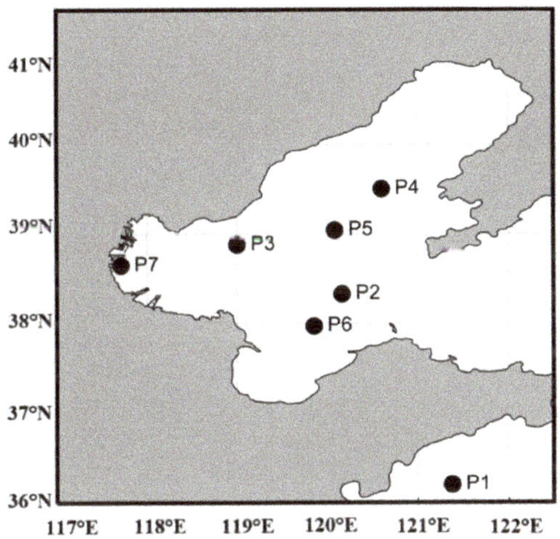

Figure 2. Position of Bohai Sea observation station.

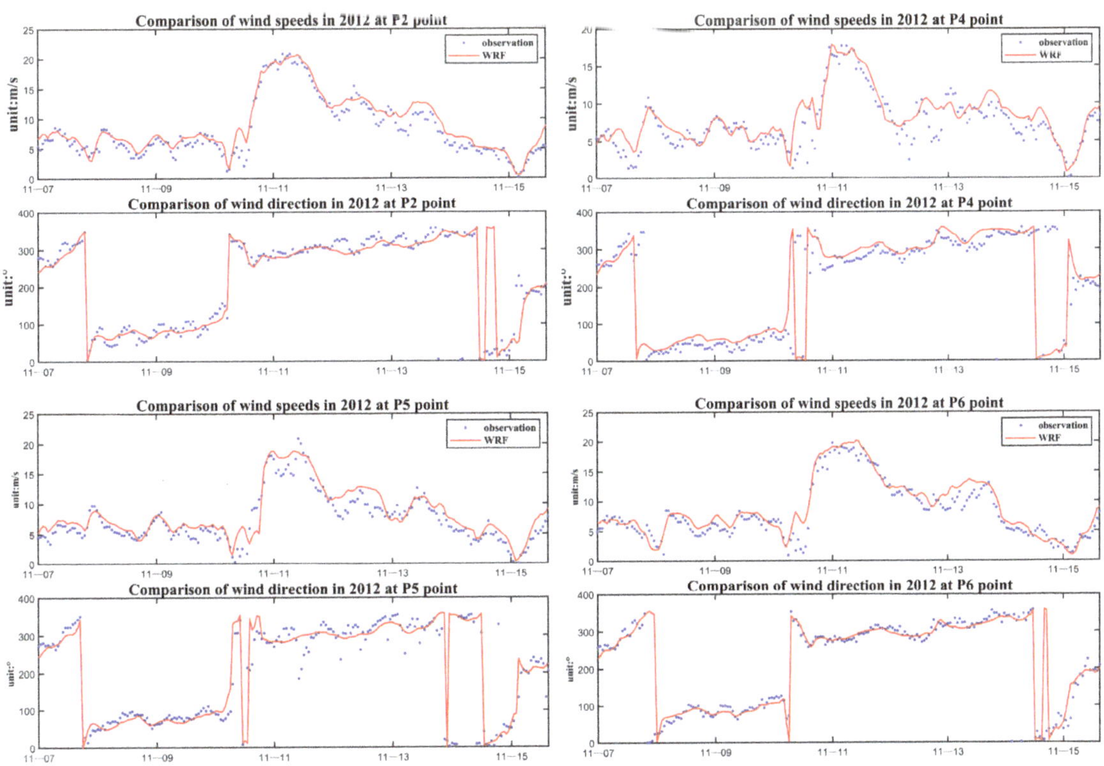

Figure 3. Comparison of wind speed and direction of "11·10" wave in 2012 (red is WRF simulation data, blue points are measured data).

Table 3. Calculation points description.

Station	Longitude	Latitude	Data Available
P1	121°22′48″	36°12′00″	2012.11.01.00–2012.11.15.23
P2	120°09′36″	38°19′48″	2012.11.01.00–2012.11.15.23
P3	119°00′00″	38°52′48″	2012.11.01.00–2012.11.15.23
P4	120°36′00″	39°30′00″	2012.11.01.00–2012.11.15.23
P5	120°04′48″	39°02′24″	2012.11.01.00–2012.11.15.23
P6	119°51′00″	37°58′12″	2012.11.01.00–2012.11.15.23

Table 4. Correlation coefficient between measured data and simulated data.

r	P2	P4	P5	P6
Wind speed	0.972	0.881	0.937	0.958
Wind direction	0.942	0.934	0.887	0.963

3. Analysis of Long-Term Characteristics of Wind Field in Bohai Sea

3.1. Temporal and Spatial Distribution Characteristics

Studying the long-term changes of wind speed and direction is conducive to the development of offshore wind energy and the orientation of wind field location. Using the WRF wind field data from 1993 to 2022, the average wind speed of the Bohai Sea in the past 30 years was analyzed, and the spatial distribution of the annual average wind speed in the past 30 years was obtained, as shown in Figure 4. On the whole, the annual average wind speed shows a decreasing trend from the middle to the surrounding areas and from the far sea to the land. The wind speed peak is above 7 m/s from the south of Liaodong Bay to the center of Bohai Sea. The sea area is greatly affected by the winter monsoon, and the annual average wind field wind direction in most sea areas is northwestward.

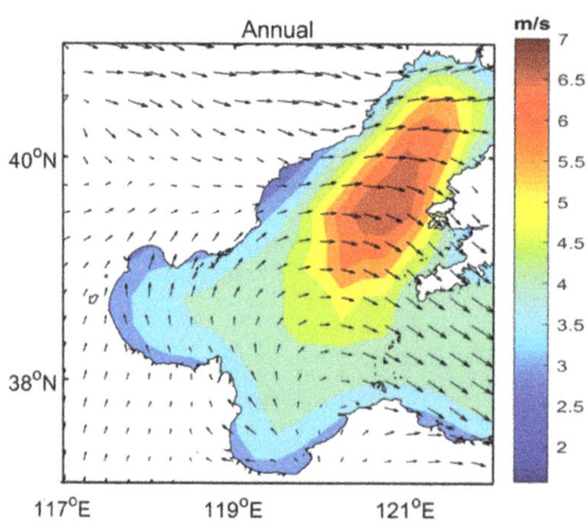

Figure 4. Spatial distribution of average wind speed and wind direction in Bohai Sea. (The arrow is the wind direction).

The wind farm data from the study area over the past 30 years are seasonally averaged to generate the seasonal mean wind field characteristics, as illustrated in Figure 5a. The results indicate a general trend of increasing wind speed from northwest to southeast, from nearshore to offshore, and from the periphery toward the central Bohai Sea. Notable seasonal variations in wind speed distribution are observed, with the highest seasonal mean wind speeds occurring in autumn and winter. In particular, winter exhibits the most pronounced peak wind speeds, exceeding 8 m/s, whereas summer records the lowest peak values, approximately 5 m/s. The wind speed distribution in spring and summer is relatively uniform, with prevailing wind directions predominantly ranging from southeast to southwest. In contrast, autumn and winter exhibit greater spatial variability in wind speed, with prevailing winds predominantly from the northwest.

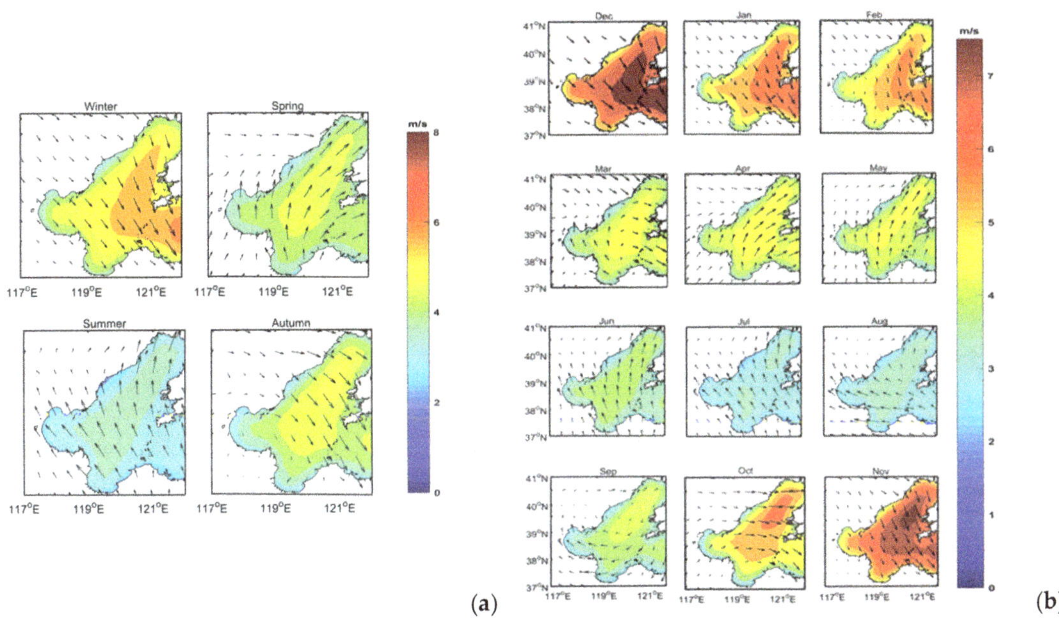

Figure 5. (**a**) The spatial distribution of seasonal average wind speed and wind direction in Bohai Sea in recent 30 years; (**b**) The spatial distribution of annual average wind speed and wind direction in Bohai Sea in recent 30 years. (The arrow is the wind direction).

In order to analyze the wind field characteristics of the Bohai Sea in greater detail, the monthly average processing was carried out in this sea area. Figure 6b is the monthly mean wind field map of the Bohai Sea. The analysis reveals that during the months of March, April, and May, the wind speed in a substantial portion of the Bohai Sea attained speeds of 4 m/s. In the months of November, December, and January, the wind speed in the southeastern region of the Bohai Sea exceeded 7 m/s, particularly in December. The predominant wind direction is from the northwest. Conversely, the monthly average wind speed in June, July, and August is minimal, with the predominant wind direction being southeast. These summer wind field characteristics have the potential to exert a significant influence on the marine ecosystem and the organization of marine fishery production activities. The maximum wind speeds recorded during different months are primarily concentrated in the south of Liaodong Bay and the central region of the Bohai Sea. This phenomenon holds significant value in the selection of offshore wind power project sites and facilitates effective risk assessment and response measures.

In order to further analyze the distribution of wind speed and wind direction in the Bohai Sea, nine representative points were selected, covering Bohai Bay, Liaodong Bay, Laizhou Bay, Bohai Strait, and central Bohai Sea. The specific location is shown in Figure 6a, which can reflect the spatial distribution characteristics of the wind field more comprehensively. Based on the wind speed and wind direction data of 30 years from 1993 to 2022, the wind rose diagram of each feature point is drawn, as shown in Figure 6b.

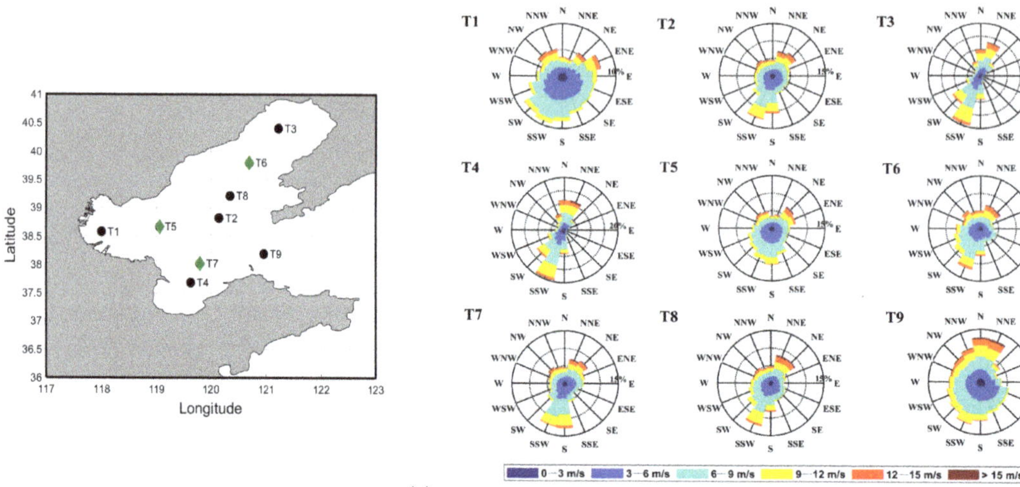

(a) (b)

Figure 6. (**a**) The distribution of stations in Bohai Sea; (**b**) the wind rose diagram (m/s) of 9 characteristic points.

As for Figure 6b, it can be concluded that the wind speed of T5 and T6 is low, and the dominant wind speed range is concentrated in 0–6 m/s. The strong wind direction is ENE. The wind speed in Laizhou Bay is mainly 3–9 m/s, the high wind speed of 6–9 m/s appears locally, and the frequency is slightly lower than that of Bohai Bay. The wind speed of T2 is higher, followed by T8, and the strong wind direction is NE and NNE. The wind speed of T9 is the highest in all regions, and the wind speed frequency and wind direction are highly consistent.

3.2. Long-Term Variation Features

Wind is one of the main driving forces of atmospheric motion. Analyzing the long-term trend of a wind field is helpful to understand the evolution law of the wind field in the Bohai Sea. Taking the Bohai Sea as the research object, four characteristic points T1, T2, T3, and T4 (Figure 7a) are selected to represent the wind field characteristics of Bohai Bay, central Bohai Sea, Liaodong Bay, and Laizhou Bay. This kind of point layout can comprehensively reflect the spatial distribution difference and long-term change trend of the wind field in the Bohai Sea.

As for Figure 7a, a general upward trend in wind speed over the 30-year period obtains, with T3 exhibiting higher speeds compared to T1, T2, and T4. The wind speed fluctuated within 0.5 m/s from 1993 to 1997, reached a 30-year high in 1998, and reached a 30-year low in 2019. Figure 7b illustrates the seasonal average trend in wind speed. The wind speed at the four characteristic points in spring fluctuated from 1993 to 1999, peaked in 2001, fluctuated from 2003 to 2013, and peaked in 2021. A notable peak in wind speed is observed in summer in 2018, marking a 30-year high. The wind speed at the characteristic points in autumn has been subject to fluctuation, reaching a peak in 1998. The winter characteristic points, conversely, reached their peak in 2010 and 2012.

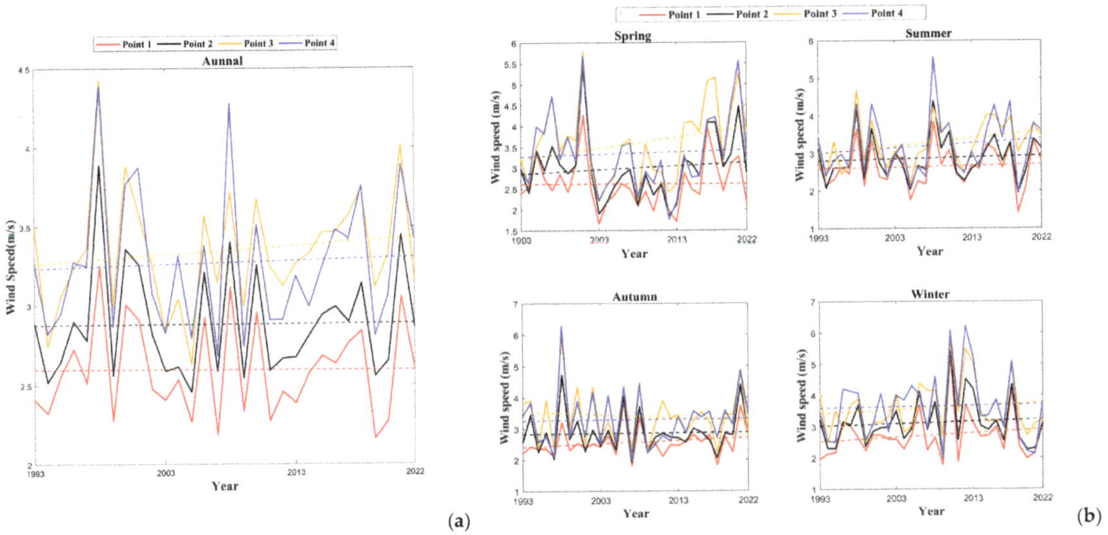

Figure 7. (**a**) Long-term trend in annual wind speed from 1993 to 2022. (**b**) Long-term trend in quarterly wind speed from 1993 to 2022.

The regression coefficients of each feature point are extracted as shown in Table 5. The enhancement of the winter monsoon may lead to the increase in winter wind speed, so the regression coefficient of winter wind speed is generally high, the change in wind speed is relatively stable and slightly increases, and the coefficient of Liaodong Bay is the largest in spring, which may be due to the significant warming enhancement of sea–land breeze circulation. The wind speed regression coefficient T3 in autumn decreased slightly.

Table 5. Linear regression coefficient of annual and seasonal average wind speed change trend.

The Representative Sea Area	Bohai Bay		In the Central Bohai Sea		Liaodong Bay		Laizhou Bay	
Feature Point	**T1**	**T1 (P)**	**T2**	**T2 (P)**	**T3**	**T3 (P)**	**T4**	**T4 (P)**
Annual average	0.0003	0.035	0.0004	0.0246	0.0059	0.0450	0.0025	0.0112
Winter	0.0133	0.0447	0.0074	0.0253	0.0121	0.0410	0.0038	0.0171
Spring	0.0002	0.0290	0.0098	0.0006	0.0265	0.0497	0.0053	0.0007
Summer	0.0044	0.0211	0.0048	0.0034	0.0215	0.0415	0.0129	0.0012
Autumn	0.0111	0.0006	0.0019	0.4031	−0.0104	0.0316	0.0021	0.0001

In order to conduct a more detailed study of the possible wind speed changes between different months and provide key information for the wind speed trend of a specific month, the linear trend in monthly average wind speed is shown in Figure 8 below.

It can be concluded from the Table 6 that in the 12 months, the wind speed is low at most points in spring and summer, while the wind speed is higher in autumn and winter, especially at the T3 and T4 points.

The regression coefficient of each feature point is extracted as shown in Table 2. From the perspective of monthly average, the wind speed change at T1 is relatively stable, and some months (May, June and December) show a downward trend, with regression coefficients of -0.0169 m/s·year^{-1}, -0.0154 m/s·year^{-1}, and -0.0105 m/s·year^{-1}. Other months show an upward trend, with the most obvious upward trend in November. The wind speed at T2 decreased in some months (May, June, and December). On the whole, the variation in wind speed in the central Bohai Sea is relatively mild, but there are obvious seasonal fluctuations. The wind speed variation at T3 attains the maximal degree, presenting remarkable seasonal characteristics throughout the year. The wind speed variation at T4 is relatively stable. However, during the summer and autumn phases, especially in June and December, it shows a significant downward trend.

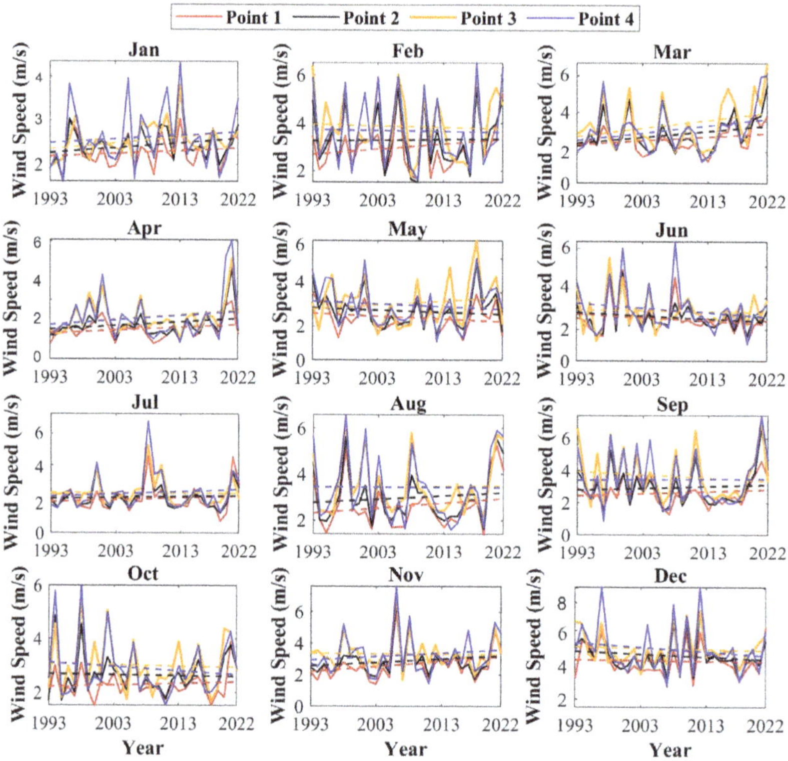

Figure 8. Long-term trend in monthly mean wind speed from 1993 to 2022.

Table 6. Linear regression coefficient of monthly mean wind speed change trend.

The Representative Sea Area	Bohai Bay		In the Central Bohai Sea		Liaodong Bay		Laizhou Bay	
Feature Point	T1	T1 (P)	T2	T2 (P)	T3	T3 (P)	T4	T4 (P)
Jan.	0.0114	0.0366	0.0182	0.0422	0.0218	0.0484	0.0161	0.0091
Feb.	0.018	0.0244	0.0084	0.0374	0.004	0.0428	0.0037	0.0123
March	0.024	0.0103	0.0349	0.0713	0.0352	0.0294	0.0382	0.0441
April	0.0258	0.0446	0.0267	0.015	0.0425	0.063	0.0308	0.0002
May	−0.0169	0.0475	−0.0098	0.0105	0.018	0.0506	−0.0044	0.0019
June	−0.0154	0.0123	−0.012	0.0388	0.0075	0.0439	−0.0158	0.002
July	0.0095	0.0043	0.0105	0.0046	−0.0005	0.0167	0.0211	0.0412
Aug.	0.0297	0.0417	0.0234	0.019	0.018	0.0087	0.0168	0.0174
Sept.	0.0136	0.0002	0.0072	0.0126	−0.0083	0.0414	0.0105	0.0003
Oct.	0.0061	0.0156	0.0047	0.0419	0.0054	0.0157	−0.0041	0.0457
Nov.	0.0333	0.0272	0.0221	0.0332	0.0007	0.04497	0.0237	0.0016
Dec.	−0.0105	0.0395	−0.0294	0.0498	−0.0162	0.0251	−0.0337	0.0434

4. Analysis of Wave Characteristics in Bohai Sea

4.1. Temporal and Spatial Distribution Characteristics of Waves

Based on the wave data of 30 years from 1993 to 2022 calculated by the SWAN model, the spatial distribution of annual average significant wave height in the Bohai Sea has obvious regional characteristics after annual average processing, as shown in Figure 9.

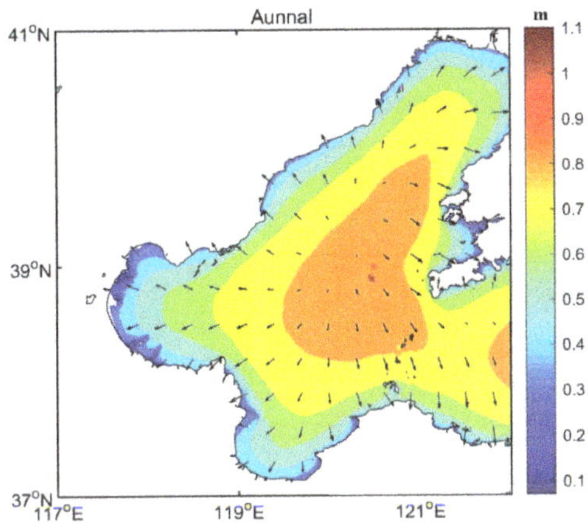

Figure 9. Spatial distribution of annual mean significant wave height and wave direction in Bohai Sea. (The arrow is the wave direction).

Regarding spatial distribution, the regions with high significant wave height are chiefly concentrated in the central sea area of the Bohai Sea, and the significant wave height is about 0.8–0.9 m. This area is far away from the land, the wind speed and wind field are relatively stable, and the wave energy accumulation is strong. Especially in winter and autumn, wind and waves are strong. The regional significant wave height is annularly distributed and decreases from the center to the periphery. In places far away from the coast, the significant wave height gradually decays to 0.2–0.4 m.

In the study area, seasonal averages of wave data spanning 30 years are calculated, yielding average wave field maps for spring, summer, autumn, and winter, as depicted in Figure 10a. The spatial distribution of monthly average significant wave height and wave direction is also shown in Figure 10b.

On the whole, the seasonal average wave distribution is similar to the annual average wave distribution, which is consistent with the trend of decreasing significant wave height from the middle to the surrounding and from the open sea to the near shore, and the maximum significant wave height appears in the central and southeastern open sea areas. Because the winter monsoon is large and lasts for a long time, it provides more energy for seawater and promotes seawater to form higher waves, so the average significant wave height in autumn and winter is higher than that in spring and autumn. In spring, the significant wave height is high, and the maximum value is about 0.8 m. It is mainly concentrated in the center of the Bohai Sea and mainly propagates to the southeast. In summer, the significant wave height decreases due to the decrease in wind force, and the wave direction propagates from southeast to southwest.

The monthly mean significant wave height and wave direction distribution in the Bohai Sea show obvious seasonal changes, which are basically consistent with the seasonal mean wind field distribution characteristics. The maximum significant wave height is concentrated in the central and southeastern offshore areas, which persists throughout the year. The monthly mean wave direction is dominated by the monsoon, and the dominant wave propagation direction in different months is periodically adjusted with the change in the seasonal wind field.

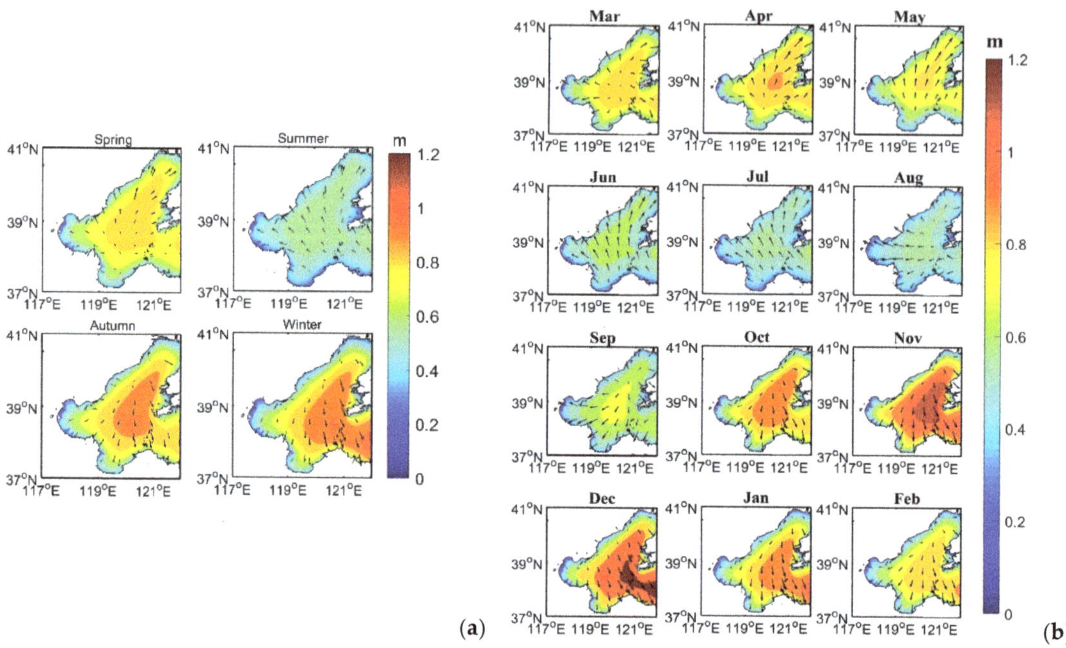

Figure 10. (**a**) Spatial distribution of seasonal mean significant wave height and wave direction in Bohai Sea. (**b**) Spatial distribution of monthly mean significant wave height and wave direction in Bohai Sea. (The arrow is the wave direction).

In this section, the nine aforementioned feature points are selected for analysis of wave distribution characteristics. In addition, the 30-year wave observation data are extracted for detailed analysis of the spatial distribution law and long-term change trend of waves in the sea area. The data extracted are then utilized to create a wave rose diagram (illustrated in Figure 11) at each feature point. This diagram offers a visual representation of the distribution characteristics of significant wave height and wave direction in the Bohai Sea over a 30-year period.

The wave directions of the Bohai Sea are mainly concentrated in the NE~ESE and SSW~W, and the wave directions of T1 and T5 are mainly concentrated in the SSW and ESE. The dominant wave directions of T4 and T7 are NE and ENE; the dominant wind waves of T3 and T6 are NE and ENE; the dominant wind waves of T2 and T8 are ENE; the dominant wind wave of T9 is W. Except for T1 and T9, which are characteristic points near the coast, the frequency of NE~NSE and SW~W in the remaining points during the 30-year statistical period is approximately the same. The strong and sub-strong wave directions of the feature points are concentrated in the SSW~W and NE~ENE.

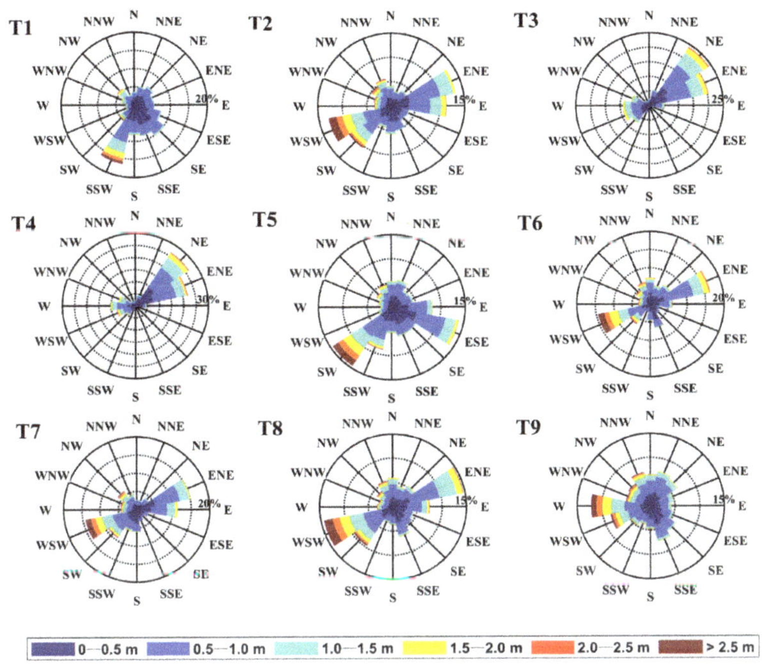

Figure 11. Wave rose diagram of 9 feature points.

4.2. Long-Term Variation Characteristics of Waves

The study of four characteristic points (T1, T2, T3, T4) in the Bohai Sea (Figure 6a) reveals an upward trend in significant wave height from 1993 to 2022 (Figure 12a). The overall trend is characterized by an upward trend in volatility. The T1 significant wave height exhibited notable fluctuations, particularly in 2007 and 2011. In contrast, the T2 significant wave height exhibited a relatively stable trend, with an approximate measurement of 1 m and a marginal upward trajectory. A notable increase in significant wave height was observed in 2017 and 2011. The variation trend in T3 wave height is analogous to that of T2, but its significant wave height is marginally higher. T4 also exhibited a significant peak, particularly in 2007 and 2011, and demonstrated a discernible upward trend over the long term.

As for Figure 12b, the long-term trend in the mean value of significant wave height varies significantly between different seasons. The four feature points demonstrate a gradual upward trend. The winter period is characterized by significant fluctuations. In contrast, the significant wave height in spring maintains relative stability, though T2 and T4 demonstrate notable fluctuations during spring 2013, with an upward trend observed. The T1 significant wave height fluctuates significantly during summer, particularly in 1996, when the peak value approaches 1.5 m. The T2 significant wave height also exhibits substantial fluctuations, with a marked upward trend, especially during the peak period of 2011. T3 displays a substantial peak in summer 2012, with a significant wave height nearing 2.5 m, and exhibits significant overall fluctuations. T4, meanwhile, exhibits significant variability, ranging from 0.5 to 2.5 m. In contrast, the fluctuation in significant wave height in autumn is minimal, and the overall trend is stable.

The regression coefficient of significant wave height change at each feature point (representing the annual significant wave height change) is extracted as shown in Table 7. In terms of the annual average, the inter-annual variability in significant wave height across the four sea areas is not consistent. T1 demonstrates a minor downward trend in significant wave height, while T2, T3, and T4 exhibit a slight upward trend.

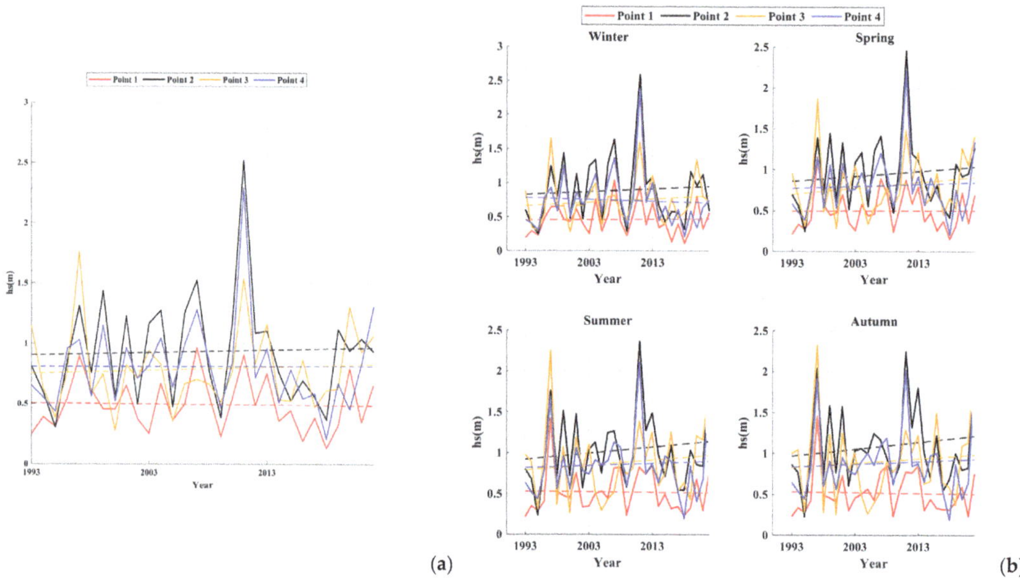

Figure 12. (**a**) Long-term trend in annual mean significant wave height from 1993 to 2022. (**b**) Long-term trend in seasonal mean significant wave height from 1993 to 2022.

Table 7. Linear regression coefficients of variation trend in annual and seasonal mean significant wave height.

The Representative Sea Area	Bohai Bay		In the Central Bohai Sea		Liaodong Bay		Laizhou Bay	
Feature Point	**T1**	**T1 (P)**	**T2**	**T2 (P)**	**T3**	**T3 (P)**	**T4**	**T4 (P)**
Annual average	0.0012	0.0326	0.0026	0.0412	0.0023	0.0333	0.0013	0.0267
Winter	0.0006	0.0465	0.0041	0.0148	0.0051	0.0335	−0.0552	0.0389
Spring	0.0002	0.0212	0.006	0.0453	0.007	0.0335	0.2012	0.0395
Summer	−0.0006	0.0028	0.0076	0.0427	0.0062	0.0008	0.0455	0.0063
Autumn	−0.001	0.0427	0.0084	0.0453	0.0053	0.0454	0.1145	0.0039

From the perspective of seasonal changes, in summer, T1 showed a downward trend, which may be related to the weakening of the summer monsoon. Other sea areas still maintained growth, with the largest increase in the central Bohai Sea. The significant wave height of T3 increases fastest in winter, which may be related to the frequent occurrence of cold waves and strong wind in winter. The increase in the middle side of the Bohai Sea in autumn reached the peak of the whole year, which may be related to the enhancement of typhoon activity in autumn.

As for Figure 13, the months of January and February are characterized by a predominance of northwesterly winds, accompanied by relatively stable significant wave heights. The spring and autumn months (April–May and September–December) correspond to the seasonal transition period. The months of September and December undergo significant fluctuations, with a notable rise in significant wave height. From June to August, the significant wave height increases slightly due to the influence of typhoon tracks. The reclamation area along the coast of Laizhou Bay has increased from 2000 to 2025, leading to an anomalous decrease in significant wave height in September. The study reveals that the change in the wave field in the Bohai Sea is regulated by air–sea interaction, topographic forcing, and human activities, which has important guiding value for ship navigation safety and coastal engineering design.

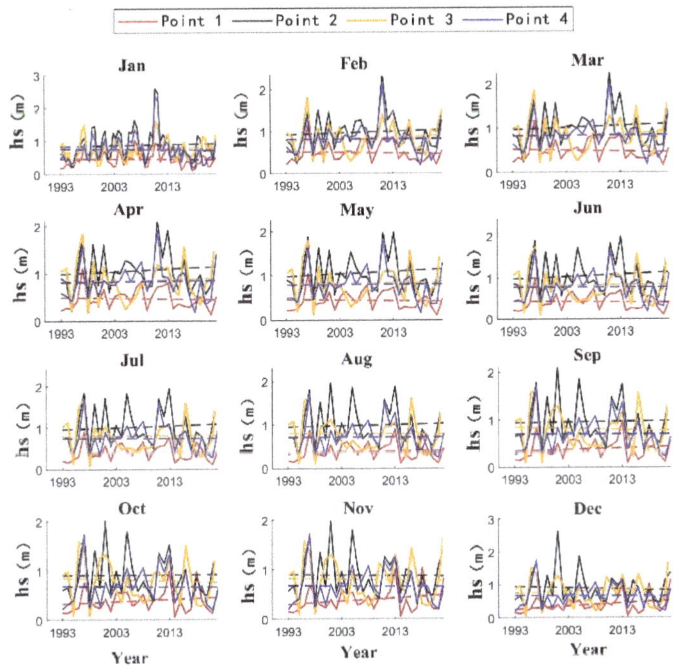

Figure 13. Monthly mean long-term trend in significant wave height from 1993 to 2022.

In order to conduct a more detailed study of the possible changes in significant wave height between different months and provide key information for the wind speed trend in a specific month, the linear change trend in monthly average significant wave height is shown in Figure 13.

As for Table 8, except that T1 showed a downward trend in February, March, April, and May, the significant wave height in other months showed an upward trend. The overall performance of T2 shows an increasing trend, and the regression coefficients in most months are positive, indicating that the significant wave height in the sea area is on the rise for a long time, especially in February–May and June–August. The change trend of T3 is more complex, with an upward trend in some months (January~July), but a downward trend in August~October, and a rebound in winter (November~December). The trend of the T4 significant wave height is stable.

Table 8. Average wind speed change trend linear regression coefficient.

The Representative Sea Area	Bohai Bay		In the Central Bohai Sea		Liaodong Bay		Laizhou Bay	
Feature Point	T1	T1 (P)	T2	T2 (P)	T3	T3 (P)	T4	T4 (P)
Jan.	0.0006	0.0372	0.0013	0.0085	0.0027	0.0409	0.0444	0.0211
Feb.	−0.0018	0.0108	0.0042	0.0417	0.0048	0.0465	0.0261	0.0447
March	−0.0021	0.0407	0.0046	0.0423	0.0038	0.0391	0.0788	0.0338
April	−0.0017	0.0193	0.0051	0.0192	0.0029	0.0402	0.0352	0.0344

Table 8. *Cont.*

The Representative Sea Area	Bohai Bay		In the Central Bohai Sea		Liaodong Bay		Laizhou Bay	
Feature Point	T1	T1 (P)	T2	T2 (P)	T3	T3 (P)	T4	T4 (P)
May	−0.0014	0.0414	0.0059	0.0417	0.0022	0.0323	−0.079	0.0454
June	0.0009	0.0333	0.0058	0.0494	0.0017	0.0417	0.0006	0.0353
July	0.0001	0.0465	0.0044	0.0485	0.0009	0.034	−0.0276	0.0251
Aug.	0.0016	0.0369	0.0027	0.0074	−0.0006	0.0476	0.0136	0.0209
Sept.	0.0031	0.0304	0.0014	0.0367	−0.0023	0.0296	−0.0006	0.0255
Oct.	0.0045	0.0348	0.0004	0.0471	−0.0031	0.0492	0.0296	0.0365
Nov.	0.0064	0.0222	−0.0011	0.0161	−0.0018	0.0372	0.0981	0.0383
Dec.	0.0095	0.0456	−0.002	0.0412	0.005	0.0359	0.1278	0.0271

5. Analysis of Gumbel Extremum Characteristics

5.1. Introduction of Gumbel Extreme Value

The Gumbel distribution is the extreme value distribution of the parameters estimated by the moment method. The probability density function of the Gumbel distribution is as follows [12]:

$$f(x) = A\exp\{-A(x-B) - \exp[-A(x-B)]\} \tag{9}$$

In the formula, A and B are undetermined parameters, which are generally obtained by the moment method [12]:

$$A = \frac{\pi}{\sqrt{6}S} \tag{10}$$

$$B = \overline{X} - 0.450053S \tag{11}$$

In the formula, 0.450053 is the Euler constant, S is the standard deviation of the sample, and x is the mean of the sample.

The Gumbel distribution is widely used in marine hydrography, especially for maximum wind speed and significant wave height derivation. The Gumbel distribution has a good fit when the data samples are large enough, and the fit of the Gumbel distribution is low when the data samples are missing or relatively small.

5.2. Distribution Characteristics of Annual Extreme Values of Wind Field

In this section, the WRF model is used to study the 30-year wind field in the Bohai Sea. The annual extreme value method is used to extract the annual maximum wind speed in the Bohai Sea for 30 years as the extreme wind speed data of the corresponding year. Subsequently, based on the extreme value type I distribution (Gumbel distribution) model, the return period wind speed in the study area is calculated [35–37]. Through statistical analysis, the return period wind speeds of 100-year, 50-year, 25-year, 10-year, five-year, and two-year return periods are calculated, and on this basis, the spatial distribution maps of wind speed fields in different return periods in the Bohai Sea are drawn in Figure 14.

As for Figure 14, the maximum wind speed value once in a year in the Bohai Sea mainly occurs in the central waters. The central waters are far away from land and are less blocked by terrain, which allows the wind to fully develop; in the western and northern waters, the wind speed during the recurrence period is relatively low, mostly concentrated between 20 m/s and 30 m/s [38]. As the recurrence period shortens, the wind speed gradually decreases, and the spatial distribution maintains the characteristics of "high in the middle and low in the surrounding areas". These areas may be affected by local climatic conditions, such as monsoon direction, ocean circulation, etc., making the extreme wind speed value relatively low. The extreme wind speed in the Laizhou Bay waters is more prominent, especially in extreme weather events. The terrain of Laizhou Bay is relatively closed and may be strongly affected by specific meteorological conditions (such as cold air moving south, typhoons, etc.), resulting in a significant increase in the extreme wind speed.

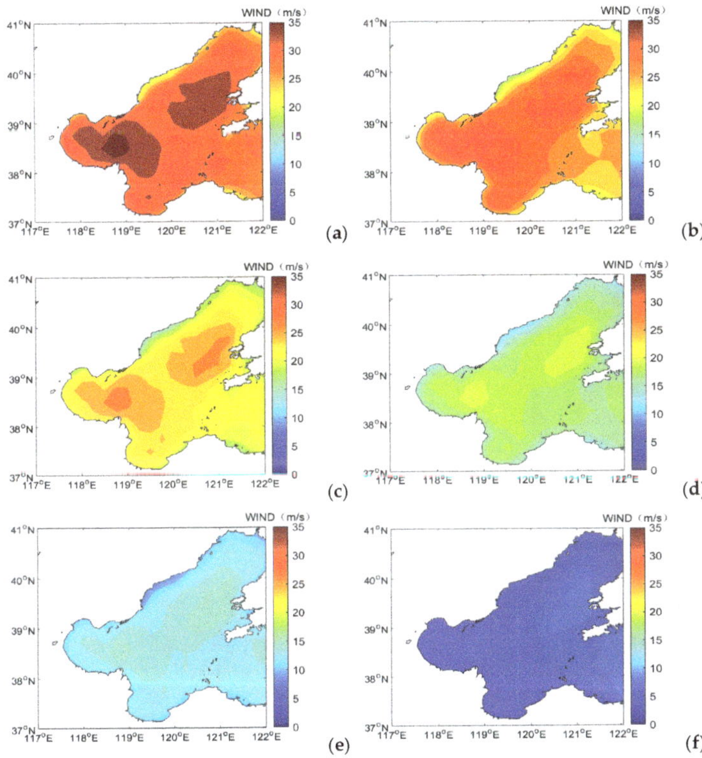

Figure 14. Spatial distribution of wind speed field in different return periods in Bohai Sea. (**a**) Once in 100 years; (**b**) once in 50 years; (**c**) once in 25 years; (**d**) once in 10 years; (**e**) once in 5 years; (**f**) once in 2 years.

As for Table 9, the extreme value of wind speed in Laizhou Bay is more prominent, especially in extreme weather events, which should be taken more seriously. Because of the differences in wind speed extremes in different regions, it is important to strengthen the prediction and preventive measures of wind speed extremes in Bohai Bay, especially in the extreme return period (100 years, 50 years). It is also important to strengthen the wind safety design of infrastructure. At the same time, Bohai Bay, central Bohai Sea, and Liaodong Bay still need to create emergency response measures in the case of a five-year and a two-year return period.

Table 9. Extreme values of wind speed at different return periods at characteristic points.

The Representative Sea Area	Bohai Bay	In the Central Bohai Sea	Liaodong Bay	Laizhou Bay
Feature Point	**T1**	**T2**	**T3**	**T4**
Once in 100 years	32.02	32.05	30.24	33.21
Once in 50 years	27.73	28.00	26.41	28.88
Once in 25 years	23.40	23.91	22.56	24.52
Once in 10 years	17.58	18.40	17.36	18.65
Once in 5 years	12.96	14.04	13.25	14.00
Once in 2 years	6.00	7.46	7.04	6.97

5.3. Distribution Characteristics of Wave Extreme Values

The wind field of the WRF model is utilized as the input wind field of the SWAN model, with the objective of forecasting the 30-year waves in the Bohai Sea. The maximum value of the significant wave height in the Bohai Sea over the past 30 years (1993–2022) is extracted as the annual extreme significant wave height. The Gumbel distribution model is then employed to calculate and analyze the significant wave height of the 100-year, 50-year, 20-year, and five-year return periods in the study area [39–42]. The wave height distribution characteristics of different return periods are obtained, and the corresponding wave field distribution map of the Bohai Sea is drawn, as shown in Figure 15.

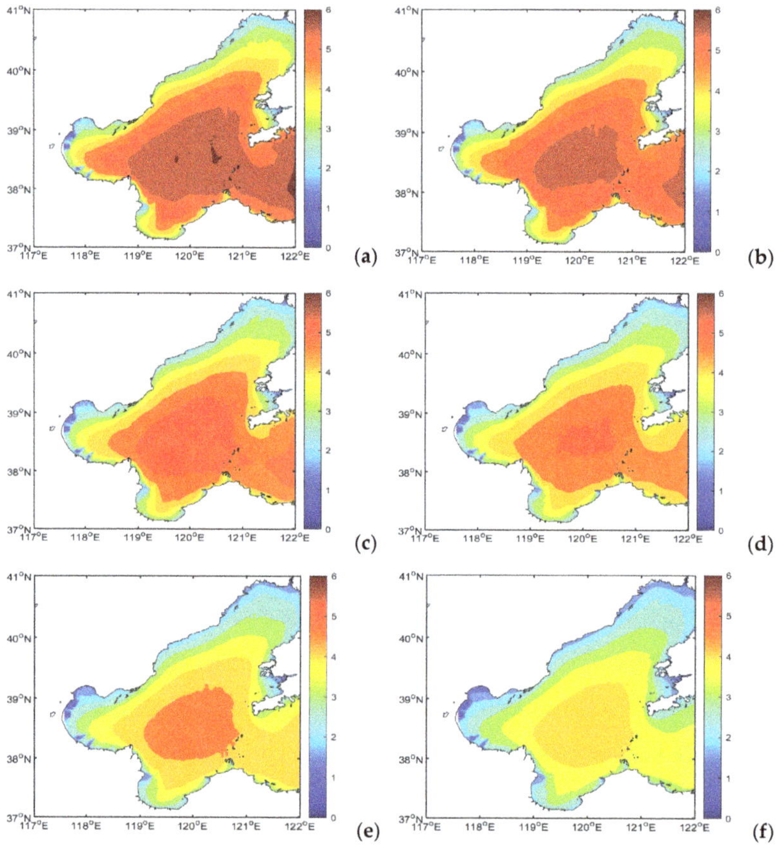

Figure 15. The spatial distribution of significant wave height of four characteristic points in different return periods in Bohai Sea. (**a**) Once in 100 years; (**b**) once in 50 years; (**c**) once in 25 years; (**d**) once in 10 years; (**e**) once in 5 years; (**f**) once in 2 years.

Figure 15 shows that the maximum significant wave height for 100-year and 50-year return periods is approximately 6 m and 5.5 m, respectively, in the central and southeastern sea areas. These regions are particularly susceptible to wind-driven processes, and the substantial water depth in these areas fosters the formation and development of waves. The maximum significant wave heights of the five-year and 10-year return periods are recorded as 5.2 m and 5.0 m, respectively. It is acknowledged that these areas may be subject to influence from local wind and wave conditions, including, but not limited to, variations in wind direction and wind size. The T2 sea area may be distinguished by distinctive seabed topography or hydrological conditions, including variations in water depth and the presence of seabed obstacles. These factors may promote the formation of waves and energy concentration, resulting in higher wave extremes. The extreme value of the T2 wave is significantly higher than that of other regions.

Table 10 shows the wave extreme values of T1, T2, T3, and T4 for different return periods. By comparison, the spatial distribution characteristics of wave extreme values in different regions and the variation law with the shortening of the return period can be observed. The wave extreme value of T2 is significantly higher than that of other regions [43]. Even at lower return periods, the wave extreme value of T2 is still 4.21 m·year^{-1}, showing that it has a significant wave advantage. The wave extreme values of T1 and T3 are relatively low and close together, and the wave extreme value of T4 is higher than that of Bohai Bay and Liaodong Bay. In the case of the lower return period, the extreme significant wave height in Laizhou Bay reaches 3.80 m annually, showing relatively strong wave characteristics.

Table 10. Wave extreme values of different return periods at characteristic points.

The Representative Sea Area	Bohai Bay	In the Central Bohai Sea	Liaodong Bay	Laizhou Bay
Feature Point	T1	T2	T3	T4
Once in 100 years	3.87	5.85	3.62	5.33
Once in 50 years	3.65	5.58	3.46	5.08
Once in 25 years	3.36	5.22	3.24	4.74
Once in 10 years	3.13	4.94	3.08	4.48
Once in 5 years	2.89	4.65	2.91	4.21
Once in 2 years	2.54	4.21	2.65	3.80

6. Conclusions

By comparing the simulated data with the measured data in order to verify the accuracy of the model, the following conclusions are drawn:

(1) With regard to the annual average wind speed, the wind speed in the Bohai Sea shows obvious seasonal differences. The wind speed in autumn and winter is greater than that in spring and summer, which is mainly due to the frequent southward movement of cold air and the influence of monsoon. In a similar fashion, the significant wave height in the Bohai Sea displays comparable seasonal characteristics. The annual mean significant wave height exhibits a gradual decrease from the middle to the periphery, from the southeast to the northwest, and from the far sea to the land.

(2) Further analysis of the wind direction data indicates that the wind direction of the Bohai Sea exhibits a significant direction. The prevalence of strong winds and sub-strong winds is observed to be concentrated in the N~ENE direction, while the normal wind direction and sub-normal wind direction are concentrated in the S~SSW direction. This distribution of wind direction exerts a direct influence on the direction of wave generation and propagation.

(3) The long-term change trend indicates that the wind speed and significant wave height in the Bohai Sea show an upward trend as a whole. This conclusion is substantiated through comprehensive analysis of wind and wave field data spanning from 1993 to 2022. This upward trend is significant on the whole, which may be related to global climate change, marine environmental warming, and other factors.

(4) The spatial distribution of wind speed and significant wave height in different return periods is consistent with the extreme wind speed and significant wave height. The maximum wind speeds of the 100-, 50-, 25-, and 10-year return periods are primarily

located in the T4 sea area, while the maximum wind speeds of the five-year and two-year return periods are situated in the T2 sea area. A similar spatial distribution of extreme significant wave height values is also evident. The maximum significant wave height that occurs once every 100 years is higher than that observed in the current T2 sea area, and the significant wave height in the nearshore area is comparatively low.

This study not only provides detailed data support for the wind and wave characteristics in the Bohai Sea but also provides an important scientific basis and decision-making reference for the design of offshore extreme conditions.

Author Contributions: H.Z.: Investigation, conceptualization, software simulation, validation, writing—original draft preparation, data processing, data visualization, data curation; Z.W.: Investigation, conceptualization, software simulation, review & editing; X.M.: Investigation, conceptualization, software simulation, review & editing. All authors have read and agreed to the published version of the manuscript.

Funding: This study was supported financially by the National Key Research and Development Program of China (2022YFC3104205), the National Natural Science Foundation of China (42377457), and the fundamental research funds for the central universities (202262005).

Data Availability Statement: The original contributions presented in this study are included in the article. Further inquiries can be directed to the corresponding author.

Acknowledgments: We would like to thank the Marine Big Data Center of the Institute for Advanced Ocean Study of the Ocean University of China for providing data conversion and storage, as well as the professional and technical services provided by Yujie Dong.

Conflicts of Interest: The authors declare no conflict of interest.

References

1. Wang, Y.; Jin, W.B.; Gao, Q. Research Progress on the Relationship between Marine Disasters and Marine Economy. *Ocean. Dev. Manag.* **2016**, *33*, 88–91+100. (In Chinese)
2. Wang, H. Analysis of Spatial-Temporal Characteristics of Wavesand Prediction of Significant Wave Height in the South China Sea. Master's Thesis, Guangdong Ocean University, Zhanjiang, China, 2021. (In Chinese)
3. Qin, N.N.; Zhong, W.; Zhao, J.Q. Statistical Characteristics of Tropical Cyclones in the South China Sea from 1949 to 2017. *Meteorol. Res. Appl.* **2019**, *40*, 1–4+17. (In Chinese)
4. Patlakas, P.; Stathopoulos, C.; Tsalis, C.; Kallos, G. Wind and wave extremes associated with tropical-like cyclones in the Mediterranean basin. *Int. J. Climatol.* **2021**, *41*, E1623–E1644. [CrossRef]
5. Tao, A.F.; Shen, Z.C.; Li, S. Research progrecs for disastrous waves in China. *Sci. Technol. Rev.* **2018**, *36*, 26–34. (In Chinese)
6. Cai, F.F. Research on the Impact of Group-Occurring Marine Disasters on the Sustainable Development of Marine Economy in China. Master's Thesis, Shandong University of Finance and Economics, Jinan, China, 2023. (In Chinese)
7. Luo, H. Research on Numberical Simulationwave in Bohai Bay by SWAN Model. Ph.D. Thesis, Tianjin University, Tianjin, China, 2012. (In Chinese)
8. Liu, Q. Research on the Spatio–Temporal Characteristics of Lower Atmospheric Ducts Across the China Seas Based on the Long–Term Simulations Using the WRF Mode. Master's Thesis, Qilu University of Technology, Jinan, China, 2024. (In Chinese)
9. Wu, Z.Y.; Jiang, C.B.; He, Z.Y.; Chen, J. Coupled atmosphere and wave model and its application in an idealized typhoon. *J. Mar. Sci.* **2019**, *37*, 9–15. (In Chinese)
10. Du, M.J.; Yi, K.; Wen, R.Q.; Zhang, Z.L.; Wang, H. Study on the applicability of different combination of physical parameterization schemes of WRF model in assessing offshore wind energy resources of China. *Mar. Forecast.* **2023**, *40*, 65–78. (In Chinese)
11. Qiao, F.; Chen, S.; Li, C.; Zhao, W.C.; Pan, Z. The study of wind, wave, current extreme parameters and climatic characters of the South China Sea. *Mar. Technol. Soc. J.* **1999**, *33*, 61–68. [CrossRef]
12. Wang, Z.; Yu, M.; Dong, S.; Wu, K.; Gong, Y. Wind and wave climate characteristics and extreme parameters in the Bay of Bengal. *Reg. Stud. Mar. Sci.* **2020**, *39*, 101403. [CrossRef]
13. Islek, F.; Yuksel, Y.; Sahin, C.; Guner, H.A. Long-term analysis of extreme wave characteristics based on the SWAN hindcasts over the Black Sea using two different wind fields. *Dyn. Atmos. Ocean.* **2021**, *94*, 101165. [CrossRef]
14. Fang, G.S. Estimation and mapping of extreme wind speeds in southeast coastal region of China. *State Key Lab. Disaster Reduct. Civ. Eng.* **2021**, *54*, 43–53. (In Chinese)

15. Soukissian, T.H.; Tsalis, C. The effect of the generalized extreme value distribution parameter estimation methods in extreme wind speed prediction. *Nat. Hazards* **2015**, *78*, 1777–1809. [CrossRef]
16. Katalinić, M.; Parunov, J. Comprehensive wind and wave statistics and extreme values for design and analysis of marine structures in the Adriatic Sea. *J. Mar. Sci. Eng.* **2021**, *9*, 522. [CrossRef]
17. Davison, A.C.; Huser, R. Statistics of extremes. *Annu. Rev. Stat. Its Appl.* **2015**, *2*, 203–235. [CrossRef]
18. Wei, Z. Forecasting wind waves in the US Atlantic Coast using an artificial neural network model: Towards an AI-based storm forecast system. *Ocean Eng.* **2021**, *237*, 109646. [CrossRef]
19. Lobeto, H.; Menendez, M.; CLosada, I.J. Future behaviour of wind wave extremes due to climate change. *Sci. Rep.* **2021**, *11*, 7869. [CrossRef]
20. Yuksel, Y.; Yuksel, Z.T.; Islek, F.; Sahin, C.; CGuner, H.A. Spatiotemporal long-term trends of wind and wave climate and extreme characteristics over the Sea of Marmara. *Ocean Eng.* **2021**, *228*, 108946. [CrossRef]
21. Qiao, W.; Song, J.; He, H.; Li, F. Application of different wind field models and wave boundary layer model to typhoon waves numerical simulation in WAVEWATCH III model. *Tellus. A Dyn. Meteorol. Oceanogr.* **2019**, *71*, 1657552. [CrossRef]
22. Hong, S.-Y.; Jeong-Ock, J.L. The WRF single-moment 6-class microphysics scheme (WSM6). *Asia-Pac. J. Atmos. Sci.* **2006**, *42*, 129–151.
23. Nikishova, A.; Kalyuzhnaya, A.; Boukhanovsky, A.; Hoekstra, A. Uncertainty quantification and sensitivity analysis applied to the wind wave model SWAN. *Environ. Model. Softw.* **2017**, *95*, 344–357. [CrossRef]
24. Nayak, S.; Bhaskaran, P.K.; Venkatesan, R.; Dasgupta, S. Modulation of local wind–waves at Kalpakkam from remote forcing effects of Southern Ocean swells. *Ocean Eng.* **2013**, *64*, 23–35. [CrossRef]
25. Wu, W.; Li, P.; Zhai, F.; Gu, Y.; Liu, Z. Evaluation of different wind resources in simulating wave height for the Bohai, Yellow, and East China Seas (BYES) with SWAN model. *Cont. Shelf Res.* **2020**, *207*, 104217. [CrossRef]
26. Lv, X.; Yuan, D.; Ma, X.; Tao, J. Wave characteristics analysis in Bohai Sea based on ECMWF wind field. *Ocean Eng.* **2014**, *91*, 159–171. [CrossRef]
27. Rusu, L.; Bernardino, M.; Guedes Soares, C. Wind and wave modelling in the Black Sea. *J. Oper. Oceanogr.* **2014**, *7*, 5–20. [CrossRef]
28. Shi, A.S. Numerical Simulation and Statistial Analysis of Typhoon Waves Along Jiangsu Province. Master's Thesis, Yangzhou University, Yangzhou, China, 2023. (In Chinese)
29. Yin, C. Tropical Cyclone Induced Storm Wave Hazard Assessmentin Hainan Island, China. Ph.D. Thesis, University of Chinese Academy of Science, Beijing, China, 2020. (In Chinese)
30. Fu, Y.C. Wave and Wave Energy Analysis of Characteristic Island Reefs in the South China Sea Based on SWAN Refined Grid. Master's Thesis, Ludong University, Yantai, China, 2022. (In Chinese)
31. Scala, P.; Manno, G.; Ingrassia, E.; Ciraolo, G. Combining Conv-LSTM and wind-wave data for enhanced sea wave forecasting in the Mediterranean Sea. *Ocean. Eng.* **2025**, *326*, 120917. [CrossRef]
32. Komen, G.J.; Hasselmann, S.; Hasselmann, K. On the existence of a fully developed wind-sea spectrum. *J. Phys. Oceanogr.* **1984**, *14*, 1271–1285. [CrossRef]
33. Hasselmann, K.; Barnett, T.P.; Bouws, E.; Carlson, H.; Cartwright, D.E.; Enke, K.; Ewing, J.A.; Gienapp, H.; Hasselmann, D.E.; Kruseman, P.; et al. *Measurements of Wind-Wave Growth and Swell Decay During The Joint North Sea Wave Project (JONSWAP)*; Deutsches Hydrographisches Institut: Hamburg, Germany, 1973; p. 95.
34. Xiao, H. Numerical Simulation of Typhoon Waves in Taiwan Strait Based on SWAN model. Master's Thesis, Fuzhou University, Fuzhou, China, 2022. (In Chinese)
35. Kang, D.; Ko, K.; Huh, J. Determination of extreme wind values using the Gumbel distribution. *Energy* **2015**, *86*, 51–58. [CrossRef]
36. Pinheiro, E.C.; Ferrari, S.L. A comparative review of generalizations of the Gumbel extreme value distribution with an application to wind speed data. *J. Stat. Comput. Simul.* **2016**, *86*, 2241–2261. [CrossRef]
37. El-Shanshoury, G.I.; Ramadan, A.A. Estimation of extreme value analysis of wind speed in the North-Western coast of Egypt. *Arab. J. Nucl. Sci. Appl.* **2012**, *45*, 265–274.
38. Dai, Y.; Qiao, L.; Xu, J.; Zhou, C.; Ding, D.; Bi, W. Estimation of extreme marine hydrodynamic variables in western Laizhou Bay. *J. Ocean Univ. China* **2015**, *14*, 425–432. [CrossRef]
39. Shao, Z.; Liang, B.; Li, H.; Lee, D. Study of sampling methods for assessment of extreme significant wave heights in the South China Sea. *Ocean Eng.* **2018**, *168*, 173–184. [CrossRef]
40. Teena, N.V.; Sanil Kumar, V.; Sudheesh, K.; Sajeev, R. Statistical analysis on extreme wave height. *Nat. Hazards* **2012**, *64*, 223–236. [CrossRef]
41. Wang, Y.X.; Liang, B.C.; Shao, Z.X. Calculation of extreme wave height in southern Yellow Sea. *J. Hohai Univ. (Nat. Sci.)* **2023**, *51*, 117–122. (In Chinese)

42. O'Grady, J.G.; Hemer, M.A.; McInnes, K.L.; Trenham, C.E.; Stephenson, A.G. Projected incremental changes to extreme wind-driven wave heights for the twenty-first century. *Sci. Rep.* **2021**, *11*, 8826. [CrossRef] [PubMed]
43. Wang, Z.F.; Wu, K.J.; Zhou, L.M.; Wu, L.Y. Wave characteristics and extreme parameters in the Bohai Sea. *China Ocean Eng.* **2012**, *26*, 341–350. [CrossRef]

Journal of
*Marine Science
and Engineering*

MDPI

Article

Variation of Wyrtki Jets Influenced by Indo-Pacific Ocean–Atmosphere Interactions

Qingfeng Feng, Jiajie Zhou, Guoqing Han and Juncheng Xie *

Marine Science and Technology College, Zhejiang Ocean University, Zhoushan 316000, China;
fengqingfeng@zjou.edu.cn (Q.F.); zhou_jiajie2022@zjou.edu.cn (J.Z.); hanguoqing@zjou.edu.cn (G.H.)
* Correspondence: xiejc@zjou.edu.cn

Abstract: As important components of the equatorial current system in the Indian Ocean, Wyrtki jets (WJs) play a significant role in distributing heat and matter in the East and West Indian Oceans. By dividing the El Niño-Southern Oscillation (ENSO) and Indian Ocean Dipole (IOD) events into several phases, we find that the spring branch exhibits positive (negative) anomalies during the El Niño (La Niña) decaying phase, while the fall branch exhibits negative (positive) anomalies during the El Niño (La Niña) developing phase. The spring and fall branches are characterized by negative (positive) anomalies under the influence of positive (negative) dipole events, and these anomalies are particularly pronounced during fall. This study systematically analyzes the characteristics of WJs under the interactions between the Indo-Pacific ocean and the atmosphere, based on the phase-locking characteristics of ENSO, and reveals the regulatory mechanisms underlying their different response patterns.

Keywords: Wyrtki jets; ENSO; IOD; ocean–atmosphere interaction

1. Introduction

The Indian Ocean is the third largest ocean in the world, and its unique geographical features make it the most significant monsoon region among the world's oceans located north of 10° S. Due to the dominant role of the Asian monsoon system, the Indian Ocean has formed a unique and diverse local climate and monsoon ocean circulation system [1,2]. The South Asian monsoon significantly impacts the process of ocean dynamics in this region. During winter, this region is controlled by the northeast monsoon, while during summer, the strong southwest monsoon dominates the dynamic changes in the ocean.

Spring (April–May) and fall (October–November) in the Northern Hemisphere serve as the transition periods of the monsoon system. During these periods, semi-annual zonal jets called "Indian Ocean equatorial jets" or "Wyrtki jets" (WJs) usually appear on the surface of the equatorial Indian Ocean. These jets have a narrow north–south range, move in an eastward direction, and are divided into spring (boreal spring) and autumn branches (boreal fall) according to the seasonal characteristics [3–6]. According to Wyrtki (1973) [3], these jets are driven by the westerly wind over the equator and mainly flow through the area between 60° E and 90° E. Since their discovery, many scholars have studied the spatiotemporal distribution characteristics and formation mechanism of WJs through observations and numerical simulations [7,8].

The formation of WJs is a direct response to the westerly wind burst in the equatorial Indian Ocean. The maximum intensity of these jets is usually located on the sea surface between 75° E and 80° E. The maximum surface velocity can exceed 1 m/s and gradually decreases with increasing depth [3,7]. Han et al. (1999) [9] identified wind drive as the main driver behind the formation of WJs. Using data on drifting buoys, Qiu et al. (2009) [10] found that WJs have an average climatological velocity of 0.50 m/s and 0.70 m/s in spring and fall, respectively, and show obvious seasonal changes that are usually manifested as the strong fall branch and the weak spring branch [11,12]. However, in certain years, there is also a phenomenon of the weak fall branch and the strong spring branch. Duan et al. (2016) [13] pointed out that the intensity of the spring branch in 2013 exceeded that of the fall branch mainly due to atmospheric intraseasonal oscillation (ISO). The intraseasonal signals of the WJs can regulate the seasonal and interannual variations of the WJs themselves, as well as larger-scale ocean–atmosphere interaction events across different timescales [14].

WJs play an important role in the dynamic changes of the upper equatorial Indian Ocean. During the eastward transport process, the sea surface height of the tropical eastern Indian Ocean increases while that of the western Indian Ocean decreases [2,9,10]. During the monsoon transition, WJs affect local wind changes along the coasts of Sumatra and Java and the Indonesian Throughflow by regulating sea surface height variations in the equatorial eastern Indian Ocean and generating eastward-propagating Kelvin waves [15]. Cao et al. (2024) [16] reveals that the intraseasonal variabilities of the WJs are stronger in spring than in fall, influencing sea-level anomalies along the southern coast of Sumatra-Java through interactions with surface wind forcing and equatorial Kelvin waves. The surface salinity of the equatorial Indian Ocean shows significant seasonal and interannual changes, and studies have shown that these changes are mainly controlled by advection transport, especially WJs, which significantly affects the salinity changes throughout the ocean [17–21]. WJs transport warm and salty upper seawater from the western equatorial Indian Ocean to the east, thereby modifying the stratification in the eastern and western Indian Oceans. This eastward transport also affects the zonal distribution of salinity and energy in the tropical Indian Ocean and changes the thermohaline structure of the upper Indian Ocean [2,22,23]. WJs affect sea level through the process of mass transport and further affect the thermocline depth along the equator [12]. These variations in the ocean's mass balance facilitate the development of sea surface temperature anomalies (SSTAs) by altering upwelling [22].

El Niño-Southern Oscillation (ENSO) is the most significant interannual variation signal in the tropical Pacific climate system. ENSO cold and warm events trigger significant anomalies in the tropical Pacific sea surface temperature (SST) and abnormal changes in ocean circulation, which have important effects on the Pacific Ocean and even on the global climate [13,24–26]. Compared with the Pacific Ocean, the tropical Indian Ocean has a smaller sea temperature variability but has equally significant zonal anomaly changes. The tropical Indian Ocean also affects the Asian climate through its interactions with the atmosphere. The Indian Ocean Dipole (IOD) is one of the dominant modes of interannual variability of SST in the tropical Indian Ocean and is the most significant factor affecting such variability besides ENSO. IOD events usually demonstrate a seesaw pattern in the temperature anomalies of the eastern and western basins of the tropical Indian Ocean. The average SSTAs in the 10° S–10° N, 50°–70° E, and 10° S–0°, 90°–110° E sea areas show inverse phase changes [27]. IOD events mainly affect the equatorial region and have significant seasonal phase-locking characteristics. Specifically, these events usually start to develop in the northern hemisphere during summer, reach a mature phase during fall, and then decline rapidly during winter [28]. Large-scale ocean circulation is mainly driven by wind or uneven density distribution. Wind stress anomalies are key factors that lead to changes in ocean circulation and trigger circulation anomalies by changing the horizontal pressure gradient within the ocean.

Previous studies have extensively examined the influence of IOD-related SSTAs on zonal winds and currents near the equator [29–36]. Similarly, the impact of ENSO on surface winds, waves, and currents in the Indian Ocean has been well established [37–39]. These studies highlight the significant role of both ENSO and IOD in driving the interannual variability of WJs [12,40–43]. Chu et al. (2023) [44] found that WJs exhibit the strongest correlation with ENSO or IOD during the same period based on lag correlation analysis. Through the Bjerknes positive feedback mechanism, ENSO events are progressively amplified, thereby strengthening the Walker circulation and affecting the winds in the equatorial Indian Ocean [45–49]. These processes, which connect surface winds in the Indian Ocean with ENSO, not only have important impacts on the formation and development of IOD but also change the interannual characteristics of WJs to some extent. At the same time, the changes in IOD events are fed back to the interannual variability of WJs. Wind-driven changes in the equatorial circulation and wind-modulated Kelvin and Rossby waves influence WJs to modulate the onset and recession processes of the IOD, thus making WJs a key element in the IOD process [12]. The interactions between the Indo-Pacific ocean and the atmosphere play a key role in this process; apart from regulating the interactions between ENSO and IOD, this ocean–atmosphere interaction also shapes the interannual variability patterns of WJs to some extent. Therefore, the feedback mechanisms of these two climate events on WJs warrant further investigation.

We examine ENSO events by dividing them into developing and decaying phases and screening for years with decaying and developing phases corresponding to the spring and fall branches, respectively. We also screen the years of IOD events in the spring and fall branches to comprehensively explore the interannual characteristics of WJs. We use multiple linear regression to strip the event impact of the observed features of the composite analysis and evaluate the specific contribution of these events to WJs. The purpose of this study is to systematically discuss the interannual characteristics of WJs and their change mechanisms to supplement the literature, to deepen our present understanding of the Indian Ocean current system, and to provide new perspectives and rationales for exploring the interactions of the Indian Ocean with the atmosphere.

The rest of this paper is organized as follows. Section 2 describes the data used in this study, while Section 3 outlines the methods employed. Section 4 analyzes the interannual variability of WJs and quantifies the relative contributions of ENSO and IOD to WJs, and explores the regulatory mechanism of this anomalous phenomenon. Section 5 summarizes and discusses the main conclusions of the study.

2. Data

To explore the interannual characteristics of WJs, the latest Simple Ocean Data Assimilation version 3.15.2 dataset was used in this study, with zonal current, wind stress (where τ_x and τ_y refers to the zonal and meridional wind stress, respectively), and sea temperature acting as the variables. SODA3 is often used in exploring low-frequency large-scale zonal currents [50–52]. The oceanic model in SODA3.15.2 switches to GFDL MOM5/Sea Ice Simulator 1 with a $1/2^\circ \times 1/2^\circ$ horizontal resolution and 50-layer vertical resolution. We computed for the monthly average temporal resolution of the data, which cover the years 1980 to 2023 [53].

To ensure consistency in the dynamic process, we used the forcing field of SODA3—the fifth generation of the European Center for Medium-Range Weather Forecasts atmospheric reanalysis dataset (ERA5)—as our wind field data. ERA5 reanalysis data provide grid products with a spatial resolution of $0.25^\circ \times 0.25^\circ$ and the atmospheric parameters of 37 atmospheric layers. We mainly used 10 m sea surface wind field data from 1980 to 2023 and computed for their monthly average temporal resolution. ERA5 benefits from the latest

Integrated Forecast System Cy41r2 and integrates the advancements in model physics, core dynamics, and data assimilation over the past decade [54].

We also used zonal current data from the Acoustic Doppler Current Profiler (ADCP) mounted on the moored buoy located at the equator (0° N, 80.5° E) within the Research Moored Array for African-Asia Australian Monsoon Analysis and Prediction (RAMA) [55]. This site has a data coverage depth of 30 m to 365 m and uses an upward-looking ADCP with an interval of 5 m and a total of 68 layers to cover the period from November 2004 to July 2019. The other site is located at 0° N, 90° E, with a data coverage depth ranging from 30 m to 400 m at 10 m intervals, consisting of a total of 38 layers, covering the period from November 2000 to November 2017. We compared the measured zonal current obtained by ADCP with that of SODA3.15.2 (Figures S1 and S2) to verify the feasibility of using these data.

3. Methods

The correlation coefficient is commonly used to measure the strength of the relationship between two variables. The formula for calculating correlation is as follows:

$$R = \frac{\sum_1^n (x_i - \overline{x})(y_i - \overline{y})}{\sqrt{\sum_1^n (x_i - \overline{x})^2 (y_i - \overline{y})^2}} \tag{1}$$

$$t = R\sqrt{\frac{n-2}{1-R^2}} \tag{2}$$

Here, R represents the correlation coefficient between the two variables x and y, n denotes the length of the variables, $\overline{x}$ and $\overline{y}$ are the mean values of the two variables, and t is the test statistic.

Given the strong interdependence between ENSO and IOD, a simple regression or composite analysis cannot effectively distinguish their independent effects [56–58]. Therefore, we used multiple regression analysis methods (including partial correlation and partial regression) for analysis. Partial correlation analysis helps separate the correlation between different predictors and the research object, thereby allowing us to clearly identify the independent relationship between the research object and each predictor [59]. The partial correlation of interdependence can be formulated as follows:

$$r_{12,3} = \frac{r_{12} - r_{13}r_{23}}{\sqrt{(1 - r_{13}^2)(1 - r_{23}^2)}} \tag{3}$$

where $r_{12,3}$ refers to the partial correlation coefficient between the two variables A1 and A2 after removing the influence of variable A3.

The partial regression coefficient is used to measure the relative contribution of each predictor after removing the linear effects of other predictors. To intuitively show the relative contribution of the predictor, we multiplied the partial regression coefficient by one standard deviation of the predictor to scale it and to compare the anomaly size driven by the typical change in the predictor [56,58]. In this study, we regarded ONI and DMI as predictors and performed multivariate regression analysis to separate their effects on zonal current and wind stress anomalies and to quantify their relative contributions. The concurrent index reconstruction method of the two climate models is as follows:

$$V(\mathrm{m}) = a \times \mathrm{ONI(m)} + b \times \mathrm{DMI(m)} + \mathrm{Res} \tag{4}$$

where V refers to zonal current (U) and wind stress anomalies (τ_x and τ_y), a denotes the partial regression coefficient of ONI after removing the DMI-related signals $(ONI|_{DMI})$, b denotes the partial regression coefficient of DMI after removing the ONI-related signals

($DMI|_{ONI}$), and m refers to month (where spring lasts from April to June, while fall lasts from October to December).

In this study, the F-test is used to examine whether the trend in event frequency changes is significant. In multiple regression analysis, the F-test is applied to determine whether the difference between the residual sum of squares and the regression sum of squares is significant. The specific calculation formula is as follows:

$$F = \frac{\sum_{i=1}^{n}\left(y_i - \widetilde{y}_i\right)^2/k}{\sum_{i=1}^{n}\left(\widetilde{y}_i - \overline{y}\right)^2/(n-k-1)} \sim F(k, n-k-1) \quad (5)$$

Here, $\widetilde{y}_i$ represents the fitted regression value, y_i is the observed value for year i, and $\overline{y}$ is the average value over n years. At a significance level of a, the test follows an F-distribution with degrees of freedom k and $n-k-1$. When $F > F_a$, the null hypothesis is rejected.

When calculating the monthly zonal current and wind stress anomalies, we also calculate the climate mean state for the entire period according to the annual cycle based on the monthly data from 1980 to 2023. The anomalies are first obtained by subtracting the climate mean state of the corresponding month from the monthly data at each grid point, and then the linear trend is removed, allowing us to extract the anomalies of the interannual changes. We use two main statistical methods for significance testing, namely, the F- and T-tests. Additionally, we use Python (version 3.10.9) for analysis and visualization.

4. Results

4.1. Spatial Distribution and Seasonal Variation Characteristics of WJs

Based on the definition of WJs in the equatorial Indian Ocean from 60° E to 90° E, 2° S to 2° N as discussed in previous studies [10,60], we regionally averaged the zonal current in the upper 15 m of the ocean in this region to further analyze the occurrence period of WJs. Figure 1 shows the distribution of zonal current from 1980 to 2023, with the horizontal axis representing the year and the vertical axis representing the month. From Figure 1a, it can be observed that WJs start to appear during the monsoon transition period (April–May and October–November) each year and may last until June and December in certain years (Figure 1a). Observations of their meridional distribution show that WJs reach their maximum strength between 70° E and 80° E before gradually weakening as they move eastward (Figure 1b,c). Additionally, We considered these persistent WJs while analyzing the characteristics of the spring and fall branches (Figure 1a) to further understand their interannual characteristics.

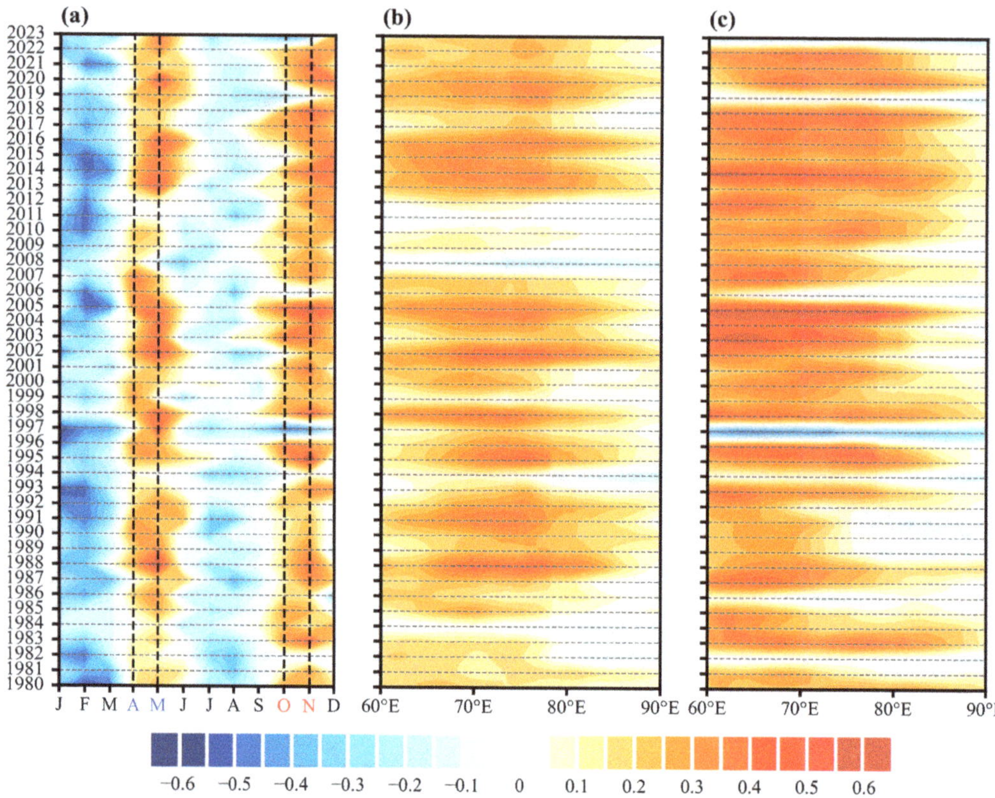

Figure 1. Spatiotemporal distribution of zonal currents in the upper layer of the equatorial Indian Ocean (shading; units: m/s). (**a**) The temporal evolution of zonal currents; and (**b**,**c**) the zonal current (upper 15 m) distribution after meridional averaging (from 2° N to 2° S) in spring (April–May) and fall (October–November), respectively.

4.2. Selection of ENSO and IOD Events

We calculated the Ocean Niño Index (ONI) and Dipole Mode Index (DMI) using sea temperature from the Simple Ocean Data Assimilation version 3.15.2 (SODA3.15.2) dataset (Figures 2 and S3). We defined the year in which the ENSO event occurred as the developing year and the following year as the decaying year [61]. We also adopted the ENSO phase division method proposed by Warner and Moum (2019) [62]. The ENSO developing years include El Niño warming and La Niña cooling, while the ENSO decaying years include El Niño cooling and La Niña warming.

These transitions include the transition to the fully perturbed (or peak) state of El Niño/La Niña and the relaxations back to the neutral state. We marked the transitions of these four phases with star-shaped dots (Figure 2a). When El Niño (or La Niña) reaches its peak, the inflection point of the gradient marks the phase transition. The months prior to this point are considered El Niño warming (or La Niña cooling), while the following months are considered El Niño cooling (or La Niña warming). Given that some ENSO events end in the same year or last for two to three years, we counted El Niño warming (or La Niña cooling) as the development phase of ENSO events and count El Niño cooling (or La Niña warming) as the decay phase (Figure 2a). Following this rule, we selected those years corresponding to the spring and fall branches in the ENSO decay and development phases, respectively (Table 1).

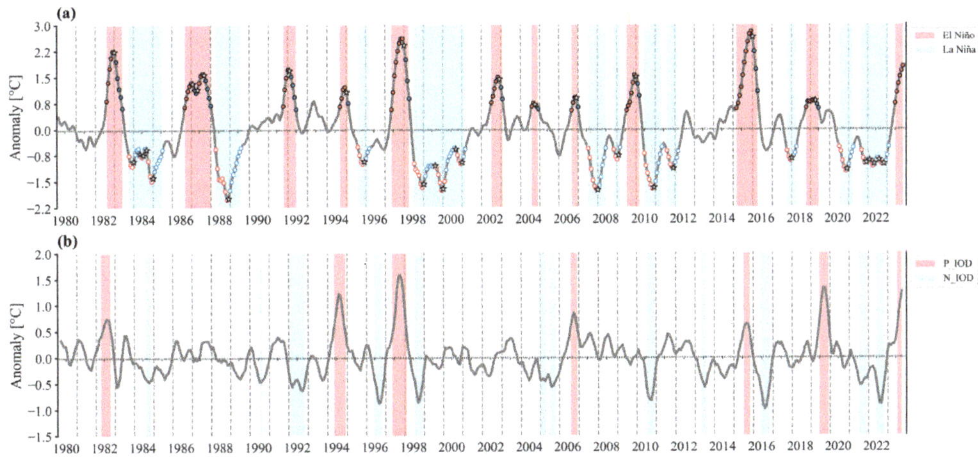

Figure 2. (**a**) ONI, where the shaded area indicates that the SSTAs exceeds ±0.5 °C for more than five consecutive months, the light red shade represents the El Niño event, and the light blue shade represents the La Niña event. The red dots correspond to the developing phase of the ENSO events (El Niño warming or La Niña cooling), the blue dots correspond to the decaying phase of the ENSO events (El Niño cooling or La Niña warming), and the white star-shaped dots represent the phase transition of the event. (**b**) DMI, where the light red and light blue shades represent pIOD and nIOD events, respectively, and the judgment criterion is that the SSTAs exceeds ±0.4 °C for more than three consecutive months.

Table 1. Years of the selected IOD and ENSO events (1980–2023).

Phase		Spring Wyrtki Jet	Fall Wyrtki Jet
ENSO developing	El Niño warming	–	1982, 1986, 1991, 1994, 1997, 2002, 2006, 2009, 2015, 2018, 2023
	La Niña cooling	–	1983, 1984, 1988, 1995, 1998, 1999, 2000, 2007, 2008, 2010, 2011, 2017, 2020, 2021, 2022
ENSO decaying	El Niño cooling	1983, 1987, 1992, 1998, 2016, 2019	–
	La Niña warming	1984, 1985, 1989, 1999, 2000, 2008, 2011, 2018, 2021, 2022	–
IOD	Positive IOD	1982, 1994, 1997	1982, 1994, 1997, 2006, 2015, 2019, 2023
	Neagtive IOD	1992, 2005, 2013, 2016	1984, 1992, 1996, 1998, 2005, 2010, 2016, 2021, 2022

We also selected those years corresponding to the spring and fall branches following the DMI phase definition. IOD shows seasonal phase-locking characteristics, that is, IOD events usually start to develop in June, reach their peak in October, and quickly disappear thereafter. We observed that in some years, positive IOD (pIOD) developed during spring, while negative IOD (nIOD) developed during spring and summer, before quickly disappearing (Figure 2b). We counted these phenomena in the years of the spring branch. Meanwhile, for IOD occurring in summer and fall, we screened out those years that coexist with the fall branch. Table 1 shows that the spring branch only appears in

ENSO decaying phases, while the fall branch only appears in ENSO developing phases. IOD primarily occurs in the fall, resulting in fewer years being statistically accounted for in the spring branch compared to the fall branch.

4.3. General Description of Interannual WJ Variations

To highlight the common characteristics of WJs under the influence of interannual events, we conducted a composite analysis based on the selected years shown in Table 1. Given that WJs mainly occur during the monsoon transition period, the spring branch appears from the northeast to the southwest monsoon during the transition period (Figures 3a,b and 4a,b). During the El Niño decaying phase, significant northeasterly wind stress anomalies are observed in the western Arabian Sea and the Bay of Bengal, while significant northwesterly wind stress anomalies are observed from 45° E to 100° E in the southern Indian Ocean. During this period, the monsoon region is abnormally weakened compared to its climatological state. At this time, the spring branch shows positive anomalies, and the significant areas are mainly located between 60° E and 80° E. During the La Niña decaying phase, the spring branch exhibits negative anomalies between 70° E and 100° E. At this time, significant southwesterly wind stress anomalies are observed over the Arabian Sea and the Bay of Bengal, while the southwestern Indian Ocean, between 40° E and 70° E, experiences notable southeasterly wind stress anomalies. Induced by the wind stress anomalies, the spring branch shows significant positive (negative) anomalies in the west (east) of the equatorial Indian Ocean in El Niño (La Niña) years. During the pIOD event, significant easterly wind stress anomalies occur near Sri Lanka, leading to negative anomalies in the spring branch. Meanwhile, during the nIOD event, significant northwesterly wind stress anomalies are observed near Sumatra, resulting in pronounced positive anomalies in the spring branch, mainly between 70° E and 100° E. The anomalies of the spring branch are more significant under nIOD than under pIOD.

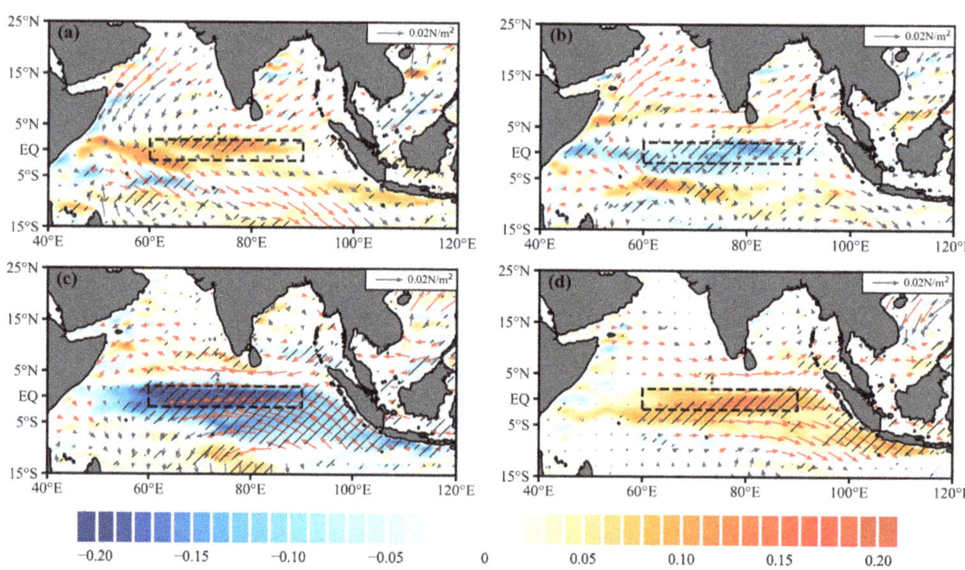

Figure 3. Composite analysis of zonal current anomalies (shading, units: m/s) and wind stress anomalies (vectors, units: N/m^2) in ENSO years. (**a**,**b**) Decaying stage of El Niño/La Niña for spring branch; (**c**,**d**) developing stage of El Niño/La Niña for fall branch. Black dotted box marks jet area, while red arrows and oblique lines indicate significant zonal wind stress and zonal current (90% significance, student's *t*-test).

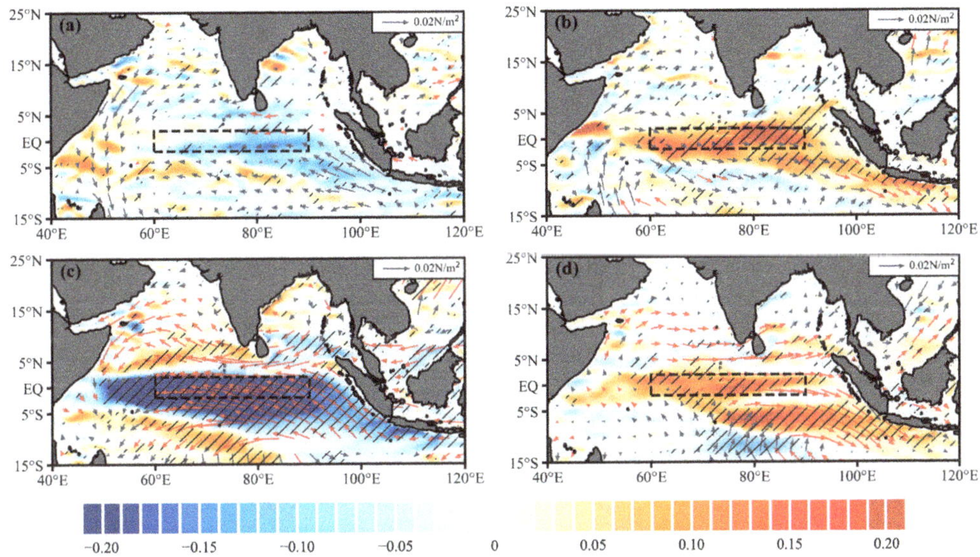

Figure 4. Composite analysis of zonal current anomalies (shading, units: m/s) and wind stress anomalies (vectors, units: N/m^2) in IOD years, (**a**,**b**) refer to the spring branch of WJs, where (**a**,**b**) are the composites for years with pIOD and nIOD, respectively; and (**c**,**d**) refer to the fall branch of WJs, where (**c**,**d**) are the composites for years with pIOD and nIOD, respectively.

The fall branch emerges during the transition from the southwest to the northeast monsoon (Figures 3c,d and 4c,d). During the El Niño developing phase, it exhibits significant negative anomalies, accompanied by pronounced southeasterly wind stress anomalies near Sumatra. These anomalies extend eastward along the equator, where strong easterly wind stress anomalies reach the central Indian Ocean. Additionally, notable northeasterly wind stress anomalies are observed over the Bay of Bengal and the Arabian Sea. During the La Niña developing phase, the pattern is reversed. The fall branch also shows significant negative anomalies during pIOD events, with wind stress anomaly patterns resembling those observed during El Niño but with greater intensity. Conversely, during nIOD events, the fall branch exhibits significant positive anomalies, with wind stress anomalies opposite to those in pIOD events. The stronger positive feedback processes in the eastern Pacific during El Niño lead to greater temperature anomaly amplitudes compared to La Niña [63]. Furthermore, the skewness characteristics of IOD events result in a greater amplitude in pIOD events than in nIOD events [64]. Consequently, the anomaly intensity of the fall branch is more pronounced during El Niño and pIOD events than during La Niña and nIOD events.

4.4. Influence of IOD and ENSO on WJ Variations

Given that ordinary composite and correlation analyses cannot easily distinguish the relationship between the spring and fall branches under the direct influence of ENSO or IOD, we calculated the partial correlation coefficients for the zonal current and wind stress anomalies using ONI and DMI, respectively, based on the screening year (Table 1) to clearly discern the independent correlation of these two branches in interannual events.

For the spring branch (Figure 5a,b), Figure 5a shows the partial correlation coefficients after removing the influence of IOD, revealing a significant positive correlation with the ENSO phase. During this period, wind stress anomalies over the Arabian Sea, the Bay of Bengal, and the southeastern Indian Ocean exhibit a strong correlation with ENSO.

These regions fall within the monsoon domain, where the ENSO decaying phase influences the wind stress anomalies, thereby modulating the spring branch anomalies. Similarly, Figure 5b presents the partial correlation coefficients after removing the influence of ENSO, showing a significant negative correlation with the IOD phase. In this case, zonal wind anomalies from Sumatra to the western Indian Ocean are strongly associated with IOD, indicating that IOD-induced equatorial wind anomalies have a direct impact on the spring branch. The key factors driving the interannual variability of the spring branch are primarily linked to the ENSO decaying phase and IOD events. For the fall branch (Figure 5c,d), the ENSO and IOD events during fall both show significant negative correlations, with the negative correlation for IOD being stronger and the negative correlation for ENSO being weaker, both passing the 90% significance test. After removing the effect of IOD, the influence of the ENSO developing phase does not cover the whole equatorial Indian Ocean (Figure 5c,d) and is rather limited to the wind stress anomalies from the Sumatra–Java coast to 78° E, which modulate the fall branch anomalies. Similarly, after removing the influence of ENSO, the IOD events influence the wind stress anomalies along the Sumatra–Java coast, extending across the entire equatorial Indian Ocean. Meanwhile, wind anomalies over the Bay of Bengal and the Arabian Sea also contribute, collectively modulating the fall branch anomalies.

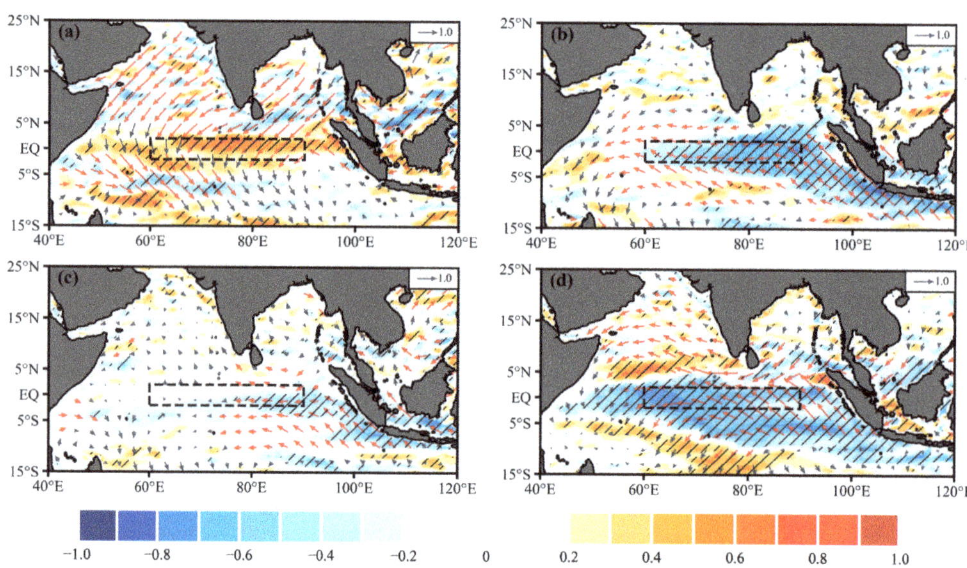

Figure 5. The spatial distribution of the partial correlation coefficients of the zonal current anomalies (shading, correlation) and wind stress anomalies (vectors, correlation) based on ONI and DMI. (**a**,**b**) represent the spring branch: (**a**) $R_{spring}\,ONI|_{DMI}$ and (**b**) $R_{spring}\,DMI|_{ONI}$; (**c**,**d**) represent the fall branch: (**c**) $R_{fall}\,ONI|_{DMI}$ and (**d**) $R_{fall}\,DMI|_{ONI}$.

The phase-locking characteristics of ENSO influence the wind over the Indian Ocean through different causal pathways. During the developing phase of ENSO, anomalies in the Walker circulation play a crucial role. In El Niño events, the descending branch of the anomalous Walker circulation is positioned over the Maritime Continent and the tropical eastern Indian Ocean, triggering anomalous easterly winds from the tropical southeastern Indian Ocean toward the western Indian Ocean. Conversely, during La Niña events, the anomalous Walker circulation converges and ascends over the Maritime Continent and the tropical eastern Indian Ocean, while its descending branch shifts to the tropical western Indian Ocean, inducing anomalous westerly winds from the tropical western Indian Ocean toward the eastern Indian Ocean [32,65,66]. During the spring of the ENSO decaying phase, persistent large-scale convective anomalies are observed over the tropical northwestern Pacific and the northern Indian Ocean. As a response to the enhancement (suppression) of convective anomalies, Rossby wave dynamics induce significant easterly (westerly) vertical wind shear over the northern Indian Ocean. This vertical wind shear, in turn, exerts a substantial modulating effect on ISO activity through tropical wave dynamics, leading to enhanced (weakened) ISO activity over the Bay of Bengal following cold (warm) ENSO events. Consequently, the strengthened (weakened) northward-propagating ISO at the initial stage triggers an earlier (delayed) onset of the Bay of Bengal summer monsoon [67]. These findings further motivate us to explore the influence of ENSO and IOD on the zonal wind stress over the equatorial Indian Ocean, as well as the role of monsoon onset timing in regulating WJs.

Therefore, we further explored the roles of ENSO and IOD in modulating the zonal wind stresses in the WJs and the equatorial Indian Ocean. We constructed regression models for spring and fall based on the cases presented in Table. 1. Results show that building a single DMI or ONI model underestimates the strength of WJs and zonal wind stress; thus, using a multiple regression model is ideal (Table 2). The total explained variance of the spring model U (τ_x) increased by 15% to 39% (25% to 34%) compared with those regression models that only consider IOD or ENSO, while the total explained variance of the fall model U (τ_x) increased by 5% to 19% (3% to 34%). Based on these improvements in explained variance, we quantified the relative contribution of interannual events to the spring and fall branches by using a partial regression approach. The partial regression coefficients and partial correlation coefficients exhibit consistent patterns in their descriptions (Figure 6), where the partial regression coefficients represent the relative contribution of predictor factors to WJs anomalies and wind stress anomalies. Following the partial regression method in Wang et al. (2021) [58], we reconstructed the anomaly fields in the synthetic analysis by using partial regression coefficients and then separately analyzed the relative contributions of ENSO and IOD to the zonal current anomalies by using partial regression fields. During El Niño (La Niña) decaying years (Figures S4 and S5), the spatial modes of the reconstructed and original fields are almost the same, only with a few errors. Therefore, using partial regression fields to separately observe the contributions of ENSO and IOD to the spring branch can guarantee an excellent interpretability. During El Niño (La Niña) decaying years, the delayed (advanced) onset of the monsoon leads to positive (negative) anomalies in the spring branch. ENSO provides the primary contribution to the spring branch, with an anomaly contribution of +0.07 m/s (−0.05 m/s). In contrast, IOD contribution is weaker, with an anomaly contribution of +0.03 m/s (no contribution). The spatial modes of the reconstructed and pristine fields in the positive (negative) IOD years (Figures S6 and S7) are also the same, with the residual field being one order of magnitude less than the reconstructed field. The anomalies of the WJs are stronger in IOD than in ENSO. During IOD events, the WJs are primarily modulated by the IOD, with ENSO playing a weaker role. During positive (negative) IOD years, IOD provides the primary contribution to the spring branch, with an anomaly contribution of +0.15 m/s (−0.11 m/s). ENSO contribution is much weaker, with an anomaly contribution of +0.02 m/s (+0.03 m/s) in both cases.

Table 2. Explained variances R^2 of the single-regression ONI or DMI model and multiple regression on variables (τ_x and U) and their differences Δ in regions 60° E–90° E in longitude and 2° S–2° N in latitude.

Model	Spring		Fall	
	τ_x	U	τ_x	U
Multiple Regression Model	0.58	0.50	0.88	0.58
Single Regression Model (ONI)	0.26	0.14	0.54	0.38
Δ	+0.32	+0.36	+0.30	+0.20
Single Regression Model (DMI)	0.34	0.32	0.86	0.55
Δ	+0.24	+0.18	+0.02	+0.03

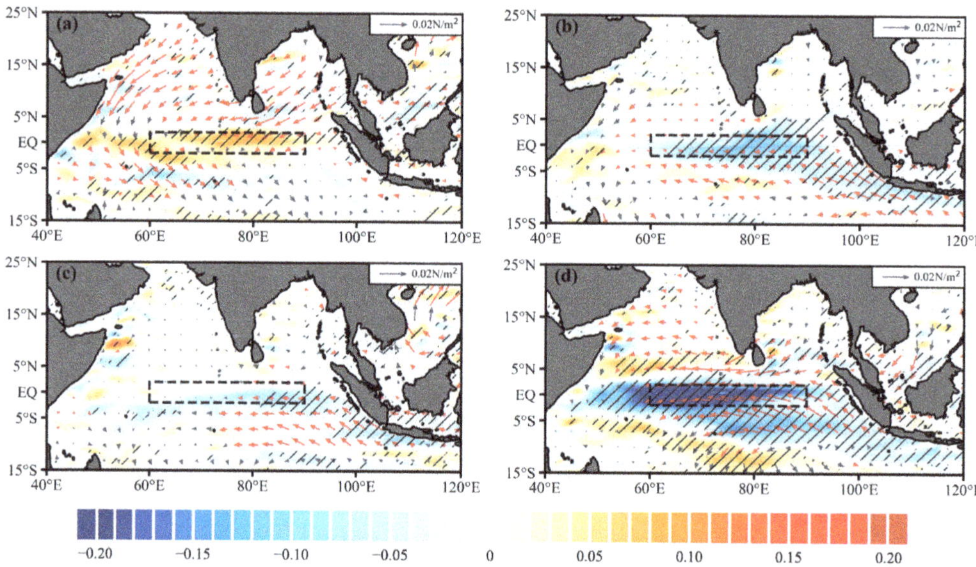

Figure 6. Spatial distribution of the partial regression coefficient zonal current anomalies (shading, units: m/s) and wind stress anomalies (vectors, units: N/m^2). (**a**,**b**) represent the spring branch: (**a**) $a_{spring}\, ONI|_{DMI}$ and (**b**) $b_{spring}\, DMI|_{ONI}$; (**c**,**d**) represent the fall branch: (**c**) $a_{fall}\, ONI|_{DMI}$ and (**d**) $b_{fall}\, DMI|_{ONI}$. The 90% significance test markings follow the same convention as in Figure 3 but are based on the F-test.

ENSO and IOD trigger eastward (westward) wind stress anomalies in the equatorial Indian Ocean and suppress (enhance) the fall branch. During the El Niño (La Niña) developing years (Figures S8 and S9), the reconstruction field works well, with IOD provides the primary contribution to the fall branch, with an anomaly contribution of −0.11 m/s (+0.06 m/s), which is greater than ENSO anomaly contribution of −0.06 m/s (+0.04 m/s). The influence of IOD also extends to as far as 60° E, while that of ENSO only extends up to 80° E. During those years with a positive (negative) IOD (Figures S10 and S11), IOD provides the primary contribution to the fall branch, with an anomaly contribution of −0.2 m/s (+0.15 m/s), which is much larger than ENSO anomaly contribution of −0.07 m/s (+0.03 m/s).

4.5. Regulatory Mechanisms of Abnormal WJs

We took the average of the WJs occurrence region and further determined the relationship of the spring and fall branches with interannual events (ENSO and IOD) through scatter relations. The spring branch exhibits positive (negative) anomalies during the El

Niño (La Niña) decaying phase and shows a significant positive correlation (0.47) with the positive and negative phases of ENSO (Figure 7a). Meanwhile, under IOD events, the spring branch exhibits negative (positive) anomalies when positive (negative) IOD occurs and demonstrates a significant negative correlation (−0.93) with the positive and negative phases of IOD (Figure 7c). During the El Niño (La Niña) developing phase, the fall branch exhibits negative (positive) anomalies and a significant negative correlation (−0.68) with ENSO (Figure 7b). In positive (negative) IOD events, the fall branch exhibits a negative (positive) anomaly and a significant negative correlation (−0.92) with IOD (Figure 7d). The positive phase anomaly of the fall branch is notably stronger than its negative phase anomaly, primarily due to the greater amplitude of pIOD compared to nIOD. Given that WJs exist in the equatorial Indian Ocean and directly affect the east–west heat distribution, the correlation with IOD is more higher than that with ENSO.

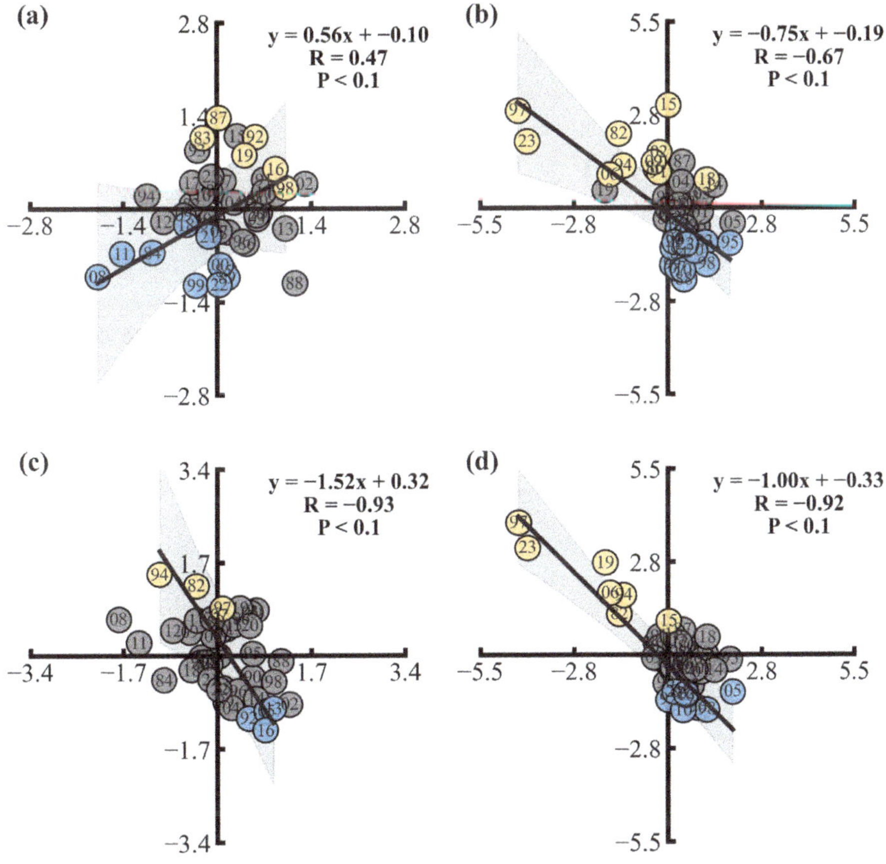

Figure 7. Scatter plots of zonal current anomalies (60° E–90° E, 2° S–2° N, upper 15 m of the equatorial Indian Ocean) versus ONI and DMI. (**a**,**c**) show the spring branch anomaly versus ENSO and IOD, respectively; and (**b**,**d**) refer to the corresponding anomalies for the fall branch. Those years with positive and negative phases are marked in yellow and blue, respectively.

Given that the responses of WJs are driven by zonal winds, we further explored the connection between zonal winds and interannual events. However, in analyzing the modulation of the spring branch by wind fields, we referred to the study by Li et al. (2022) [68], which found that the coordinated transition of equatorial westerlies and the southwest monsoon after monsoon onset plays a crucial role in regulating WJ intensity. A later (earlier) monsoon onset leads to a stronger (weaker) spring branch. Moreover, in the year following a cold (warm) ENSO event, convection over the northwestern Pacific is significantly enhanced (suppressed), creating favorable (unfavorable) conditions for the development of ISO over the Bay of Bengal. This, in turn, leads to an earlier (delayed) monsoon onset [67]. Based on this mechanism, we selected the monsoon regions of the Indian Ocean (the Bay of Bengal and the Arabian Sea) and the significantly affected areas in the southwestern Indian Ocean identified in the composite analysis. We then examined the zonal and meridional wind anomalies and marked interannual events with scatter points (Figure 8a–f). The wind anomalies reveal that during the decaying phase of El Niño (La Niña), both zonal and meridional winds in this region exhibit significant weakening (strengthening), indicating a delayed (advanced) monsoon onset. This delay (advance) in monsoon onset causes the transition of equatorial westerlies to lag (lead), resulting in a stronger (weaker) spring branch anomaly. As the primary driving force of surface currents in the Indian Ocean, the spring branch exhibits a pronounced response to the monsoon onset process. From Figure 8h, the spring zonal wind anomaly demonstrates a significant negative correlation (−0.83) with IOD events.

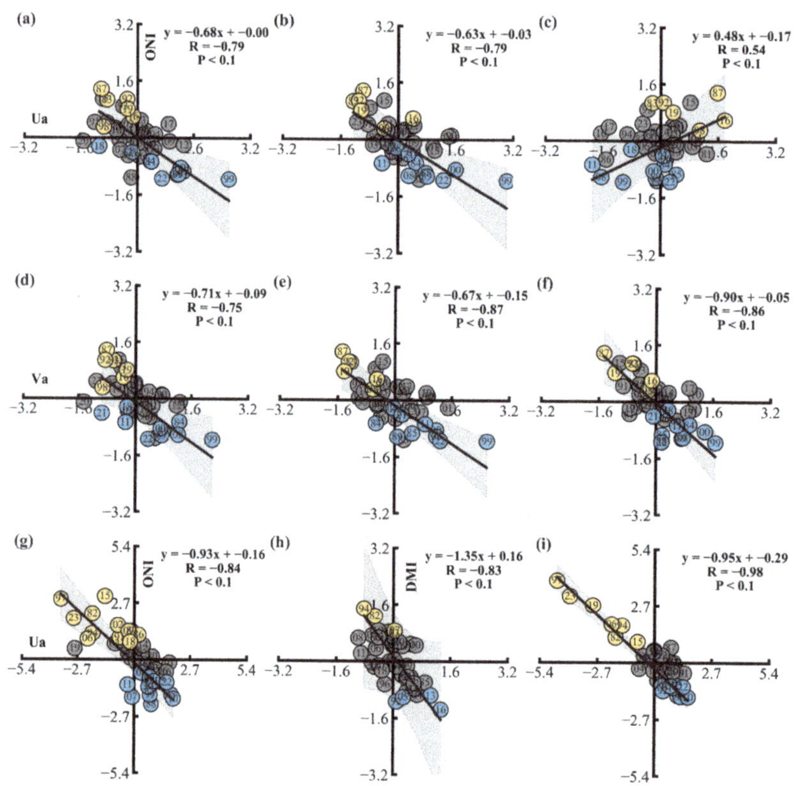

Figure 8. Scatter plots of wind anomalies. (**a–c**) show the zonal wind anomalies in the Bay of Bengal (80° E–100° E, 2° N–15° N), the Arabian Sea (40° E–80° E, 2° N–15° N), and the southwestern Indian Ocean (40° E–80° E, 2° S–12° S), respectively. (**d–f**) represent the meridional wind anomalies in the same regions as (**a–c**) during spring. (**g**) shows the zonal wind anomalies over the equatorial eastern Indian Ocean (80° E–100° E, 5° S–5° N) during fall. (**h**,**i**) depict the zonal wind anomalies over the equatorial Indian Ocean (40° E–100° E, 5° S–5° N) during spring and fall, respectively. Those years with positive and negative phases are marked in yellow and blue, respectively.

As the dominant mode of the ocean–atmosphere coupled system on interannual timescales, ENSO directly influences tropical atmospheric circulation. During the developing phase of ENSO, persistent zonal wind anomalies over the equatorial Pacific modulate the thermal contrast between the eastern and western Pacific. Through the Bjerknes positive feedback mechanism, these wind anomalies gradually amplify SSTAs while adjusting the intensity of the Walker circulation, ultimately inducing significant equatorial zonal wind anomalies over the Indian Ocean. To examine this effect, we selected the significant zonal wind region (80° E–100° E, 5° S–5° N) identified in Figure 5c. During the development of El Niño (La Niña), there is a clear suppression (enhancement) of westerly winds, leading to negative (positive) zonal wind anomalies (Figure 8g), showing a significant negative correlation (−0.84). Under the influence of strong interannual events, the zonal wind anomalies over the equator become more pronounced. When such anomalies occur, WJs exhibit corresponding anomalous signals, further confirming that their formation is driven by wind forcing. During IOD events (Figure 8i), equatorial wind anomalies show a significant negative correlation (−0.98) with the IOD phase, and the anomalous variations of WJs in both spring and fall exhibit a consistent pattern. This further verifies that wind anomalies induced by interannual events serve as the primary driver of WJ interannual variability.

5. Conclusions and Discussion

During the monsoon transition, WJs can cover the entire central equatorial Indian Ocean in the zonal direction, thus significantly affecting the east–west heat and salinity transport throughout the Indian Ocean [19–21]. When the El Niño decaying (La Niña developing) phase or the nIOD phase leads to the intensification of the spring (fall) branch, the anomalously strong eastward jet transports more warm water to the equatorial eastern Indian Ocean, significantly increasing the upper-ocean heat content in this region and creating favorable conditions for the enhancement of local atmospheric convection. These convective changes not only affect the local ecological environment and economic activities but also influence the monsoon system by modulating the Hadley circulation, thereby extending their impact across the entire East and South Asian monsoon regions. Conversely, when the La Niña decaying (El Niño developing) phase or the pIOD phase leads to the weakening of the spring (fall) branch, the positive thermal anomalies in the eastern Indian Ocean dissipate, leading to a corresponding reduction in their regulatory effect on local atmospheric circulation [14,50]. At the same time, when WJs reach the eastern boundary, they trigger Rossby waves that propagate downward and westward, affecting the strength of the Equatorial Undercurrent and playing a crucial role in regulating basin-scale material and energy distribution [69,70]. By affecting the sea-level anomalies on both sides of the Indian Ocean, it has a significant impact on the sea surface temperature changes and upwelling activities in the eastern Indian Ocean, making it have an important impact on the sea surface height, upper ocean heat content, sea surface temperature, salinity, and atmospheric convection activities in the equatorial Indian Ocean and even the northern Indian Ocean, making it a key regulatory factor in global climate variability [14,16,71].

In this study, we analyzed the interannual variability of WJs under the interactions of the Indo-Pacific ocean and the atmosphere along with their corresponding mechanisms by using ocean reanalysis data (SODA3.15.2) from 1980 to 2023. WJs are mainly active in the 60° E~90° E, 2° S~2° N region, reaching a peak intensity at 70° E~80° E and decaying from west to east. Its climatic state is that the spring branch is generally weaker than the fall branch. We selected the corresponding years through the four phases defined by Warner and Moum (2019) [62], and we filtered the spring and fall branches corresponding to the ENSO decaying and ENSO developing phases. We eventually identified the effects of ENSO and IOD on WJs via composite analysis and multiple regression, respectively.

Results show that ENSO regulates the interannual variability of the spring and fall branches of WJs through different mechanisms. Due to the seasonal phase-locking effect of ENSO, WJs exhibit distinct response characteristics: the spring branch shows positive (negative) anomalies during the El Niño (La Niña) decaying phases, while the fall branch exhibits negative (positive) anomalies during the El Niño (La Niña) developing phases. Meanwhile, IOD directly drives equatorial wind anomalies through local air–sea feedback, with positive (negative) IOD events leading to negative (positive) anomalies in both the spring and fall branches. The stronger amplitude of positive IOD events results in more pronounced anomalies in the fall branch. We used partial correlation to separate the relationship between ENSO and IOD, and results show that the spring branch has a significant positive correlation with ENSO and a significant negative correlation with IOD. For the fall branch, both ENSO and IOD exhibit significant negative correlations. The variance explained by the combined influence of ENSO and IOD on WJs is significantly higher than that of either factor alone. Multiple regression models indicate that ENSO is the primary driver of interannual variability in the spring branch, while IOD is the dominant factor for the fall branch. The mechanisms underlying interannual anomalies differ between the spring and fall branches. For the spring branch, ENSO regulates its intensity indirectly by influencing monsoon onset timing: in El Niño (La Niña) decaying years, delayed (advanced) monsoon onset leads to a lagged (advanced) shift in equatorial westerlies, thereby strengthening (weakening) the spring branch. In contrast, for the fall branch, ENSO modulates zonal currents by altering the position of the descending branch of the Walker circulation, but its impact is relatively limited, primarily affecting the central equatorial Indian Ocean. In comparison, IOD influences WJs by inducing anomalous zonal winds across the entire equatorial Indian Ocean through its basin-wide mechanisms.

This study focuses on characterizing the interannual response of WJs and their regulation mechanisms under the interactions between the Indo-Pacific ocean and the atmosphere. Previous studies have suggested that the interannual variability of the spring branch is primarily controlled by intraseasonal wind variations and cannot be explained by interannual forcing events such as ENSO and IOD [50,72]. However, our findings demonstrate that the spring branch exhibits significant interannual variability, primarily regulated by ENSO through its influence on monsoon onset timing, which indirectly modulates the spring branch [67,68]. Due to the influence of intraseasonal effects and the limitations of data temporal resolution, the direct modulation of spring branch interannual characteristics by equatorial zonal wind anomalies cannot be accurately captured. Therefore, understanding the role of the spring branch and its low-frequency variability in the Indian Ocean dipole zonal model [41], monsoon variability, and ENSO–monsoon interactions should be further explored [66,73,74]. The mechanism of ENSO–monsoon interactions on the interannual variability of WJs in this study also warrants further investigation.

In the context of global warming, the frequency and intensity of El Niño events may increase [75,76], decrease [77], or depend on changes in the mean climate state [78]. Accordingly, ENSO characteristics may also change in the future [79], which subsequently changes the interannual characteristics of WJs. ENSO plays a key role not only in the occurrence and development of IOD [80] but also in the interannual variability of WJs. Therefore, the feedback mechanisms of the Indo-Pacific interannual events on WJs should be systematically investigated to enhance the current understanding of ocean dynamics, ecosystems, and climate effects in the tropical Indian Ocean.

Supplementary Materials: The following supporting information can be downloaded at: https://www.mdpi.com/article/10.3390/jmse13040691/s1, Figure S1: Time-depth distribution of zonal current anomaly (shading; units: m/s). (a) RAMA moorings at 0°, 80.5°E; (b) SODA3.15.2 at 0°, 80.5° E; (c) Time evolution of depth-integrated zonal current anomalies within the depth range of

40–80 m, Figure S2: Time-depth distribution of zonal current anomaly (shading; units: m/s). (a) RAMA moorings at 0°, 90°E; (b) SODA3.15.2 at 0°, 90° E; (c) Time evolution of depth-integrated zonal current anomalies within the depth range of 40–80 m, Figure S3: Comparison of NOAA's Oceanic Niño Index (ONI) (a) and the Dipole Mode Index (DMI) (b), Figure S4: (a) Observed, (b), reconstructed, (c), residual, (d) partial regression maps of zonal current anomalies (m/s, shading) & wind stress anomalies (N/m^2, vectors) onto $ONI|_{DMI}$ (i.e., ONI with the influences of DMI removed), and (e), $DMI|_{ONI}$ during El Niño decaying events happen in spring situation. Observations are the ensembles of El Niño decaying events. Only zonal wind stress and zonal current values that are statistically significant at the 90% confidence level are shown for (b), (d) & (e). The red-outlined boxes indicate the key regions of Wyrtki jets, Figure S5: (a) Observed, (b), reconstructed, (c), residual, (d) partial regression maps of zonal current anomalies (m/s, shading) & wind stress anomalies (N/m^2, vectors) onto $ONI|_{DMI}$, and (e), $DMI|_{ONI}$ during La Niña decaying events happen in spring situation, Figure S6: (a) Observed, (b), reconstructed, (c), residual, (d) partial regression maps of zonal current anomalies (m/s, shading) & wind stress anomalies (N/m^2, vectors) onto $ONI|_{DMI}$, and (e), $DMI|_{ONI}$ during pIOD events happen in spring situation, Figure S7: (a) Observed, (b), reconstructed, (c), residual, (d) partial regression maps of zonal current anomalies (m/s, shading) & wind stress anomalies (N/m^2, vectors) onto $ONI|_{DMI}$, and (e), $DMI|_{ONI}$ during nIOD events happen in spring situation, Figure S8: (a) Observed, (b), reconstructed, (c), residual, (d) partial regression maps of zonal current anomalies (m/s, shading) & wind stress anomalies (N/m^2, vectors) onto $ONI|_{DMI}$ (i.e., ONI with the influences of DMI removed), and (e), $DMI|_{ONI}$ during El Niño developing events happen in fall situation. Observations are the ensembles of El Niño developing events. Only zonal wind stress and zonal current values that are statistically significant at the 90% confidence level are shown for (b), (d) & (e). The red-outlined boxes indicate the key regions of Wyrtki jets, Figure S9: (a) Observed, (b), reconstructed, (c), residual, (d) partial regression maps of zonal current anomalies (m/s, shading) & wind stress anomalies (N/m^2, vectors) onto $ONI|_{DMI}$, and (e), $DMI|_{ONI}$ during La Niña developing events happen in fall situation, Figure S10: (a) Observed, (b), reconstructed, (c), residual, (d) partial regression maps of zonal current anomalies (m/s, shading) & wind stress anomalies (N/m^2, vectors) onto $ONI|_{DMI}$, and (e), $DMI|_{ONI}$ during pIOD events happen in fall situation, Figure S11: (a) Observed, (b), reconstructed, (c), residual, (d) partial regression maps of zonal current anomalies (m/s, shading) & wind stress anomalies (N/m^2, vectors) onto $ONI|_{DMI}$, and (e), $DMI|_{ONI}$ during nIOD events happen in fall situation.

Author Contributions: Conceptualization, Q.F., J.Z. and G.H.; funding acquisition, G.H. and J.X.; methodology, Q.F. and J.X.; project administration, Q.F. and J.Z.; validation, G.H.; visualization, G.H. and J.X.; writing—original draft, Q.F., J.Z. and J.X. Q.F. and J.Z. contributed equally to this paper. All authors have read the submitted version of the manuscript.

Funding: This study was supported by the National Natural Science Foundation of China Project (No. 42206005 and 42206004), and the National Undergraduate Training Program for Innovation and Entrepreneurship (No. 202210340017).

Data Availability Statement: The SODA3 data can be accessed from the Department of Atmospheric and Oceanic Science, University of Maryland (http://dsrs.atmos.umd.edu/DATA/soda3.15.2/REGRIDED/ocean/, accessed on 10 April 2024). The ECMWF ERA5 data are accessible at the Copernicus Climate Change Service (C3S) Climate Date Store (https://cds.climate.copernicus.eu/datasets/reanalysis-era5-single-levels-monthly-means?tab=overview, accessed on 16 April 2024). The RAMA data can be accessed from the Department of Pacific Marine Environmental Laboratory, National Oceanic and Atmospheric Administration (https://www.pmel.noaa.gov/tao/drupal/disdel/, accessed on 21 November 2024).

Acknowledgments: The computations for this study were performed using a system featuring a CPU of i5-11400F and 32 GB of RAM, which supported further data processing and visualization.

Conflicts of Interest: The authors declare no conflicts of interest.

Abbreviations

ENSO	El Niño-Southern Oscillation
IOD	Indian Ocean Dipole
ONI	Ocean Niño Index
DMI	Dipole Mode Index
pIOD	positive IOD
nIOD	negative IOD
WJs	Wyrtki jets
ISO	Intraseasonal oscillation
SSTAs	Sea surface temperature anomalies

References

1. Schott, F.A.; McCreary, J.P. The Monsoon Circulation of the Indian Ocean. *Prog. Oceanogr.* **2001**, *51*, 1–123. [CrossRef]
2. Schott, F.A.; Xie, S.-P.; McCreary, J.P., Jr. Indian Ocean Circulation and Climate Variability. *Rev. Geophys.* **2009**, *47*, RG1002. [CrossRef]
3. Wyrtki, K. An Equatorial Jet in the Indian Ocean. *Science* **1973**, *181*, 262–264. [CrossRef]
4. Molinari, R.L.; Olson, D.; Reverdin, G. Surface Current Distributions in the Tropical Indian Ocean Derived from Compilations of Surface Buoy Trajectories. *J. Geophys. Res. Ocean.* **1990**, *95*, 7217–7238. [CrossRef]
5. Reppin, J.; Schott, F.A.; Fischer, J.; Quadfasel, D. Equatorial Currents and Transports in the Upper Central Indian Ocean: Annual Cycle and Interannual Variability. *J. Geophys. Res. Ocean.* **1999**, *104*, 15495–15514. [CrossRef]
6. Yuan, D.; Han, W. Roles of Equatorial Waves and Western Boundary Reflection in the Seasonal Circulation of the Equatorial Indian Ocean. *J. Phys. Oceanogr.* **2006**, *36*, 930–944. [CrossRef]
7. O'Brien, J.J.; Hurlburt, H.E. Equatorial Jet in the Indian Ocean: Theory. *Science* **1974**, *184*, 1075–1077. [CrossRef]
8. Nagura, M.; McPhaden, M.J. Wyrtki Jet Dynamics: Seasonal Variability. *J. Geophys. Res. Ocean.* **2010**, *115*. [CrossRef]
9. Han, W.; McCreary, J.P.; Anderson, D.L.T.; Mariano, A.J. Dynamics of the Eastern Surface Jets in the Equatorial Indian Ocean. *J. Phys. Oceanogr.* **1999**, *29*, 2191–2209. [CrossRef]
10. Qiu, Y.; Li, L.; Yu, W. Behavior of the Wyrtki Jet Observed with Surface Drifting Buoys and Satellite Altimeter. *Geophys. Res. Lett.* **2009**, *36*, L18607. [CrossRef]
11. Jensen, T.G. Equatorial Variability and Resonance in a Wind-Driven Indian Ocean Model. *J. Geophys. Res. Ocean.* **1993**, *98*, 22533–22552. [CrossRef]
12. McPhaden, M.J.; Wang, Y.; Ravichandran, M. Volume Transports of the Wyrtki Jets and Their Relationship to the Indian Ocean Dipole. *J. Geophys. Res. Ocean.* **2015**, *120*, 5302–5317. [CrossRef]
13. Duan, Y.; Liu, L.; Han, G.; Liu, H.; Yu, W.; Yang, G.; Wang, H.; Wang, H.; Liu, Y.; Zahid; et al. Anomalous Behaviors of Wyrtki Jets in the Equatorial Indian Ocean during 2013. *Sci. Rep.* **2016**, *6*, 29688. [CrossRef]
14. Cao, G.; Xu, T.; Wei, Z. Research progress on intraseasonal variability of Wyrtki jet. *Prog. Geophys.* **2024**, *39*, 1293–1303. [CrossRef]
15. Shinoda, T.; Han, W.; Metzger, E.J.; Hurlburt, H.E. Seasonal Variation of the Indonesian Throughflow in Makassar Strait. *J. Phys. Oceanogr.* **2012**, *42*, 1099–1123. [CrossRef]
16. Cao, G.; Xu, T.; Wei, Z. Seasonal Differences of Wyrtki Jet Intraseasonal Variabilities. *Front. Mar. Sci.* **2024**, *11*, 1517779. [CrossRef]
17. Zhang, Y.; Du, Y. Seasonal Variability of Salinity Budget and Water Exchange in the Northern Indian Ocean from HYCOM Assimilation. *Chin. J. Ocean. Limnol.* **2012**, *30*, 1082–1092. [CrossRef]
18. Zhang, Y.; Du, Y.; Zheng, S.; Yang, Y.; Cheng, X. Impact of Indian Ocean Dipole on the Salinity Budget in the Equatorial Indian Ocean. *J. Geophys. Res. Ocean.* **2013**, *118*, 4911–4923. [CrossRef]
19. Zhang, Y.; Du, Y.; Zhang, Y.; Gao, S. Asymmetry of Upper Ocean Salinity Response to the Indian Ocean Dipole Events as Seen from ECCO Simulation. *Acta Oceanol. Sin.* **2016**, *35*, 42–49. [CrossRef]
20. Wang, J. Observational Bifurcation of Wyrtki Jets and Its Influence on the Salinity Balance in the Eastern Indian Ocean. *Atmos. Ocean. Sci. Lett.* **2017**, *10*, 36–43.
21. Xie, C.; Ding, R.; Xuan, J.; Huang, D. Interannual Variations in Salt Flux at 80°E Section of the Equatorial Indian Ocean. *Sci. China Earth Sci.* **2023**, *66*, 2142–2161. [CrossRef]
22. Murtugudde, R.; McCreary, J.P., Jr.; Busalacchi, A.J. Oceanic Processes Associated with Anomalous Events in the Indian Ocean with Relevance to 1997–1998. *J. Geophys. Res. Ocean.* **2000**, *105*, 3295–3306. [CrossRef]
23. Masson, S.; Delecluse, P.; Boulanger, J.-P.; Menkes, C. A Model Study of the Seasonal Variability and Formation Mechanisms of the Barrier Layer in the Eastern Equatorial Indian Ocean. *J. Geophys. Res. Ocean.* **2002**, *107*, SRF-18-1–SRF 18-20. [CrossRef]

24. McPhaden, M.J.; Zebiak, S.E.; Glantz, M.H. ENSO as an Integrating Concept in Earth Science. *Science* **2006**, *314*, 1740–1745. [CrossRef]
25. Guan, C.; Chen, Y.; Wang, F. Seasonal Variability of Zonal Heat Advection in the Mixed Layer of the Tropical Pacific. *Chin. J. Ocean. Limnol.* **2013**, *31*, 1356–1367. [CrossRef]
26. Guan, C.; Hu, S.; McPhaden, M.J.; Wang, F.; Gao, S.; Hou, Y. Dipole Structure of Mixed Layer Salinity in Response to El Niño-La Niña Asymmetry in the Tropical Pacific. *Geophys. Res. Lett.* **2019**, *46*, 12165–12172. [CrossRef]
27. Saji, N.H.; Goswami, B.N.; Vinayachandran, P.N.; Yamagata, T. A Dipole Mode in the Tropical Indian Ocean. *Nature* **1999**, *401*, 360–363. [CrossRef]
28. Chowdary, J.S.; Gnanaseelan, C. Basin-Wide Warming of the Indian Ocean during El Niño and Indian Ocean Dipole Years. *Int. J. Climatol.* **2007**, *27*, 1421–1438. [CrossRef]
29. Webster, P.J.; Moore, A.M.; Loschnigg, J.P.; Leben, R.R. Coupled Ocean–Atmosphere Dynamics in the Indian Ocean during 1997–1998. *Nature* **1999**, *401*, 356–360. [CrossRef]
30. Feng, M.; Meyers, G.; Wijffels, S. Interannual Upper Ocean Variability in the Tropical Indian Ocean. *Geophys. Res. Lett.* **2001**, *28*, 4151–4154. [CrossRef]
31. Rao, S.A.; Behera, S.K.; Masumoto, Y.; Yamagata, T. Interannual Subsurface Variability in the Tropical Indian Ocean with a Special Emphasis on the Indian Ocean Dipole. *Deep. Sea Res. Part. II Top. Stud. Oceanogr.* **2002**, *49*, 1549–1572. [CrossRef]
32. Saji, N.H.; Yamagata, T. Possible Impacts of Indian Ocean Dipole Mode Events on Global Climate. *Clim. Res.* **2003**, *25*, 151–169. [CrossRef]
33. Feng, M.; Meyers, G. Interannual Variability in the Tropical Indian Ocean: A Two-Year Time-Scale of Indian Ocean Dipole. *Deep. Sea Res. Part. II Top. Stud. Oceanogr.* **2003**, *50*, 2263–2284. [CrossRef]
34. McPhaden, M.J.; Nagura, M. Indian Ocean Dipole Interpreted in Terms of Recharge Oscillator Theory. *Clim. Dyn.* **2014**, *42*, 1569–1586. [CrossRef]
35. Nyadjro, E.S.; McPhaden, M.J. Variability of Zonal Currents in the Eastern Equatorial Indian Ocean on Seasonal to Interannual Time Scales. *J. Geophys. Res. Ocean.* **2014**, *119*, 7969–7986. [CrossRef]
36. Yuan, D.; Liu, H. Long-Wave Dynamics of Sea Level Variations during Indian Ocean Dipole Events. *J. Phys. Oceanogr.* **2009**, *39*, 1115–1132. [CrossRef]
37. Chambers, D.P.; Tapley, B.D.; Stewart, R.H. Anomalous Warming in the Indian Ocean Coincident with El Niño. *J. Geophys. Res. Ocean.* **1999**, *104*, 3035–3047. [CrossRef]
38. Le Blanc, J.-L.; Boulanger, J.-P. Propagation and Reflection of Long Equatorial Waves in the Indian Ocean from TOPEX/POSEIDON Data during the 1993–1998 Period. *Clim. Dyn.* **2001**, *17*, 547–557. [CrossRef]
39. Huang, B.; Kinter III, J.L. Interannual Variability in the Tropical Indian Ocean. *J. Geophys. Res. Ocean.* **2002**, *107*, 20-1–20-26. [CrossRef]
40. Gnanaseelan, C.; Deshpande, A.; McPhaden, M.J. Impact of Indian Ocean Dipole and El Niño/Southern Oscillation Wind-Forcing on the Wyrtki Jets. *J. Geophys. Res. Ocean.* **2012**, *117*, C8. [CrossRef]
41. Joseph, S.; Wallcraft, A.J.; Jensen, T.G.; Ravichandran, M.; Shenoi, S.S.C.; Nayak, S. Weakening of Spring Wyrtki Jets in the Indian Ocean during 2006–2011. *J. Geophys. Res. Ocean.* **2012**, *117*, C4. [CrossRef]
42. Wu, Y.; Liu, L.; Zhang, X.; Duan, Y.; Yang, G.; Yang, Y.; Zahid; Oloo, P.; Sagero, P. Different Impacts from Various El Niño Events on Wyrtki Jets in Boreal Autumn Season. *Pure Appl. Geophys.* **2018**, *175*, 4567–4577. [CrossRef]
43. Deng, K.; Cheng, X.; Feng, T.; Ma, T.; Duan, W.; Chen, J. Interannual Variability of the Spring Wyrtki Jet. *J. Ocean. Limnol.* **2021**, *39*, 26–44. [CrossRef]
44. Chu, X.; Han, W.; Zhang, L.; Chen, G. Effects of Climate Modes on Interannual Variability of the Equatorial Currents in the Indian Ocean. *Clim. Dyn.* **2023**, *60*, 3681–3694. [CrossRef]
45. Ueda, H.; Matsumoto, J. A Possible Triggering Process of East-West Asymmetric Anomalies over the Indian Ocean in Relation to 1997/98 El Niño. *J. Meteorol. Soc. Japan. Ser. II* **2000**, *78*, 803–818. [CrossRef]
46. Hendon, H.H. Indonesian Rainfall Variability: Impacts of ENSO and Local Air–Sea Interaction. *J. Clim.* **2003**, *16*, 1775–1790. [CrossRef]
47. Lau, N.-C.; Nath, M.J. Atmosphere–Ocean Variations in the Indo-Pacific Sector during ENSO Episodes. *J. Clim.* **2003**, *16*, 3–20. [CrossRef]
48. Tokinaga, H.; Tanimoto, Y. Seasonal Transition of SST Anomalies in the Tropical Indian Ocean during El Niño and Indian Ocean Dipole Years. *J. Meteorol. Soc. Japan. Ser. II* **2004**, *82*, 1007–1018. [CrossRef]
49. Wang, B.; Wu, R.; Li, T. Atmosphere–Warm Ocean Interaction and Its Impacts on Asian–Australian Monsoon Variation. *J. Clim.* **2003**, *16*, 1195–1211. [CrossRef]
50. Deshpande, A.; Gnanaseelan, C.; Chowdary, J.S.; Rahul, S. Interannual Spring Wyrtki Jet Variability and Its Regional Impacts. *Dyn. Atmos. Ocean.* **2017**, *78*, 26–37. [CrossRef]

51. Gnanaseelan, C.; Deshpande, A. Equatorial Indian Ocean Subsurface Current Variability in an Ocean General Circulation Model. *Clim. Dyn.* **2018**, *50*, 1705–1717. [CrossRef]
52. Zhang, Y.; Guan, Y.P.; Huang, R.X. 3D Structure of Striations in the North Pacific. *J. Phys. Oceanogr.* **2021**, *51*, 3651–3662. [CrossRef]
53. Carton, J.A.; Chepurin, G.A.; Chen, L. SODA3: A New Ocean Climate Reanalysis. *J. Clim.* **2018**, *31*, 6967–6983. [CrossRef]
54. Naseef, T.M.; Kumar, V.S. Climatology and Trends of the Indian Ocean Surface Waves Based on 39-Year Long ERA5 Reanalysis Data. *Int. J. Climatol.* **2020**, *40*, 979–1006. [CrossRef]
55. McPhaden, M.J.; Meyers, G.; Ando, K.; Masumoto, Y.; Murty, V.S.N.; Ravichandran, M.; Syamsudin, F.; Vialard, J.; Yu, L.; Yu, W. RAMA: The Research Moored Array for African–Asian–Australian Monsoon Analysis and Prediction*. *Bull. Am. Meteorol. Soc.* **2009**, *90*, 459–480. [CrossRef]
56. Cai, W.; van Rensch, P.; Cowan, T.; Hendon, H.H. Teleconnection Pathways of ENSO and the IOD and the Mechanisms for Impacts on Australian Rainfall. *J. Clim.* **2011**, *24*, 3910–3923. [CrossRef]
57. Ham, Y.-G.; Kug, J.-S.; Park, J.-Y. Two Distinct Roles of Atlantic SSTs in ENSO Variability: North Tropical Atlantic SST and Atlantic Niño. *Geophys. Res. Lett.* **2013**, *40*, 4012–4017. [CrossRef]
58. Wang, J.-Z.; Wang, C. Joint Boost to Super El Niño from the Indian and Atlantic Oceans. *J. Clim.* **2021**, *34*, 4937–4954. [CrossRef]
59. Ashok, K.; Behera, S.K.; Rao, S.A.; Weng, H.; Yamagata, T. El Niño Modoki and Its Possible Teleconnection. *J. Geophys. Res.* **2007**, *112*, C11007. [CrossRef]
60. Hastenrath, S.; Greischar, L. The Monsoonal Current Regimes of the Tropical Indian Ocean: Observed Surface Flow Fields and Their Geostrophic and Wind-Driven Components. *J. Geophys. Res. Ocean.* **1991**, *96*, 12619–12633. [CrossRef]
61. Rasmusson, E.M.; Carpenter, T.H. Variations in Tropical Sea Surface Temperature and Surface Wind Fields Associated with the Southern Oscillation/El Niño. *Mon. Weather. Rev.* **1982**, *110*, 354–384. [CrossRef]
62. Warner, S.J.; Moum, J.N. Feedback of Mixing to ENSO Phase Change. *Geophys. Res. Lett.* **2019**, *46*, 13920–13927. [CrossRef]
63. Guan, C.; McPhaden, M.J.; Wang, F.; Hu, S. Quantifying the Role of Oceanic Feedbacks on ENSO Asymmetry. *Geophys. Res. Lett.* **2019**, *46*, 2140–2148. [CrossRef]
64. Cai, W.; Zheng, X.-T.; Weller, E.; Collins, M.; Cowan, T.; Lengaigne, M.; Yu, W.; Yamagata, T. Projected Response of the Indian Ocean Dipole to Greenhouse Warming. *Nat. Geosci.* **2013**, *6*, 999–1007. [CrossRef]
65. Krishnamurthy, V.; Kirtman, B.P. Variability of the Indian Ocean: Relation to Monsoon and ENSO. *Q. J. R. Meteorol. Soc.* **2003**, *129*, 1623–1646. [CrossRef]
66. Annamalai, H.; Xie, S.P.; McCreary, J.P.; Murtugudde, R. Impact of Indian Ocean Sea Surface Temperature on Developing El Niño. *J. Clim.* **2005**, *18*, 302–319. [CrossRef]
67. Li, K.; Liu, Y.; Li, Z.; Yang, Y.; Feng, L.; Khokiattiwong, S.; Yu, W.; Liu, S. Impacts of ENSO on the Bay of Bengal Summer Monsoon Onset via Modulating the Intraseasonal Oscillation. *Geophys. Res. Lett.* **2018**, *45*, 5220–5228. [CrossRef]
68. Li, K.; Yin, Y.; Yang, Y.; Liu, Y.; Yu, W. Dynamic Response of the Spring Wyrtki Jet to the Monsoon Onset Over the Bay of Bengal. *Geophys. Res. Lett.* **2022**, *49*, e2022GL101435. [CrossRef]
69. Huang, K.; Wang, D.; Chen, G.; Nagura, M.; Han, W.; McPhaden, M.J.; Feng, M.; Chen, J.; Wu, Y.; Zhang, X.; et al. Intensification and Dynamics of the Westward Equatorial Undercurrent During the Summers of 1998 and 2016 in the Indian Ocean. *Geophys. Res. Lett.* **2022**, *49*, e2022GL100168. [CrossRef]
70. Srinivas, G.; Amol, P.; Mukherjee, A. Influence of the Extreme Indian Ocean Dipole 2019 on the Equatorial Indian Ocean Circulation. *Clim Dyn* **2024**, *62*, 7111–7125. [CrossRef]
71. Masumoto, Y.; Hase, H.; Kuroda, Y.; Matsuura, H.; Takeuchi, K. Intraseasonal Variability in the Upper Layer Currents Observed in the Eastern Equatorial Indian Ocean. *Geophys. Res. Lett.* **2005**, *32*, 2. [CrossRef]
72. Prerna, S.; Chatterjee, A.; Mukherjee, A.; Ravichandran, M.; Shenoi, S.S.C. Wyrtki Jets: Role of Intraseasonal Forcing. *J. Earth Syst. Sci.* **2019**, *128*, 21. [CrossRef]
73. Wu, R.; Kirtman, B.P. Impacts of the Indian Ocean on the Indian Summer Monsoon–ENSO Relationship. *J. Clim.* **2004**, *17*, 3037–3054. [CrossRef]
74. Krishnan, R.; Ayantika, D.C.; Kumar, V.; Pokhrel, S. The Long-Lived Monsoon Depressions of 2006 and Their Linkage with the Indian Ocean Dipole. *Int. J. Climatol.* **2011**, *31*, 1334–1352. [CrossRef]
75. Roeckner, E.; Oberhuber, J.M.; Bacher, A.; Christoph, M.; Kirchner, I. ENSO Variability and Atmospheric Response in a Global Coupled Atmosphere-Ocean GCM. *Clim. Dyn.* **1996**, *12*, 737–754. [CrossRef]
76. Cai, W.; Borlace, S.; Lengaigne, M.; van Rensch, P.; Collins, M.; Vecchi, G.; Timmermann, A.; Santoso, A.; McPhaden, M.J.; Wu, L.; et al. Increasing Frequency of Extreme El Niño Events Due to Greenhouse Warming. *Nat. Clim. Change* **2014**, *4*, 111–116. [CrossRef]
77. Marjani, S.; Alizadeh-Choobari, O.; Irannejad, P. Frequency of Extreme El Niño and La Niña Events under Global Warming. *Clim. Dyn.* **2019**, *53*, 5799–5813. [CrossRef]
78. Wang, B.; Luo, X.; Yang, Y.-M.; Sun, W.; Cane, M.A.; Cai, W.; Yeh, S.-W.; Liu, J. Historical Change of El Niño Properties Sheds Light on Future Changes of Extreme El Niño. *Proc. Natl. Acad. Sci.* **2019**, *116*, 22512–22517. [CrossRef]

79. Yang, S.; Li, Z.; Yu, J.-Y.; Hu, X.; Dong, W.; He, S. El Niño–Southern Oscillation and Its Impact in the Changing Climate. *Natl. Sci. Rev.* **2018**, *5*, 840–857. [CrossRef]
80. Stuecker, M.F.; Timmermann, A.; Jin, F.-F.; Chikamoto, Y.; Zhang, W.; Wittenberg, A.T.; Widiasih, E.; Zhao, S. Revisiting ENSO/Indian Ocean Dipole Phase Relationships. *Geophys. Res. Lett.* **2017**, *44*, 2481–2492. [CrossRef]

Journal of
Marine Science
and Engineering

MDPI

Article

The Impact of Air–Sea Flux Parameterization Methods on Simulating Storm Surges and Ocean Surface Currents

Li Cai [1], Bin Wang [1,*], Wenqian Wang [1] and Xingru Feng [2]

[1] POWERCHINA Huadong Engineering Corporation Limited, Hangzhou 310014, China; cai_l@hdec.com (L.C.); wang_wq3@hdec.com (W.W.)
[2] Key Laboratory of Ocean Observation and Forecasting, Key Laboratory of Ocean Circulation and Waves, Institute of Oceanology, Chinese Academy of Sciences, Qingdao 266071, China; fengxingru07@qdio.ac.cn
* Correspondence: binwangdut@outlook.com

Abstract: As the primary driver of energy transfer between atmospheric and oceanic systems, the air–sea momentum flux fundamentally governs coupled model dynamics through its regulation of wind stress partitioning. Given the complexity of the physical processes involved, simplified representations of these interactions are widely adopted to balance computational efficiency and physical fidelity. This systematic evaluation of five wind stress parameterizations reveals scheme-dependent variability in momentum partitioning efficiency, particularly under typhoon conditions. Our results quantify how the wind stress drag coefficient's formulation alters atmosphere–ocean feedback, with wave-state aware schemes exhibiting superior surge prediction accuracy compared to wind-speed-dependent approaches. Specifically, a larger wind stress drag coefficient leads to increased atmospheric bottom stress and sea surface stress, resulting in weaker winds and larger sea surface currents and storm surges. These findings provide actionable guidelines into the performance and sensitivity of various air–sea coupled models and offer useful suggestions for improving operational marine forecasting systems.

Keywords: air–sea interaction; parameterization; coupled models; storm surges; surface currents

1. Introduction

The air–sea momentum flux holds a crucial position in air–sea coupled models. It serves as a key factor in mediating the energy and momentum transfer between the ocean and the atmosphere, which is fundamental for driving various oceanic and atmospheric processes. This flux is essential for driving ocean currents, influencing sea surface temperature, and modulating the intensity and track of tropical cyclones [1]. The inclusion of wave models in coupled atmosphere–ocean systems has been shown to significantly improve the representation of air–sea interactions, particularly during extreme events such as storms and typhoons [2–4]. For instance, studies have demonstrated that wave coupling can reduce wind speed and enhance agreement with observations during high-wind conditions [5,6]. Additionally, the interaction between waves and the ocean can lead to changes in SST and mixed layer depth, which further influence atmospheric conditions [7,8]. However, the processes involved in air–sea interaction are highly complex, encompassing the exchange of heat, momentum, water vapor, and other factors. For example, the presence of swell waves can significantly alter air–sea momentum fluxes by modifying surface stress and turbulence properties [9]. Direct numerical simulations of these processes are often computationally intensive and challenging due to the wide range of spatial and temporal scales involved.

Consequently, parameterization methods have been developed to represent these processes in a simplified manner, enhancing the efficiency and feasibility of numerical modeling [1].

Accurate parameterization of the air–sea momentum flux is essential because it directly governs the transfer of momentum from the atmosphere to the ocean. This transfer involves two primary pathways: the immediate transfer to surface waves and the direct transfer to ocean currents through surface friction. The momentum that drives ocean currents originates from two sources: the dissipation of wave energy and the direct influence of wind [10,11]. The momentum transfer from the atmosphere, commonly referred to as wind stress, is described by Equation (1) [12]:

$$\tau = \rho U_*^2 = \rho C_d U_{10}^2 \tag{1}$$

here, τ denotes the wind stress, ρ represents the air density, U_* is the friction velocity, U_{10} is the wind speed measured at a height of 10 m above the sea surface, and C_d is the wind stress drag coefficient.

Theoretically, C_d is governed by the dimensionless roughness length z_0 under neutral atmospheric conditions ($\Psi = 0$), as articulated by the Monin–Obukhov similarity theory [13,14]:

$$C_d^{1/2} = \frac{C_{dn}^{1/2}}{1 - \frac{C_d^{1/2}}{\kappa}\Psi};\ C_{dn} = \frac{\kappa^2}{[\ln(\frac{10}{z_0})]^2} \tag{2}$$

where κ (=0.41) denotes the von Kármán constant. This inverse-square logarithmic relationship reveals that even minor perturbations in z_0 disproportionately amplify uncertainties in C_d. Such sensitivity underscores the necessity of physically consistent z_0 parameterizations to resolve nonlinear feedbacks between wave-state-dependent roughness and momentum transfer efficiency. The accurate quantification of the air–sea momentum flux can be fundamentally reduced to determining the sea surface roughness length (z_0), which encapsulates microscale hydrodynamic processes that govern wind stress generation.

Over the years, laboratory experiments and observational studies have revealed that the wind stress drag coefficient is influenced by three key factors: (1) wave age [15–18], with studies demonstrating its role in modulating drag coefficient through air–sea momentum partitioning; (2) wave steepness [19–21], where empirical relationships between steepness and drag coefficient were established; and (3) sea surface roughness [12,17,22–25], with research quantifying how microscale roughness elements alter turbulent stress. Recent studies have also highlighted the importance of incorporating wave effects into air–sea momentum flux parameterizations, particularly under high-wind conditions [2–4,26]. For instance, Ref. [2] demonstrated that including wave effects in coupled atmosphere–ocean models can significantly improve the representation of air–sea interactions during extreme events.

However, the parameterization schemes for the air–sea momentum flux vary significantly under different oceanic and climatic conditions, which introduces complexity and uncertainty into simulations. For instance, under high-wind-speed conditions, the parameterization of the sea surface wind stress drag coefficient may deviate substantially from the actual conditions. Observations have indicated that when wind speed exceeds a certain threshold, the drag coefficient decreases rather than increases with further increases in wind speed [27–30]. This behavior has been attributed to the formation of sea spray and whitecaps, which can reduce the effective roughness of the sea surface [13,29]. Therefore, optimizing the parameterization schemes of the wind stress drag coefficient under diverse conditions to enhance simulation accuracy remains a significant challenge in current research.

Ref. [11] carried out a study to assess five distinct parameterization methods for air–sea momentum flux. Their research focused on two typhoon scenarios and provided valuable insights into the performance of these methods. Their results indicated that the typhoon's track and minimum sea level pressure showed relatively low sensitivity to the choice of parameterization schemes. However, significant differences were observed in the spatial distribution of the wind stress drag coefficient and its variation with wind speed across the various methods. Notably, the parameterization method that incorporated sea spray effects demonstrated better agreement with the maximum wind speeds observed, particularly at higher wind speeds. Additionally, the air–sea momentum flux is a critical factor in the feedback mechanisms between the atmosphere and ocean during typhoons, influencing sea surface temperature (SST), ocean mixed layer depth, and ocean currents [29,31].

This study aims to evaluate the impact of air–sea momentum flux parameterization schemes on storm surge and ocean current simulations using a coupled atmosphere–wave–ocean model. Unlike previous studies focusing on isolated factors (e.g., wave age or wind speed), our work integrates multiple influences (sea spray, regional variations, and high-wind dynamics) to provide a holistic understanding of air–sea interactions. By comparing five schemes under typhoon conditions, we identify optimal parameterizations for coastal hazard prediction and demonstrate the necessity of incorporating sea spray and wave-state feedbacks into coupled models. The structure of this paper is organized as follows: Section 2 offers a comprehensive description of the model, model configuration, and the design of numerical experiments. Section 3 examines the influence of the drag coefficient (C_d) on the simulation outcomes. Section 4 provides a detailed discussion of the findings, and the key conclusions are summarized in Section 5.

2. Model and Methods

2.1. Coupled Model Configuration

This section discusses the impact of different air–sea momentum flux parameterization techniques on atmospheric and oceanic simulation outcomes, utilizing a coupled atmosphere–wave–ocean model. Ref. [11] have already provided a detailed introduction to the model and parameterization methods. This paper will briefly introduce the relevant model settings, and for more details, please refer to their article.

This research utilizes the COAWST model system, version 3.1, developed by Ref. [32] and further improved in Ref. [33], which integrates three core components:

(1) Atmosphere model: The Advanced Weather Research and Forecasting (WRF; [34] Model, version 3.6, configured with a horizontal resolution of $1/12° \times 1/12°$ and 31 vertical sigma layers. The model's domain spans 15° S–45° N and 99° E–135° E, initialized using the NCEP Final Operational Global Analysis (FNL) dataset (https://rda.ucar.edu/ (accessed on 1 March 2020), 0.25° resolution, 6 hourly intervals).

(2) Ocean model: The Regional Ocean Modeling System (ROMS), with the same horizontal resolution as WRF but 30 s-coordinate layers in the vertical. Boundary conditions are derived from the ETOPO1 dataset, which is provided by the National Geophysical Data Center (NGDC) (https://www.ncei.noaa.gov/ (accessed on 1 March 2020)).

(3) Wave model: the Simulating Waves Nearshore (SWAN) model operates on a coincident grid with identical bathymetric inputs to ROMS, ensuring spatial consistency across coupled components.

2.2. Air–Sea Momentum Flux Parameterization Schemes

Five air–sea momentum flux parameterization methods are compared, focusing on the wind stress drag coefficient (C_d) formulation. The key schemes include:

(1). WRF–ROMS—without explicit wave effect:

$$z_0 = \frac{0.018}{g}u_*^2 \tag{3}$$

(2). WRF–ROMS–SWAN1—considering wave height and wave steepness [20]:

$$z_0 = 1200H_s(\frac{H_s}{L_p})^{4.5} \tag{4}$$

(3). WRF–ROMS–SWAN2—considering wave height and wave age [18]:

$$z_0 = 3.35H_s(\frac{u_*}{C_p})^{3.4} \tag{5}$$

(4). WRF–ROMS–SWAN3—considering wavelength and wave age [17]:

$$z_0 = \frac{25}{\pi}L_p(\frac{u_*}{C_p})^{4.5} \tag{6}$$

(5). WRF–ROMS–SWAN4—considering wave age and sea spray [12]:

$$\frac{gz_0}{u_*^2} = \begin{cases} (0.085\beta_*^{3/2})^{1-\frac{1}{\omega}}[0.03\beta_* exp(-0.14\beta_*)]^{\frac{1}{\omega}}, & 0.35 < \beta_* < 35 \\ 17.6^{1-\frac{1}{\omega}}(0.008)^{\frac{1}{\omega}}, & \beta_* > 35 \end{cases} \tag{7}$$

here, u_* denotes the friction velocity, g represents gravitational acceleration, H_s corresponds to the significant wave height, and spectral wave characteristics are quantified through C_p (the phase speed of the spectral peak) and L_p (the wavelength of the spectral peak); β_* is defined as $C_p/u_*, \omega = min(1, a_{rc}/\kappa u_*)$, in which a_{rc} represents the equilibrium settling velocity of marine spray droplets, empirically determined as $0.64\ m/s$ through aerosol dynamic measurements.

2.3. Numerical Experiments

This investigation employs numerical simulations to examine two 2016 typhoon events (Haima and Nida), initialized with three-day forecast data. Our analysis concentrates on outputs generated by an advanced air–sea coupled modeling system. The modeling framework builds upon established coupling methodologies, with comprehensive validation procedures documented in prior research [11].

3. Results

3.1. Surface Currents

The parameterization of sea surface roughness plays a crucial role in modulating the momentum exchange between the atmosphere and the ocean, which, in turn, impacts the ocean's dynamic processes, with surface currents being of particular significance. This section examines the simulated surface currents. When a typhoon is located over the open ocean, the sea surface currents near the radius of maximum wind speed are predominantly influenced by surface winds and exhibit minimal interaction with coastlines or tidal currents. The intense winds associated with a typhoon exert wind stress on the sea surface, facilitating the transfer of momentum from the atmosphere to the ocean and driving ocean surface currents [13]. Moreover, the low-pressure center of a typhoon generates a pressure gradient force that induces water to flow from regions of higher pressure (outside the typhoon) toward regions of lower pressure (inside the typhoon). Consequently, ocean surface currents converge toward the typhoon's center, particularly in the vicinity of the maximum wind speed radius. This convergence can result in localized enhancements in surface current speeds.

To conduct the analysis, the simulation results of Typhoon Haima at the 51st hour were chosen. The corresponding simulated surface currents are presented in Figure 1, which provides a visual representation of the current patterns. Consistent with typhoon-induced current asymmetry, rightward track regions exhibited 15–20% higher surface current velocities compared to leftward sectors (Figure 1). This observation aligns with the asymmetric characteristics of typhoon-induced currents, where the right side of the typhoon's path usually experiences stronger wind forcing, leading to enhanced surface currents.

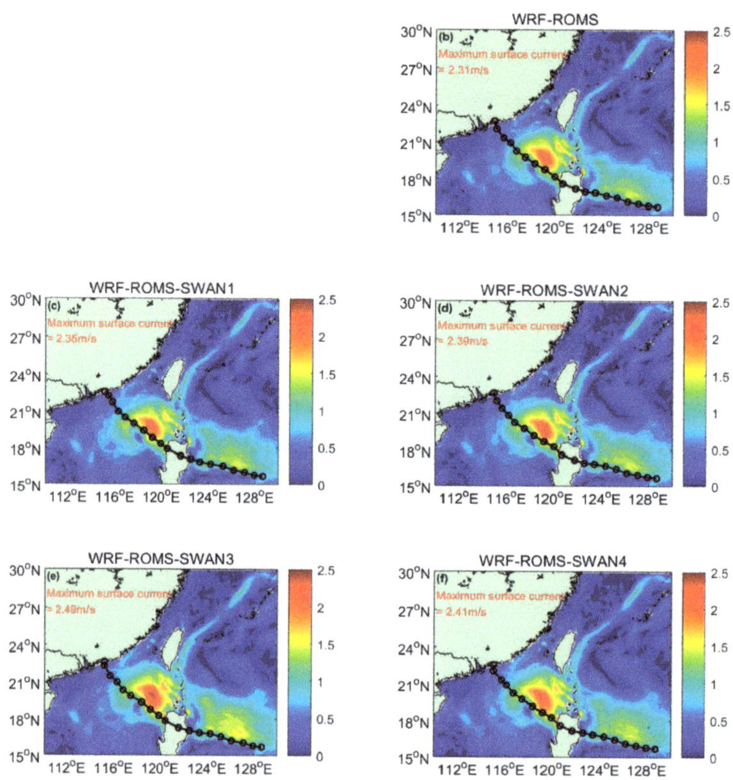

Figure 1. Surface currents at the time of 51st hour of the simulation for Typhoon Haima. The values of the simulated maximum surface current velocity for the different numerical experiments are also shown in the corresponding subplot.

When comparing the simulation results across different experiments, the maximum ocean surface currents simulated by each experiment were 2.31 m/s, 2.35 m/s, 2.39 m/s, 2.49 m/s, and 2.41 m/s, corresponding to the WRF-ROMS and WRF-ROMS-SWAN experiments 1 through 4, respectively. Although the maximum surface currents simulated by each group of experiments are quite similar, the WRF-ROMS-SWAN3 experiment demonstrates the broadest spatial influence, suggesting that this particular parameterization scheme may be more effective in capturing the extent of the surface currents. This could be attributed to a more accurate depiction of wind stress and its impact on the ocean surface.

Notably, the simulated wind speed in the WRF-ROMS-SWAN3 experiment is significantly weaker than that in the other experiments at the 51st hour of simulation (as shown in Figure 4 of the work of Ref. [11]. This raises an interesting point: despite the weaker wind speeds, the WRF-ROMS-SWAN3 experiment produces stronger surface currents. The inverse relationship between simulated wind speeds and surface current magnitudes in WRF-ROMS-SWAN3 reveals enhanced momentum transfer efficiency through its wave-state-dependent C_d formulation.

To elucidate this phenomenon, the C_d and sea surface wind stress at the corresponding time were also examined. The spatial distributions of C_d in different experiments are illustrated in Figure 9 of the work of Ref. [11], where significant differences are observed. In the WRF-ROMS experiment, C_d is mainly correlated with wind speed; hence, its spatial distributions are highly similar. In contrast, in the WRF-ROMS-SWAN experiments 1 to 4, the spatial distribution of C_d is influenced not only by wind speed but also by the wave field. Notably, the C_d value in the WRF-ROMS-SWAN3 experiment is significantly higher than that in the other experiments, which may contribute to its enhanced performance in simulating surface currents despite lower wind speeds.

Wind stress was computed using the wind speed and C_d data according to Equation (1) during the simulation, and the wind stress distribution is illustrated in Figure 2. The results revealed that the wind stress in the WRF-ROMS-SWAN3 experiment was the highest among all experiments, attributed to its larger C_d value, despite having a lower wind speed compared to the other experiments. This indicates that the greater sea surface roughness in the WRF-ROMS-SWAN3 experiment led to increased bottom stress in the atmospheric model, which subsequently weakened the wind field relative to the other experiments. However, the sea surface wind stress was also higher than in the other experiments, thereby driving stronger sea surface currents.

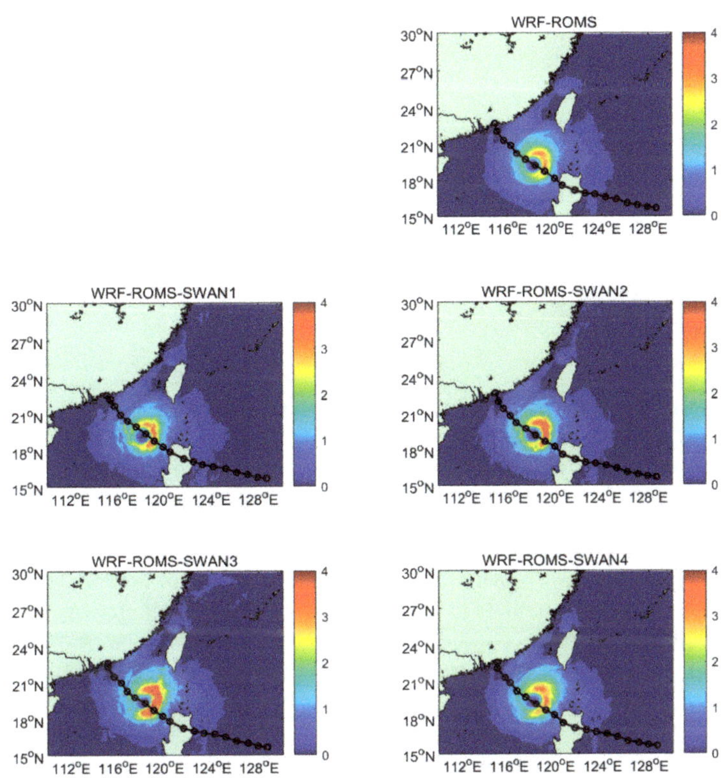

Figure 2. Sea surface wind stress at the time of 51st hour of the simulation for Typhoon Haima.

The comparison of surface currents in different experiments, along with the analysis of C_d and sea surface wind stress, provides valuable insights into the performance and sensitivity of various air–sea coupled models. It highlights the importance of optimizing the parameterization schemes of C_d and understanding the complex feedback mechanisms between the atmosphere and ocean. The results of this research provide valuable guidance for enhancing the operational marine forecasting system and promoting the development of more integrated models that are capable of predicting storm surges and sea level anomalies.

For Typhoon Nida, the intensity is not as strong as that of Typhoon Haima, and consequently, the surface currents caused by Typhoon Nida are also relatively smaller. However, the conclusion remains consistent: the results from the WRF-ROMS-SWAN3 experiment show the largest surface currents, with a maximum value of 2.19 m/s (Figure 3). We also analyzed the C_d and wind stresss, which were found to be the largest in the WRF-ROMS-SWAN3 experiment as well (For brevity the figures are not shown here).

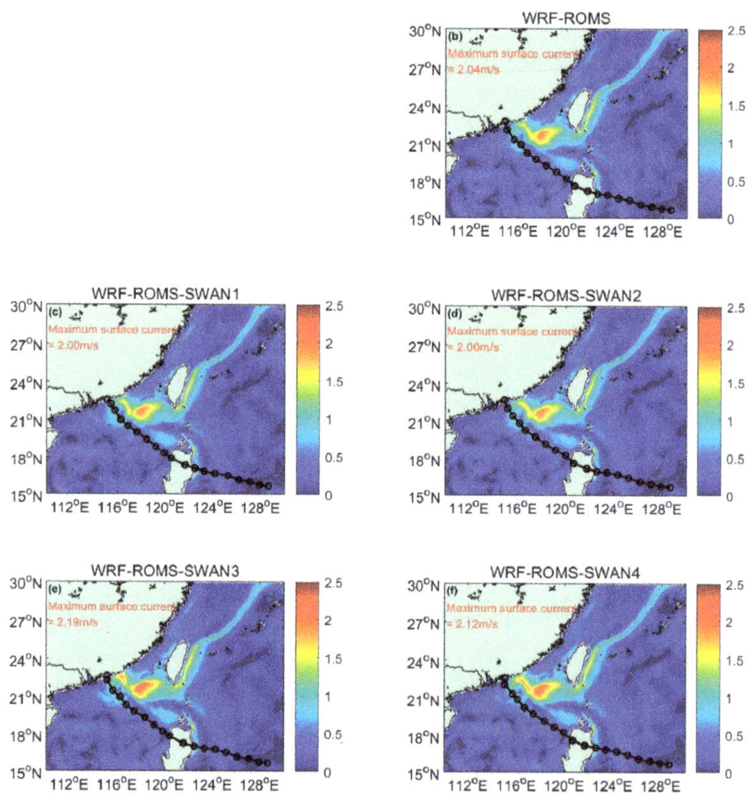

Figure 3. Surface currents at the time of 27th hour of the simulation for Typhoon Nida.

Although Typhoon Nida was less intense, the surface current and wave field characteristics it induced were similar to those of Typhoon Haima, particularly in terms of the higher flow velocity on the right side of the typhoon's track. The WRF-ROMS-SWAN3 experiment demonstrated good performance in simulating the surface currents of both typhoons, indicating that the parameterization scheme used in this experiment is highly applicable across typhoons of different intensities. These findings help to improve and validate air–sea coupled models, enhancing the simulation and prediction capabilities for oceanic dynamic processes induced by typhoons.

3.2. Storm Surge

Another important ocean dynamic process during a typhoon is the storm surge, which can lead to significant economic losses and casualties. For instance, during Typhoon Haima, the storm surge disaster resulted in an economic loss of approximately USD 1.93 billion. The maximum storm surge observed occurred on the right flank of the typhoon's landfall location. To elucidate the influence of different sea surface roughness parameterization schemes on the simulation of storm surges, Figure 4 presents the simulated maximum storm surges for various experiments. These simulations are crucial for understanding and predicting the potential damage of storm surges, which can affect a wide range of sectors, including agriculture, aquaculture, shipping, and infrastructure. Accurate simulations can help in developing more effective disaster prevention and mitigation strategies to reduce the economic and social impacts of such natural disasters.

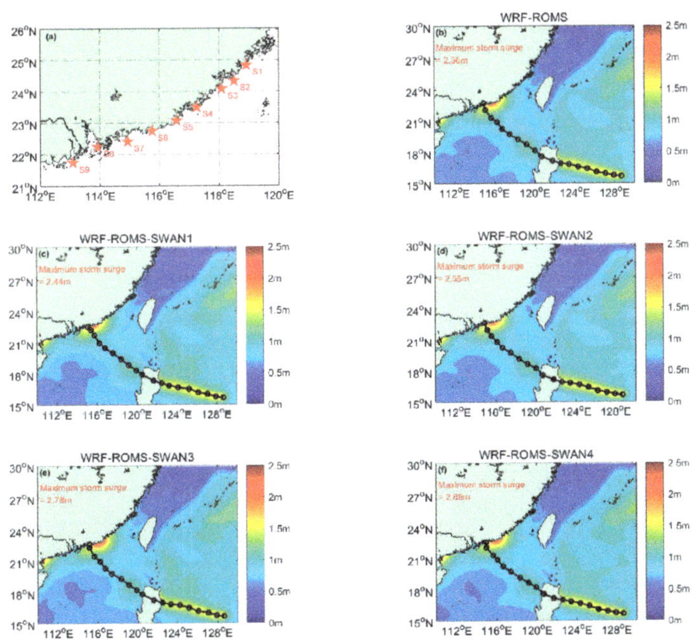

Figure 4. Simulated maximum storm surge distribution during Typhoon Haima for five numerical experiments (**a–f**). The red stars represent the nine stations selected for analyzing the storm surge time series.

As shown in Figure 4, the simulated maximum storm surge value differs among the experiments. In the WRF-ROMS-SWAN1 experiment, it reaches 2.44 m, while in the WRF-ROMS-SWAN3 experiment, it reaches 2.78 m. The significant storm surge predominantly occurred along the coastal regions situated to the right of the typhoon's trajectory. This is attributed to the onshore wind direction in this region during the typhoon, which pushes seawater towards the coast, thereby increasing the sea level [35].

To provide a more detailed analysis of the storm surge processes, Figure 5 illustrates the storm surge time series recorded at nine strategically chosen stations during Typhoon Haima. These stations were selected based on their locations in the coastal regions where the storm surge impact was most significant, as shown in Figure 4a. Specifically, stations S1 to S5 are located on the right flank of the typhoon's path, stations S6 and S7 are positioned near the typhoon center, and stations S8 and S9 are on the left flank of the typhoon.

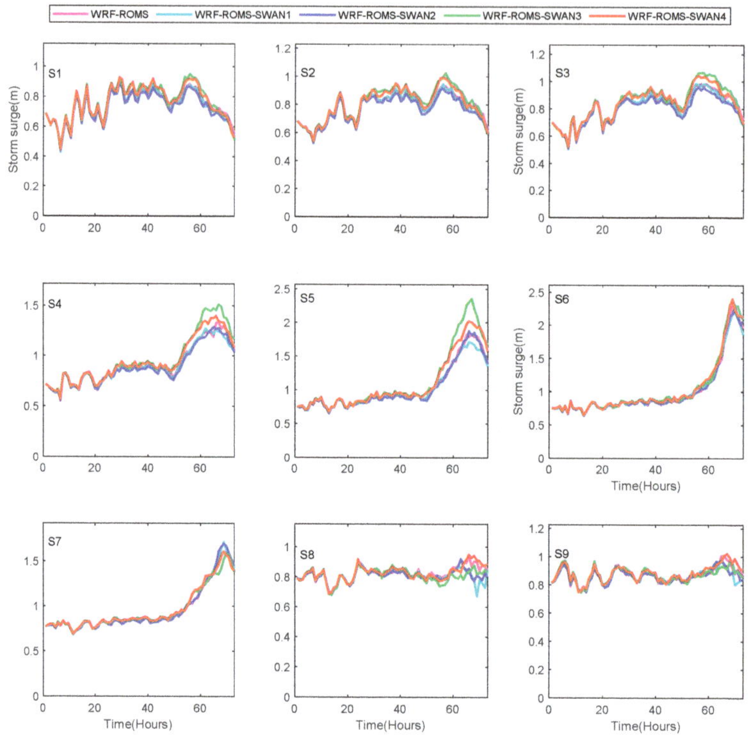

Figure 5. Storm surge time series during Typhoon Haima at coastal stations; see Figure 4a for station locations.

Through a comparative analysis of the storm surge simulations from the five experiments (as shown in Figures 4 and 5), it is apparent that at stations S1 to S5, the storm surge simulated by the WRF-ROMS-SWAN3 experiment is significantly higher than that from the other experiments. This difference is likely attributable to the influence of wind stress, which is a key factor in determining storm surge height. In contrast, the results from stations S6 and S7, which are located near the typhoon center, show relatively consistent storm surge magnitudes across all experiments. On the left side of the typhoon's path, a process of storm surge reduction may occur, characterized by a decrease in sea level. This phenomenon is explained by the strong winds near the typhoon's center, which push seawater in the direction of the typhoon's movement, leading to a relative decrease in the sea level on the left side.

Figure 6 presents time series data of wind speed, C_d, wind stress, and sea level pressure to illustrate the differences in storm surge outcomes among various experiments. For clarity, the results are shown only for three representative stations. The figure indicates that the wind speed in the WRF-ROMS-SWAN3 experiment was generally lower compared to that of other experiments at these stations. However, the wind stress in WRF-ROMS-SWAN3 was higher, primarily due to the larger C_d values in this experiment. This higher wind stress explains why the storm surge simulated by WRF-ROMS-SWAN3 was generally the highest during Typhoon Haima.

At station S8, the maximum simulated storm surge in the WRF-ROMS-SWAN3 experiment was lower than in other experiments. This is attributed to the station's location on the left side of the typhoon track and its proximity to the landfall point. Here, the impact of sea level pressure on the storm surge is substantial, overshadowing the influence of wind stress. Specifically, the sea level pressure can counteract the wind stress, resulting in a decrease in storm surge height.

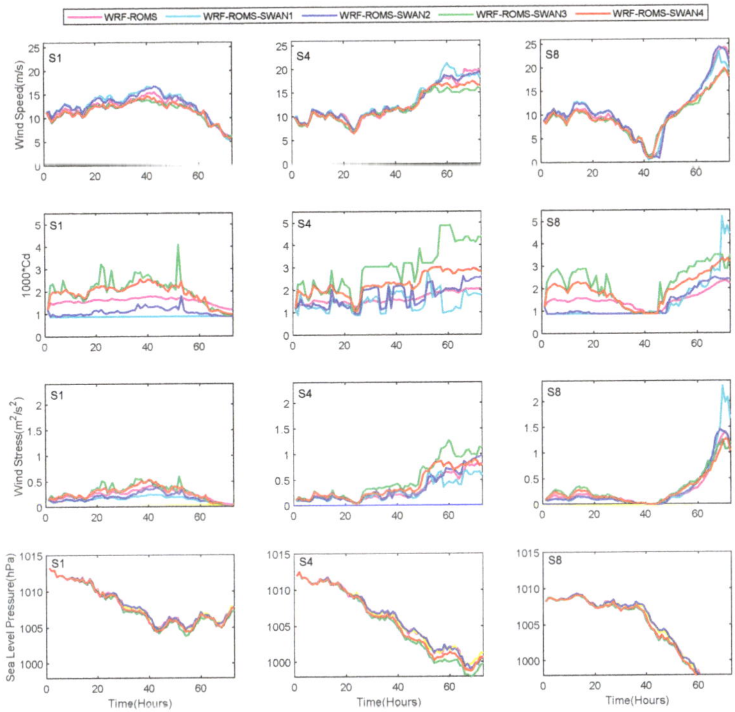

Figure 6. Time series of wind speed, C_d, wind stress and sea level pressure at stations S1, S4 and S8 along the coast during typhoon Haima.

These results highlight the necessity of incorporating both wind stress and sea level pressure into storm surge modeling, particularly in regions near the typhoon's landfall point. The precise depiction of these factors is essential for enhancing the accuracy of storm surge predictions, thereby supporting more effective disaster preparedness and mitigation efforts.

For Typhoon Nida, the simulation results from different models exhibit some discrepancies. The WRF-ROMS-SWAN1 model simulates a maximum storm surge of 2.35 m, whereas the WRF-ROMS-SWAN4 model simulates a maximum storm surge of 2.66 m (as shown in Figure 7). These differences may be associated with various factors, including model configuration, parameterization methods, and initial and boundary conditions. For example, under different coupling modes, the models have different accuracies in simulating the path and intensity of the typhoon, which, in turn, affects the simulation results of the storm surge. Overall, a difference of 0.31 m is relatively significant in the simulation of storm surges, indicating that different model configurations have a fairly noticeable impact on the simulation results. Additionally, since the maximum wind speed of Typhoon Nida is higher than that of Typhoon Haima, the resulting maximum storm surge is also larger. We used the same nine stations to analyze the specific storm process. From the specific storm surge process (Figure 8), the conclusion is consistent with Haima; that is, for S1–S5 stations, the maximum storm surge simulated by WRF-ROMS-SWAN3 is the largest.

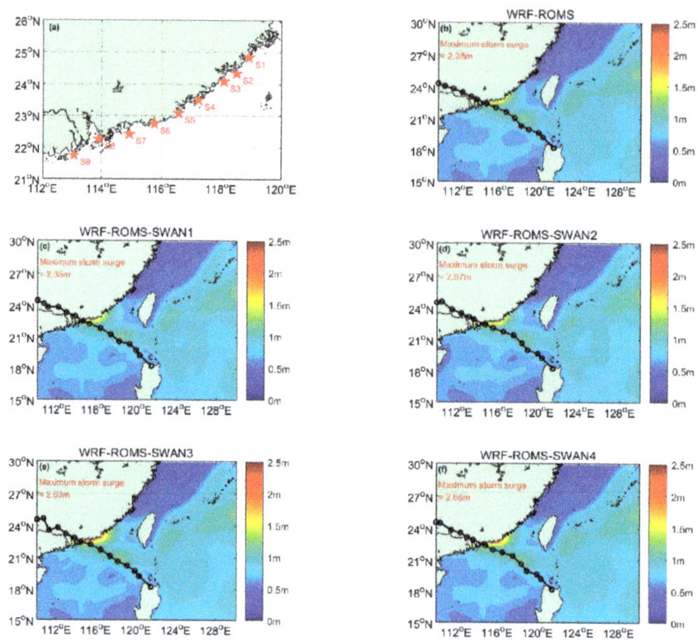

Figure 7. Simulated maximum storm surge distribution during Typhoon Nida for five numerical experiments (**a–f**). The red stars represent the nine stations selected for analyzing the storm surge time series.

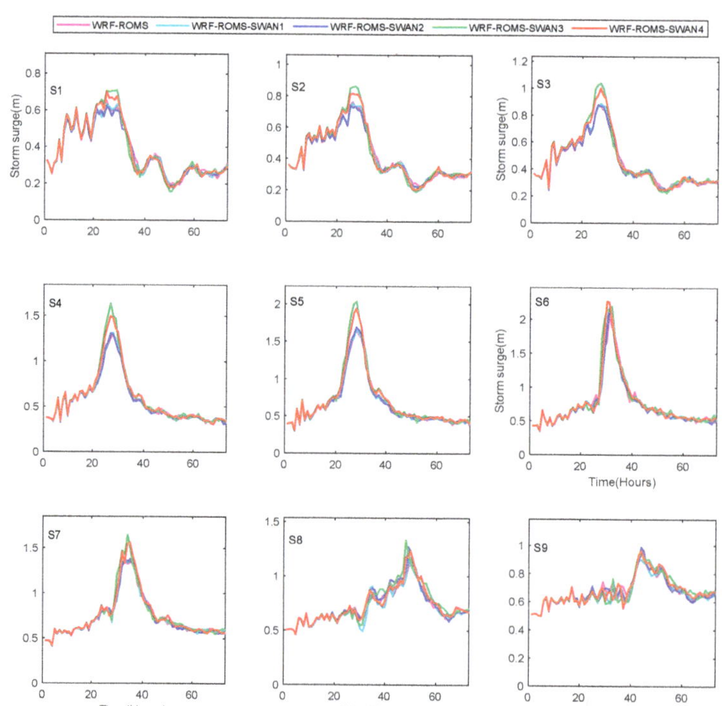

Figure 8. Storm surge time series during Typhoon Nida at coastal stations; see Figure 7a for station locations.

We also presented the charts of wind speed, C_d, wind stress, and sea surface pressure for three stations (S1, S4, and S8) during Typhoon Nida (Figure 9). It is evident that, for most of the time at these stations, the wind speed in the WRF-ROMS-SWAN3 experiment is the lowest compared to other experiments. However, due to the higher C_d values in WRF-ROMS-SWAN3, the wind stress in this experiment is greater than that in the other experiments. This is why the storm surge of WRF-ROMS-SWAN3 is the largest for most of the time. However, the wind speed, C_d, and wind stress of WRF-ROMS-SWAN4 are quite similar to those of WRF-ROMS-SWAN3, and at times, they even surpass the results of WRF-ROMS-SWAN3. This leads to WRF-ROMS-SWAN4 simulating the largest storm surge. At station S8, the sea surface pressure is notably lower than at stations S1 and S4, and this has a significant impact. Therefore, the simulation results from all experimental groups at station S8 are relatively consistent.

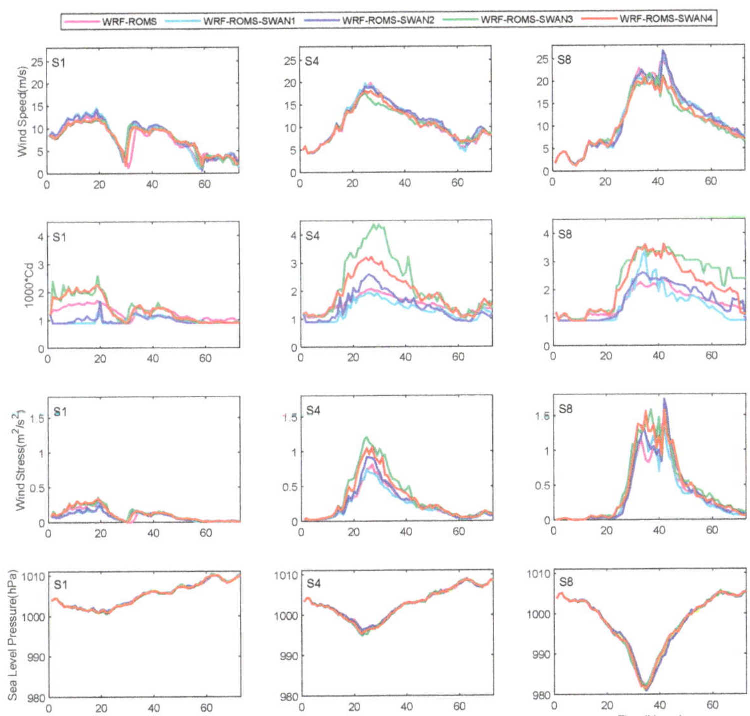

Figure 9. Time series of wind speed, C_d, wind stress and sea level pressure at stations S1, S4 and S8 along the coast during typhoon Nida.

4. Discussion

The WRF-ROMS-SWAN3 experiment shows excellent performance in simulating surface currents and storm surges. This can be mainly ascribed to its distinctive way of representing the wind stress drag coefficient (C_d), which is closely related to wave dynamics. The coupling between them plays a crucial role in enhancing the simulation accuracy. In traditional parameterizations (e.g., WRF-ROMS), C_d is primarily governed by wind speed, leading to a simplified linear relationship that fails to capture the saturation of momentum transfer under high-wind conditions [29]. In contrast, the WRF-ROMS-SWAN3 experiment incorporates wave-state dependencies, where C_d is modulated by both wind speed and wave age. This dual dependency enhances the realism of sea surface roughness calculations, as waves with longer fetch or higher steepness increase surface roughness [20], thereby amplifying wind stress even under lower wind speeds. This mechanism explains why WRF-ROMS-SWAN3 generated stronger surface currents despite weaker simulated winds (Figure 6): the elevated C_d compensated for reduced wind momentum by enhancing the efficiency of energy transfer from the atmosphere to the ocean.

The asymmetry in surface current intensity (stronger currents on the typhoon's right flank) further highlights the role of wave–current interactions. During typhoon passage, wind–wave alignment on the right side generates younger, steeper waves, which increase local roughness and C_d [17]. These transient wave states intensify wind stress in specific regions, driving stronger convergence of surface currents toward the typhoon center. Conversely, on the left flank, wave decoupling from wind direction reduces roughness, leading to weaker stress and current velocities. This spatial heterogeneity underscores the necessity of integrating wave dynamics into coupled models to resolve momentum transfer gradients.

For storm surges, the dominance of wind stress over sea level pressure at coastal stations (S1–S5) arises from the combined effects of onshore winds and wave setup. The WRF-ROMS-SWAN3 experiment's higher C_d amplified wind stress, which directly enhanced water mass transport toward the coast. However, near the typhoon center (stations S6 and S7), the balance between wind stress and pressure gradient forces minimized inter-experiment differences. Here, the low-pressure-induced suction effect dominated surge dynamics, masking variations caused by C_d parameterizations. This dichotomy illustrates how localized feedback mechanisms—wind stress in shallow coastal zones versus pressure gradients in deep offshore regions—govern surge magnitude.

The discrepancies between Typhoons Haima and Nida further emphasize environmental dependencies. Nida's higher maximum wind speeds likely induced stronger wave breaking and sea spray, which nonlinearly modulate C_d [28]. In such conditions, parameterizations that neglect spray-mediated momentum dissipation (e.g., WRF-ROMS-SWAN1) overestimate C_d, whereas schemes accounting for spray effects (e.g., WRF-ROMS-SWAN3) better replicate the observed saturation of drag coefficients [12]. This explains why WRF-ROMS-SWAN3 maintained accuracy across both typhoons, despite differences in intensity.

Ultimately, the sensitivity of simulations to C_d parameterizations stems from the interplay between atmospheric forcing, wave-state variability, and ocean response. Models that decouple these components (e.g., WRF-ROMS) oversimplify momentum transfer pathways, while coupled frameworks (e.g., WRF-ROMS-SWAN) resolve critical feedback loops. Future efforts should prioritize dynamic C_d schemes that adapt to real-time wave and spray conditions, particularly in coastal zones where surge predictions hinge on precise wind stress estimates.

5. Conclusions

The research results highlight the significant influence of air–sea momentum flux parameterization, especially the representation of the wind stress drag coefficient (C_d), on the accuracy of coupled atmosphere–wave–ocean models when simulating typhoon-induced surface currents and storm surges. The superior performance of the WRF-ROMS-SWAN3 experiment stems from its explicit incorporation of wave-state dependencies into C_d, which dynamically accounts for wavelength and wave age. This approach enhances sea surface roughness calculations, enabling more efficient momentum transfer from the atmosphere to the ocean, even under high-wind conditions where traditional wind-speed-dependent schemes (e.g., WRF-ROMS) fail to capture saturation effects. The resultant amplification of wind stress compensates for weaker simulated winds, driving stronger surface currents and elevated storm surges in coastal regions.

The spatial asymmetry in surface currents—stronger flows on the typhoon's right flank—is mechanistically linked to wave–current interactions. Younger, steeper waves on the right side increase local roughness, intensifying wind stress and current convergence toward the typhoon center. Conversely, reduced coupling between wind and waves on the left flank diminishes stress and flow velocities. For storm surges, wind stress dominates in shallow coastal zones (e.g., stations S1–S5), where elevated C_d enhances onshore water transport, while pressure gradients prevail near the typhoon center (e.g., stations S6 and S7), minimizing parameterization-induced discrepancies.

The variation in model performance across typhoons (Haima vs. Nida) underscores the environmental sensitivity of C_d. Under extreme winds (e.g., Typhoon Nida), schemes neglecting sea-spray-mediated momentum dissipation (e.g., WRF-ROMS-SWAN1) overestimate drag, whereas wave-coupled frameworks (e.g., WRF-ROMS-SWAN3) better replicate observed saturation trends. This highlights the necessity of adaptive C_d parameterizations that integrate real-time wave and spray dynamics, particularly for high-intensity typhoons.

To advance operational marine forecasting, future efforts should prioritize the development of dynamically coupled models that resolve atmosphere–wave–ocean feedbacks, with a focus on coastal regions where surge prediction accuracy hinges on precise wind stress estimates. Additionally, expanding validation datasets to include diverse typhoon intensities and oceanic conditions will further refine parameterization schemes, ultimately enhancing the reliability of storm surge and surface current forecasts.

Author Contributions: Conceptualization, B.W.; Data Curation, X.F.; Formal Analysis, W.W.; Methodology, L.C.; Writing—Original Draft, L.C.; Writing—Review and Editing, B.W. All authors have read and agreed to the published version of the manuscript.

Funding: This work was financially supported by the National Key R&D Program of China (2023YFB4203200).

Institutional Review Board Statement: Not applicable.

Informed Consent Statement: Not applicable.

Data Availability Statement: The raw data supporting the conclusions of this article will be made available by the authors on request.

Conflicts of Interest: Authors Li Cai, Bin Wang and Wenqian Wang are employed by POWERCHINA Huadong Engineering Corporation Limited, others declare no conflict of interest.

References

1. Bhattacharya, T.; Chakraborty, K.; Anthoor, S.; Ghoshal, P.K. An assessment of air-sea CO_2 flux parameterizations during tropical cyclones in the Bay of Bengal. *Dyn. Atmos. Oceans* **2023**, *103*, 101390. [CrossRef]
2. Gröger, M.; Dieterich, C.; Haapala, J.; Ho-Hagemann, H.T.M.; Hagemann, S.; Jakacki, J.; May, W.; Meier, H.E.M.; Miller, P.A.; Rutgersson, A.; et al. Coupled regional earth system modeling in the Baltic Sea region. *Earth Syst. Dynam.* **2021**, *12*, 939–973. [CrossRef]
3. Wahle, K.; Staneva, J.; Koch, W.; Fenoglio-Marc, L.; Ho-Hagemann, H.T.M.; Stanev, E.V. An atmosphere–wave regional coupled model: Improving predictions of wave heights and surface winds in the southern north Sea. *Ocean Sci.* **2017**, *13*, 289–301. [CrossRef]
4. Wiese, A.; Stanev, E.; Koch, W.; Behrens, A.; Geyer, B.; Staneva, J. The impact of the two-way coupling between wind wave and atmospheric models on the lower atmosphere over the north Sea. *Atmosphere* **2019**, *10*, 386. [CrossRef]
5. Wiese, A.; Staneva, J.; Ho-Hagemann, H.T.M.; Grayek, S.; Koch, W.; Schrum, C. Internal model variability of ensemble simulations with a regional coupled wave-atmosphere model GCOAST. *Front. Mar. Sci.* **2020**, *7*, 596843. [CrossRef]
6. Varlas, G.; Spyrou, C.; Papadopoulos, A.; Korres, G.; Katsafados, P. One-year assessment of the CHAOS two-way coupled atmosphere-ocean wave modelling system over the Mediterranean and black seas. *Mediterr. Mar. Sci.* **2020**, *21*, 372–385. [CrossRef]

7. Breivik, Ø.; Mogensen, K.; Bidlot, J.R.; Balmaseda, M.A.; Janssen, P.A.E.M. Surface wave effects in the NEMO ocean model: Forced and coupled experiments. *J. Geophys. Res. Oceans* **2015**, *120*, 2973–2992. [CrossRef]
8. Alari, V.; Staneva, J.; Breivik, Ø.; Bidlot, J.R.; Mogensen, K.; Janssen, P. Surface wave effects on water temperature in the Baltic Sea: Simulations with the coupled NEMO-WAM model. *Ocean Dyn.* **2016**, *66*, 917–930. [CrossRef]
9. Wu, L.; Sahlée, E.; Nilsson, E.; Rutgersson, A. A review of surface swell waves and their role in air–sea interactions. *Ocean Modell.* **2024**, *190*, 102397. [CrossRef]
10. Fan, Y.; Ginis, I.; Hara, T. Momentum flux budget across the air–sea interface under uniform and tropical cyclone winds. *J. Phys. Oceanogr.* **2010**, *40*, 2221–2242. [CrossRef]
11. Feng, X.; Sun, J.; Yang, D.; Yin, B.; Gao, G.; Wan, W. Effect of drag coefficient parameterizations on air–sea coupled simulations: A case study for Typhoons Haima and Nida in 2016. *J. Atmos. Oceanic Technol.* **2021**, *38*, 977–993. [CrossRef]
12. Liu, B.; Guan, C.; Xie, L. The wave state and sea spray related parameterization of wind stress applicable from low to extreme winds. *J. Geophys. Res.* **2012**, *117*, C00J22. [CrossRef]
13. Fairall, C.W.; Bradley, E.F.; Hare, J.E.; Grachev, A.A.; Edson, J.B. Bulk parameterization of air-sea fluxes for tropical ocean-global atmosphere coupled-ocean-atmosphere response experiment. *J. Geophys. Res. Oceans* **1996**, *101*, 3747–3764. [CrossRef]
14. Grachev, A.A.; Fairall, C.W.; Larsen, S.E. On the determination of the neutral drag coefficient in the convective boundary layer. *Bound.-Layer Meteorol.* **1998**, *86*, 257–278. [CrossRef]
15. Stewart, R.W. The air-sea momentum exchange. *Bound.-Layer Meteorol.* **1974**, *6*, 151–167. [CrossRef]
16. Toba, Y.; Iida, N.; Kawamura, H.; Ebuchi, N.; Jones, I.S.F. Wave dependence of sea-surface wind stress. *J. Phys. Oceanogr.* **1990**, *20*, 705–721. [CrossRef]
17. Oost, W.A.; Komen, G.J.; Jacobs, C.M.J.; Van Oort, C. New evidence for a relation between wind stress and wave age from measurements during ASGAMAGE. *Bound.-Layer Meteorol.* **2002**, *103*, 409–438. [CrossRef]
18. Drennan, W.M.; Taylor, P.K.; Yelland, M.J. Parameterizing the sea surface roughness. *J. Phys. Oceanogr.* **2005**, *35*, 835–848. [CrossRef]
19. Anctil, F.; Donelan, M.A. Air-water momentum flux observations over shoaling waves. *J. Phys. Oceanogr.* **1996**, *26*, 1344–1353. [CrossRef]
20. Taylor, P.K.; Yelland, M.J. The dependence of sea surface roughness on the height and steepness of the waves. *J. Phys. Oceanogr.* **2001**, *31*, 572–590. [CrossRef]
21. Guan, C.; Xie, L. On the linear parameterization of drag coefficient over sea surface. *J. Phys. Oceanogr.* **2004**, *34*, 2847–2851. [CrossRef]
22. Smith, S.D.; Anderson, R.J.; Oost, W.A.; Kraan, C.; Maat, N.; De Cosmo, J. Sea surface wind stress and drag coefficients: The HEXOS results. *Bound.-Layer Meteorol.* **1992**, *60*, 109–142. [CrossRef]
23. Johnson, H.K.; Højstrup, J.; Vested, H.J.; Larsen, S.E. On the dependence of sea surface roughness on wind waves. *J. Phys. Oceanogr.* **1998**, *28*, 1702–1716. [CrossRef]
24. Drennan, W.M.; Graber, H.C.; Hauser, D.; Quentin, C. On the wave age dependence of wind stress over pure wind seas. *J. Geophys. Res.* **2003**, *108*, 8062. [CrossRef]
25. Gao, Z.; Wang, Q.; Wang, S. An alternative approach to sea surface aerodynamic roughness. *J. Geophys. Res.* **2006**, *111*, D22108. [CrossRef]
26. Hu, Y.; Shao, W.; Wang, X.; Zuo, J.; Jiang, X. Analysis of wave breaking on synthetic aperture radar at C-band during tropical cyclones. *Geo-Spat. Inf. Sci.* **2024**, *27*, 2109–2122. [CrossRef]
27. Alamaro, M. Wind Wave Tank for Experimental Investigation of Momentum and Enthalpy Transfer from the Ocean Surface at High Wind Speed. Master's Thesis, Massachusetts Institute of Technology, Cambridge, MA, USA, 2001.
28. Alamaro, M.; Emanuel, K.; Colton, J.; McGillis, W.; Edson, J. Experimental investigation of air–sea transfer of momentum and enthalpy at high wind speed. In Proceedings of the 25th Conference on Hurricanes and Tropical Meteorology, San Diego, CA, USA, 29 April–3 May 2002; pp. 667–668.
29. Powell, M.D.; Vickery, P.J.; Reinhold, T.A. Reduced drag coefficient for high wind speeds in tropical cyclones. *Nature* **2003**, *422*, 279–283. [CrossRef]
30. Lin, S.; Sheng, J. Revisiting dependences of the drag coefficient at the sea surface on wind speed and sea state. *Cont. Shelf Res.* **2020**, *207*, 104188. [CrossRef]
31. Businger, J.A.; Wyngaard, J.C.; Izumi, Y.; Bradley, E.F. Flux-profile relationships in the atmospheric boundary layer. *J. Atmos. Sci.* **1971**, *28*, 181–189. [CrossRef]
32. Skamarock, W.C.; Klemp, J.B.; Dudhia, J.; Gill, D.O.; Barker, D.M.; Wang, W.; Powers, J.G. *A Description of the Advanced Research WRF Version 2*; National Center for Atmospheric Research: Boulder, CO, USA, 2005; p. 88. [CrossRef]
33. Warner, J.C.; Sherwood, C.R.; Signell, R.P.; Harris, C.K.; Arango, H.G. Development of a three-dimensional, regional, coupled wave, current, and sediment-transport model. *Comput. Geosci.* **2008**, *34*, 1284–1306. [CrossRef]

34. Warner, J.C.; Armstrong, B.; He, R.; Zambon, J.B. Development of a Coupled Ocean–Atmosphere–Wave–Sediment Transport (COAWST) modeling system. *Ocean Modell.* **2010**, *35*, 230–244. [CrossRef]
35. Leung, N.-C.; Chow, C.-K.; Lau, D.-S.; Lam, C.-C.; Chan, P.-W. WRF-ROMS-SWAN Coupled Model Simulation Study: Effect of Atmosphere–Ocean Coupling on Sea Level Predictions Under Tropical Cyclone and Northeast Monsoon Conditions in Hong Kong. *Atmosphere* **2024**, *15*, 1242. [CrossRef]

Journal of
Marine Science and Engineering

MDPI

Article

Impact of Silted Coastal Port Engineering Construction on Marine Dynamic Environment: A Case Study of Binhai Port

Xiaolong Deng [1], Zhifeng Wang [1] and Xin Ma [2,*]

[1] College of Engineering, Ocean University of China, Qingdao 266100, China; dengxiaolong@stu.ouc.edu.cn (X.D.); wzf1984@ouc.edu.cn (Z.W.)
[2] Physical Oceanography Laboratory, Ocean University of China, Qingdao 266100, China
* Correspondence: maxin@ouc.edu.cn

Abstract: Siltation around the harbour entrance poses significant challenges to the navigational safety and operational stability of coastal ports. Previous research has predominantly focused on sedimentation mechanisms in sandy coastal environments, while studies on silt-muddy coasts remain scarce. This paper investigates the causes of siltation around the entrance of Binhai Port in Jiangsu Province, China, utilising field observation data and a two-dimensional tidal current numerical model, with emphasis on hydrodynamic variations and sediment dynamics. Observations reveal that tidal currents induce sediment deposition in the outer harbour entrance area, whereas pronounced scouring occurs near breakwater heads. During extreme weather events, such as Typhoons Lekima (2019) and Muifa (2022), combined wind–wave interactions markedly intensified sediment transport and accumulation, particularly amplifying siltation at the entrance, with deposition thicknesses reaching 0.5 m and 1.0 m, respectively. The study elucidates erosion–deposition patterns under combined tidal, wave, and wind forces, identifying two critical mechanisms: (1) net sediment transport directionality driven by tidal asymmetry, and (2) a lagged dynamic sedimentary response during sediment migration. Notably, the entrance zone, functioning as a critical conduit for water– sediment exchange, exhibits the highest siltation levels, forming a key bottleneck for navigational capacity. The insights gleaned from this study are instrumental in understanding the morphodynamic processes triggered by artificial structures in silt-muddy coastal systems, thereby providing a valuable reference point for the sustainable planning and management of ports.

Keywords: hydrodynamic environment; sediment environment; characteristic analysis; extreme weather events; silt-muddy coast

1. Introduction

Coastal regions, endowed with abundant natural resources and strategic geographical advantages, have emerged as pivotal zones for global competition and development. Against the backdrop of global warming, sea-level rise, and intensified human activities [1–3], geomorphological evolution in deltaic regions has garnered widespread attention [4,5]. Ports, as quintessential manifestations of human activity and critical hubs for global trade, exert profound positive impacts on coastal urban development. The positive impacts of port-driven economies on coastal urban development include stimulated local economic growth, enhanced urban competitiveness, promoted employment and population mobility, and improved infrastructure. Notable examples such as the Port of Shanghai and the Port of Chattogram exemplify the pivotal role of ports in regional economies [6].

Ports contribute to coastal development through multifaceted mechanisms, particularly as engines for job creation and local economic advancement [7–9]. Approximately 80% of global trade relies on maritime transport, underscoring the indispensability of ports as components of international commerce [10,11]. Simultaneously, ports function as fundamental nodes and potential bottlenecks in global trade networks, facilitating sustainable trade exchanges [12]. However, the construction of ports gives rise to environmental and ecological challenges, particularly in the management of sediment and the effects of hydrodynamics. Despite the potential for further research in this area, the existing literature on nearshore hydrodynamic fields, suspended sediment dynamics, and the geomorphological evolution of subaqueous deltas induced by port infrastructure is limited. The existing studies have classified abandoned estuaries into three layers, with progressively weaker hydrodynamic forces from upper to lower strata [13]. Morphodynamic modelling further reveals that the intensity of the local tidal range governs deltaic morphology [14]. The results of the simulations indicate that post-construction flow velocities stabilise at a distance of 2 km from the port, while the tidal prism decreases by approximately 0.07% [15]. A wave–current coupled hydrodynamic model applied to the Zhuanghe Fishing Port project demonstrates minimal post-construction alterations to large-scale flow fields, with impacts confined to localised areas near the port [16]. Numerical simulations of hydrodynamic interactions at Cha Am Pier highlight sediment accumulation on updrift breakwaters, emphasising the criticality of understanding hydrodynamic conditions for effective downdrift erosion management [17].

In the context of numerous artificially excavated ports, the phenomenon of channel shallowing due to siltation is pervasive, with some ports experiencing sudden siltation events triggered by typhoons, storm waves, or cold surges [18]. Studies have demonstrated that localised channel engineering alters submarine topography, leading to the destabilisation of underwater slopes [19]. A notable example is Jiangsu Binhai Port, where severe recurrent siltation necessitates continuous maintenance dredging to ensure navigational safety. The challenges posed by high water content in dredged sediments and the operational complexities involved have resulted in low efficiency and elevated maintenance costs [20]. Post-construction siltation persists in harbour basins and navigation channels, even triggering erosional scour pits near the northern breakwater head, posing a threat to channel stability and breakwater integrity. The recent acceleration in coastal erosion following the northward shift in the Yellow River's course has amplified the hydrodynamic and sediment transport impacts of Binhai Port's development, positioning it as a critical research focus [21].

The aim of this study is to analyse the causes of excessive entrance siltation at Binhai Port, Yancheng, Jiangsu Province, through a morphodynamic process lens. To this end, a two-dimensional tidal current numerical model was developed to investigate post-construction hydrodynamic alterations, with the dual objectives of providing technical support for future dredging operations and offering novel insights into siltation mechanisms in silt-muddy coastal navigation channels.

2. Study Area and Model Validation

2.1. Study Area and Model Setup

The Binhai Port area is located in Binhai County, Yancheng City, Jiangsu Province, on the erosive silty sand coastline of the abandoned Yellow River Delta, as shown in Figure 1a. The coastline of this region juts out into the sea and is characterised by strong nearshore wave dynamics, active sediment movement and a complex underwater topography with significant underwater slope erosion. The area has a turning point at the northern tip of the abandoned Yellow River estuary, where the direction of the coastline changes from northwest–southeast (NW-SE) to north–south (N-S). The terrain is generally flat, gently sloping from southwest to northeast towards the Yellow Sea, with ground elevations ranging from 2 to 3 m. The southern area has slightly higher terrain, forming a ridge-like landform. The coastline is subject to continuous erosion. Since the implementation of coastal protection works in the 1970s, the erosion rate has slowed down considerably, while the scour rate on the shoals has increased significantly [19,22]. Recent research suggests that the remaining nearshore section of the −15 m isobath along the Jiangsu coast offers favourable conditions for deepwater port development. According to nautical charts provided by the China Maritime Safety Administration (2014 and 2016 data), the water depth distribution in the Binhai Port area is shown in Figure 2. The −5 m, −10 m, and −15 m isobaths are approximately 0.9 km, 1.9 km, and 3.8 km from the coastline, respectively [19], providing suitable deep water conditions for the development of Binhai Port. The construction of Binhai Port started in July 2009, with the construction of the north and south breakwaters completed by June 2011. The 100,000-tonne multipurpose terminal officially started operations in October 2014. The breakwaters of the port are designed in a zigzag pattern, consisting of a northern breakwater 4880 m long and a southern breakwater 1985 m long. The depth of the water at the head of the northern breakwater is 14 m, while the depth of the southern breakwater is 10 m. The entrance width is 800 m and the distance between the two breakwaters is 2.3 km. These structures provide essential protection for the harbour and significantly influence sediment transport and the dynamic equilibrium of the surrounding marine environment. The wide study area for this model extends from 33.7° to 35.8° N and 119.7° to 122.45° E, while the narrow study area extends from 34.2° to 34.4° N and 120.15° to 120.45° E, as shown in Figure 1b. The simulation time frame was set from 1 February 2021 to 1 April 2021. The reliability of the model is supported by comparisons with measured data from validation points, the distribution and locations of which are shown in Figure 1b. One observation point and eight validation points (as shown in Table 1) were selected to compare the observed results with the simulated data for tidal levels, currents, and suspended sediment concentrations.

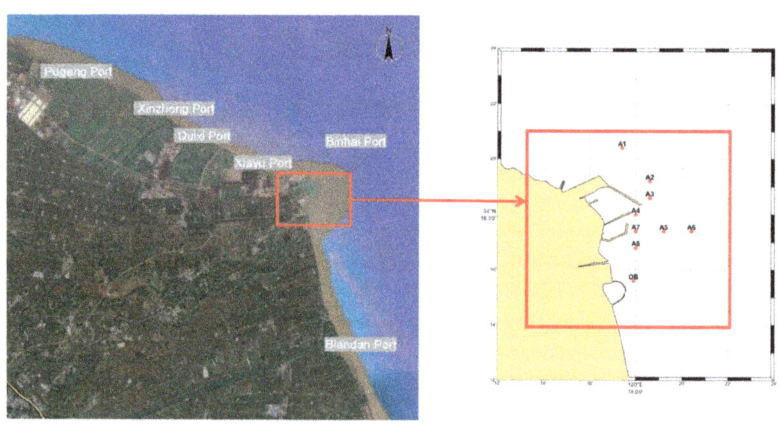

(a) (b)

Figure 1. (**a**) Geographical location of the study area; (**b**) distribution of observation points and eight characteristic points. It should be noted that the shaded area in panel (**b**) represents the study region. (Note: The geographic location map is sourced from Google Earth.)

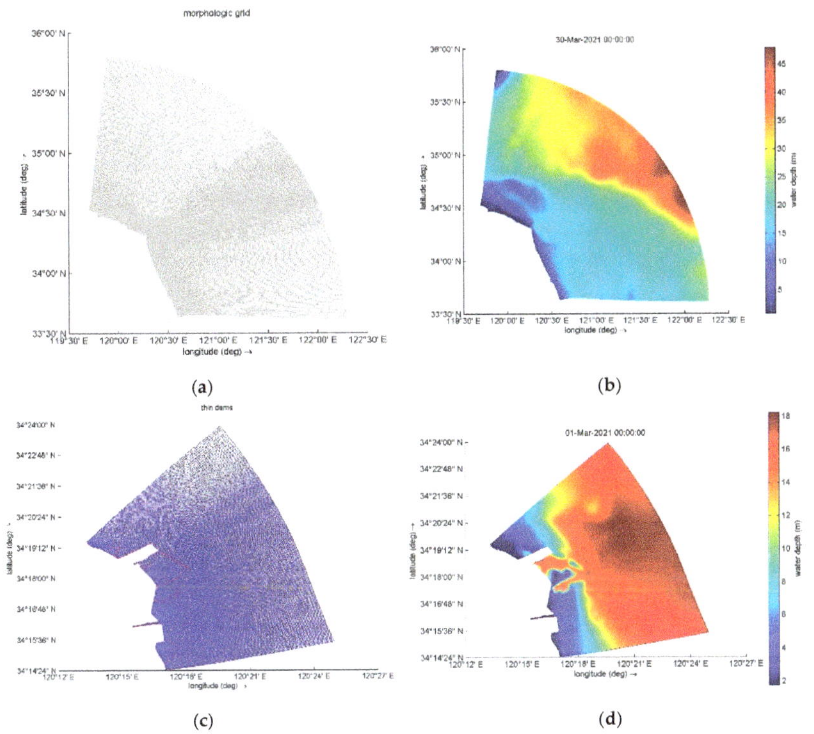

Figure 2. The grid and bathymetry of the study area (**a**,**b**) and the grid and bathymetry of the small-scale model (**c**,**d**) are presented.

Table 1. Coordinates of observation data points.

Station	Longitude (°E)	Latitude (°N)	Observation Instruments	Measured Parameters
OB (Observation Point)	120.30	34.26	KELLER Pressure Tide Gauge	Tidal Level
A1 (Validation Point)	120.29	34.34	Acoustic Doppler Current Profiler (ADCP)/OBS-3A Turbidimeter RBR Concerto Turbidimeter	Velocity/ Direction/ Sediment Concentration
A2 (Validation Point)	120.31	34.32	Acoustic Doppler Current Profiler (ADCP)/OBS-3A Turbidimeter RBR Concerto Turbidimeter	Velocity/ Direction/ Sediment Concentration
A3 (Validation Point)	120.31	34.31	Acoustic Doppler Current Profiler (ADCP)/OBS-3A Turbidimeter RBR Concerto Turbidimeter	Velocity/ Direction/ Sediment Concentration
A4 (Validation Point)	120.30	34.30	Acoustic Doppler Current Profiler (ADCP)/OBS-3A Turbidimeter RBR Concerto Turbidimeter	Velocity/ Direction/ Sediment Concentration
A5 (Validation Point)	120.32	34.29	Acoustic Doppler Current Profiler (ADCP)/OBS-3A Turbidimeter RBR Concerto Turbidimeter	Velocity/ Direction/ Sediment Concentration
A6 (Validation Point)	120.34	34.29	Acoustic Doppler Current Profiler (ADCP)/OBS-3A Turbidimeter RBR Concerto Turbidimeter	Velocity/ Direction/ Sediment Concentration
A7 (Validation Point)	120.30	34.29	Acoustic Doppler Current Profiler (ADCP)/OBS-3A Turbidimeter RBR Concerto Turbidimeter	Velocity/ Direction/ Sediment Concentration
A8 (Validation Point)	120.30	34.28	Acoustic Doppler Current Profiler (ADCP)/OBS-3A Turbidimeter RBR Concerto Turbidimeter	Velocity/ Direction/ Sediment Concentration

Note: The ocean current observation was conducted using an acoustic Doppler current profiler produced by Nortek, a company based in Norway.

Prior to the start of the observations, the Acoustic Doppler Current Profiler was set to begin measurements on the hour, with measurements taken every 10 min. Each measurement session lasted 100 s with a continuous sampling frequency of 1 Hz. The average of each session was taken as the observation value. The pressure tide gauge was also configured to start observations on the hour, with measurements taken every 10 min. Each session lasted 60 s at a sampling frequency of 1 Hz and the average value was also taken as the observation value. Simultaneously with the turbidity measurements, seawater samples were taken from the surface at a depth of 0.6H and from the bottom every 2 h. The sediment concentration in the water samples was determined by the filtration and drying method.

The predominant wind direction in this region is from southeastern (SE) directions, followed by south-southeast (SSE), with strong winds predominantly originating from north (N) directions. The frequency of winds from various directions is demonstrated in Figure 3. Wind speed and direction measurements were performed using the DEM6 Portable Three-Cup Anemometer.

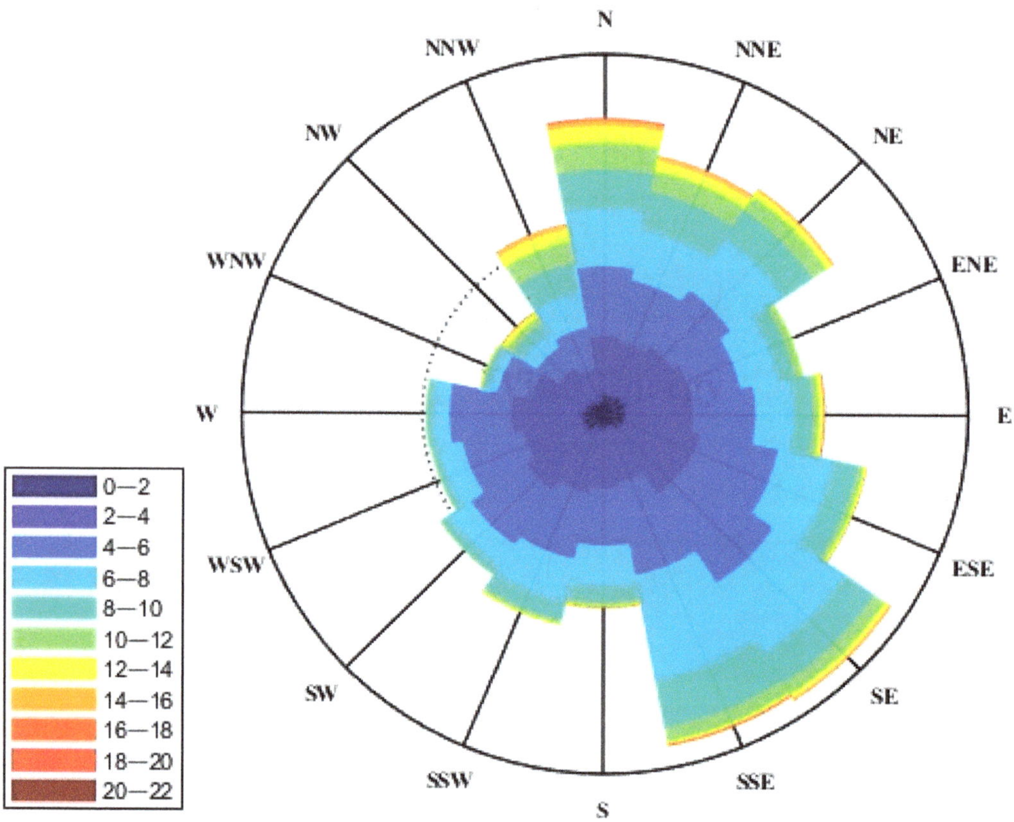

Figure 3. Rose diagrams of wind in the study area. (Note: The data are sourced from the "Hydrological Observation Report for the Optimization Plan of Coastal Port Layout", April 2021.)

This study employs the Delft3D-Flow module for numerical modelling, with the computational grid constructed using the Grid-RFFGRID module comprising 225,319 cells. Bathymetric data were interpolated via the Grid-QUICKIN method, with a large-scale orthogonal curvilinear grid configuration as illustrated in Figure 2a.

The sediment sources in the study area were categorised into three types based on the existing literature: offshore sediment supply, terrestrial input, and localised sediment recirculation. The bathymetric inputs were derived from field measurements [23], with the minimum water depth set to 47 m and subjected to smoothing procedures. Figure 2b presents the simulated bathymetric distribution. To streamline boundary condition implementation, nested grid techniques were applied, where large-scale grid data provided boundary constraints for finer-resolution subdomains (see spatial domains in Figure 2c,d). The model domain features three open boundaries and one closed land boundary. The tidal forcing at the open boundaries was implemented using harmonic constants from 13 tidal constituents extracted from the TPXO9 global tidal database following the methodology of Fu et al. [24]. This approach has been demonstrated to demonstrate superior accuracy in Chinese coastal waters. A cold-start initialisation approach was adopted to ensure numerical stability during model spin-up.

Sediment Parameter Settings: According to the General Report on Ocean Investigation and Evaluation of Jiangsu Province [25], there is a significant spatial heterogeneity in the sediment distribution in the coastal areas of Jiangsu Province. The median particle size of sediments in the Yellow River Estuary ranges from 0.01 to 0.08 mm, indicating a relative size distribution predominantly composed of fine particles; hence, the sediment parameters in the model exhibit spatial heterogeneity. Based on the relationship between suspended sediment particle size and settling velocity determined through physical model experiments, the settling velocity was set between 0.01 and 1.4 mm/s. Utilising an empirical formula for particle size and erosion rate, this study set the erosion rate at 2×10^{-5} to 5×10^{-5} kg/m^2/s, with critical erosion stress established at 0.6 to 0.12 N/m^2, and a constant critical deposition stress of 0.05 N/m^2. Sensitivity tests determined a time step of 0.5 min. The sediment concentration distribution in this region exhibits a characteristic of higher concentrations nearshore, decreasing significantly offshore. The average sediment concentrations near the -5 m and -10 m depth contours are approximately 0.98 and 0.58 kg/m^2, respectively, while in deeper waters beyond -20 m, it falls below 0.1 kg/m^2. Therefore, the sediment concentration boundary conditions for the south and northwest open boundaries were interpolated linearly from 1.0 kg/m^2 on the landward side to the offshore northeastern boundary, where the sediment concentration was set to zero. For the closed boundary, a non-penetrable condition was applied, meaning the normal sediment concentration was zero.

Wave Parameter Settings: The wind field data utilised in the SWAN model was derived from ERA5, with the spatial extent of the model for the Jiangsu coastal waters spanning from 33.7° to 35.8° N and 119.7° to 122.45° E. The computational time step was set to one hour, with a storage time step also of one hour, and the breaking wave parameter was set to 0.86. During the simulation process, interactions between waves, as well as wave dissipation caused by whitecapping, bottom friction, and wave breaking, were considered.

2.2. Validation of Tidal Currents

As demonstrated in Figure 4, a comparison was made between the simulated and observed water levels at the designated station. The root mean square error (RMSE) is 0.1452 m, and the correlation coefficient is greater than 0.87, indicating the reliability of the model for further simulation work. Furthermore, as demonstrated in Table 2, the primary tidal harmonic constants of the tidal level simulation, in conjunction with the observed data and their discrepancies, are presented. The table shows that the largest amplitude error occurs for the M4 tidal component, with a difference of 10.4 cm, while the amplitude error for the O1 tidal component is 0 cm. With regard to phase lag, the M4 tidal component exhibits the greatest discrepancy, with a difference of 28.0°, while the O1 component demonstrates the least phase lag error, with a difference of only 0.3°. These comparative results demonstrate that the model performs well in simulating the tidal current field.

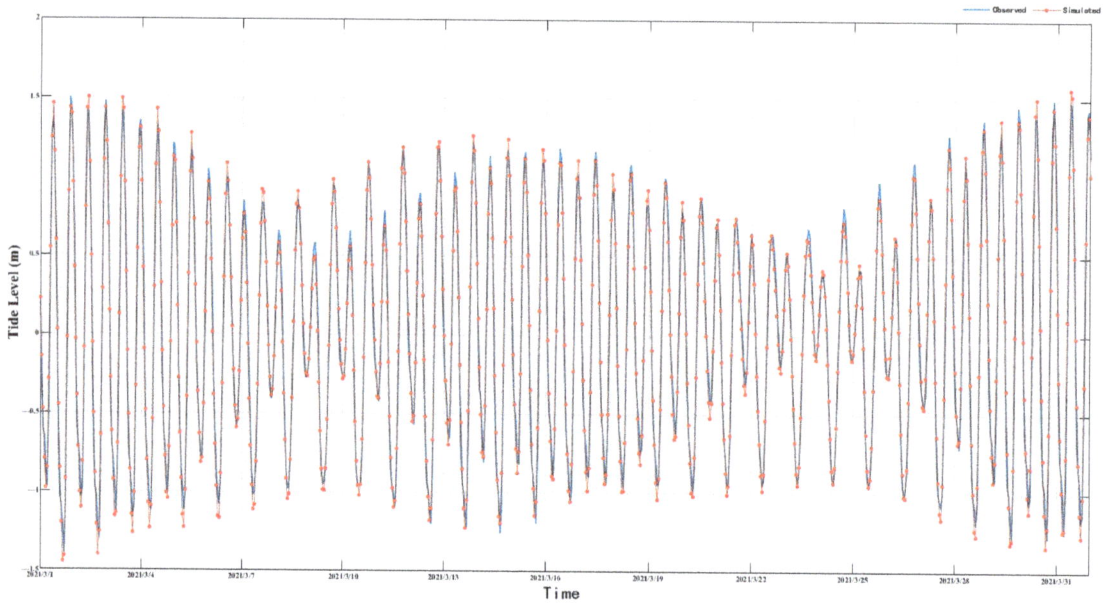

Figure 4. Tidal validation chart.

Table 2. Comparison of harmonic constants at tide gauge stations.

Station	Tidal Constituent	Observed		Simulated		Amplitude (cm)	Phase Lag (°)
		Phase Lag Difference Amplitude (cm)	Phase Lag (°)	Phase Lag Difference Amplitude (cm)	Phase Lag (°)		
OB	Q1	2.3	329.2	2.6	327.4	−0.2	1.8
	O1	22.3	329.4	22.3	329.1	0.0	0.3
	K1	20.2	37.9	18.9	36.2	1.3	1.8
	N2	12.1	208.1	16.5	208.6	−4.4	−0.5
	M2	83.3	223.0	88.1	224.2	−4.8	−1.2
	S2	28.2	290.8	31.0	293.3	−2.8	−2.5
	M4	23.1	11.4	12.7	39.4	10.4	−28.0
	MS4	19.8	86.1	13.5	108.4	6.3	−22.3

In order to enhance the precision of the model, a comparative analysis was undertaken between the observed and simulated values of flow velocity, flow direction, and sediment concentration at stations A1–A8. As demonstrated in Figure 5, the simulated flow velocities and directions exhibit a strong alignment with the observed data. The hydrodynamic numerical model that has been developed demonstrates a high degree of accuracy in replicating the flow field distribution, thereby providing reliable support for the analysis of the hydrodynamic characteristics of the coastal area.

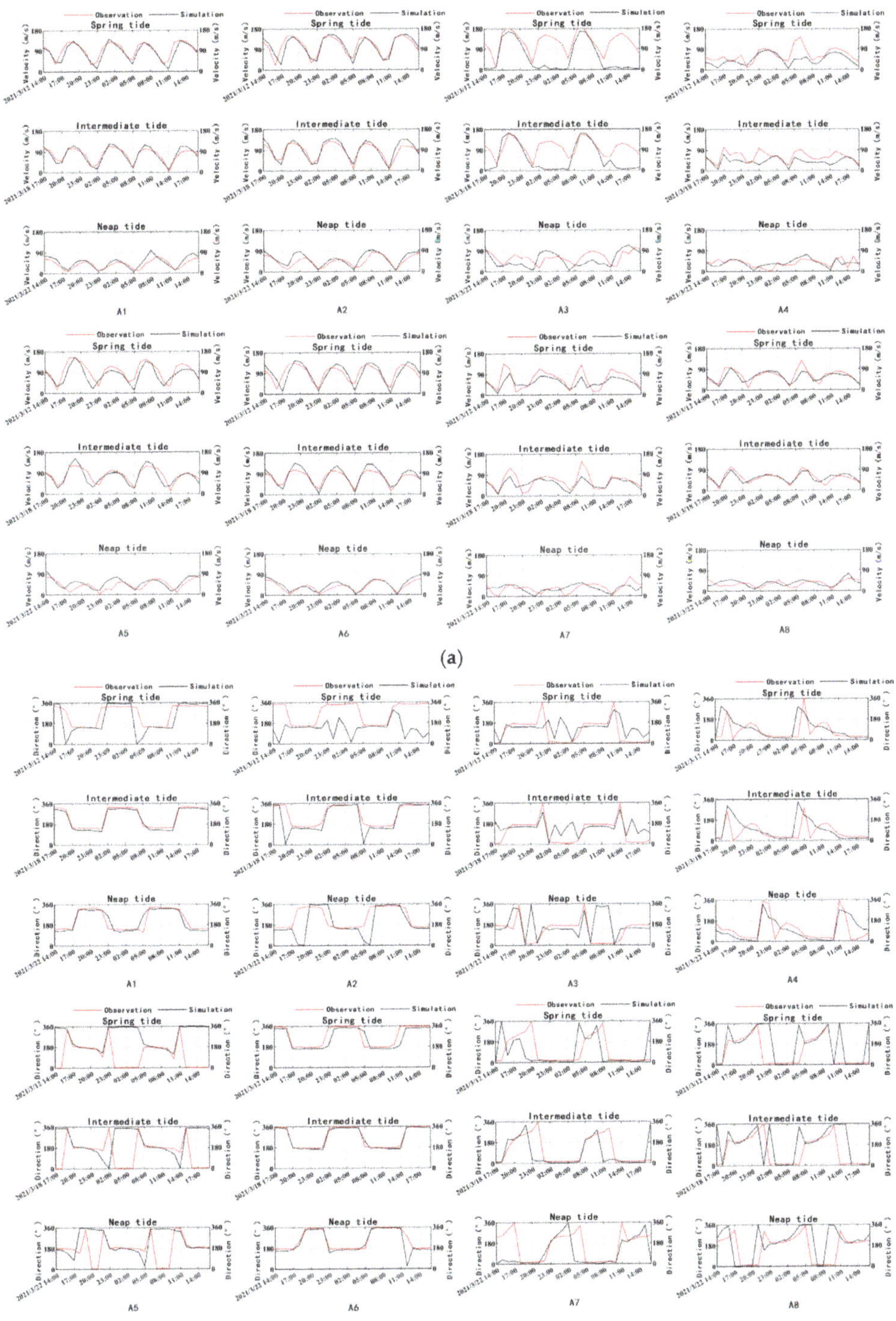

Figure 5. Validation of flow velocity and direction. (**a**) Comparison between simulated and observed flow velocities; (**b**) comparison between simulated and observed flow directions.

2.3. Wave Model Validation

The SWAN wave model was forced by wind fields derived from the Weather Research and Forecasting (WRF) model. The computational domain spanned 33.7° N–35.8° N, 119.7° E–122.45° E, with a 1 h computational time step and equivalent output interval. Depth-induced wave breaking was parameterized using a breaker index of 0.86. The simulations incorporated key wave dissipation mechanisms, including bottom friction, and depth-induced breaking, and nonlinear wave–wave interactions. Table 3: Geographic coordinates of the wave observation stations along the Jiangsu coast. Figure 6a shows the location map.

Table 3. Geographic coordinates of monitoring stations.

Station Name	Longitude	Latitude
Jiangsu Binhai Statio	120.2983° E	34.26° N

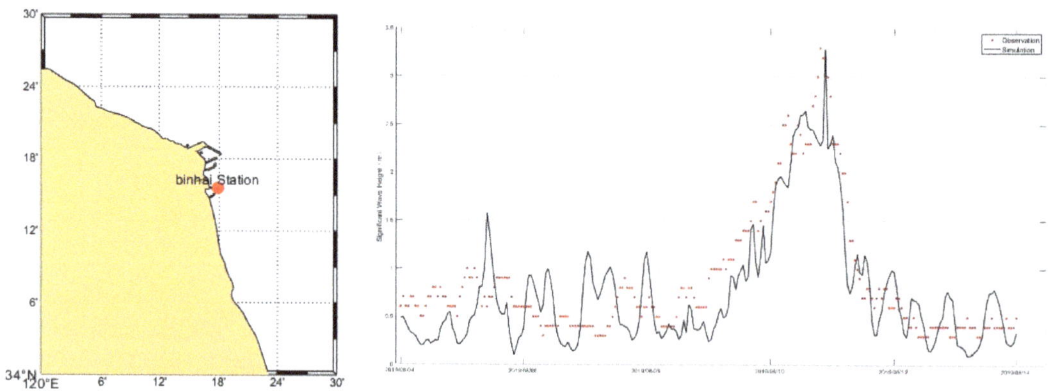

(**a**) (**b**)

Figure 6. Wave height validation chart. (**a**) Schematic of the observation station's geographic location. (**b**) Comparison of simulated and measured wave heights in August 2019.

Validation against in situ wave observations from Jiangsu Binhai Station (4–14 August 2019); Figure 6b demonstrates strong agreement between the third-generation wave numerical model outputs and field measurements. Comparative analysis with annual wave buoy data reveals that the model achieves a wave height prediction error of <15% under routine sea conditions, with key statistical metrics including a correlation coefficient (R) of 0.89, root mean square error (RMSE) of 0.23 m, and mean absolute error (MAE) of 0.18 m. During the extreme wave event induced by Typhoon Lekima (2019), the model successfully captured the evolution of the maximum significant wave height (3.3 m), with a phase error of <1 h. This high-fidelity performance validates both the applicability of the model's parameterization schemes in the shallow waters of northern Jiangsu and the reliability of the observational system. The results provide robust technical and data-driven support for coastal engineering design, storm surge forecasting, and marine environmental studies.

2.4. Suspended Sediment and Wave Validation

The sediment field was calculated from 1 March 2021 to 1 April 2021. The validation period for suspended sediment during spring tides was from 14:00 on 12 March 2021 to 17:00 on 13 March 2021, and for neap tides from 14:00 on 22 March 2021 to 19:00 on 23 March 2021. The validation process entailed the utilisation of stations A1 through A8, which were employed to validate sediment concentrations under both pure tidal flow and the combined effects of wave and current interactions. The validation results are illustrated in the figure below:

As demonstrated in Figure 7, the simulated suspended sediment concentrations (SSCs) under combined wave–current interactions demonstrate a strong correlation with measured values across all the stations. However, the SSC variation exhibits a dual-peak pattern within a tidal cycle, contrasting with the quadruple-peak characteristics of tidal current velocities. The enhancement of wave height has been observed to result in increased sediment resuspension, leading to elevated SSC levels. Conversely, a reduction in wave height has been shown to result in a decline in SSC, although this response is less pronounced due to the influence of delayed settling effects, which are dictated by factors such as sediment grain size and hydrodynamic inertia. During late ebb phases, a significant decrease in SSC has been documented, attributable to the reduction in flow velocity, which concomitantly diminishes the sediment-carrying capacity. Given the non-negligible wave effects in the study area and the strong consistency between the simulated and observed SSC under wave–current coupling (Figures 7 and 8), the adopted model incorporating wave–current interactions proves robust for capturing sediment dynamics in this silt-muddy coastal environment.

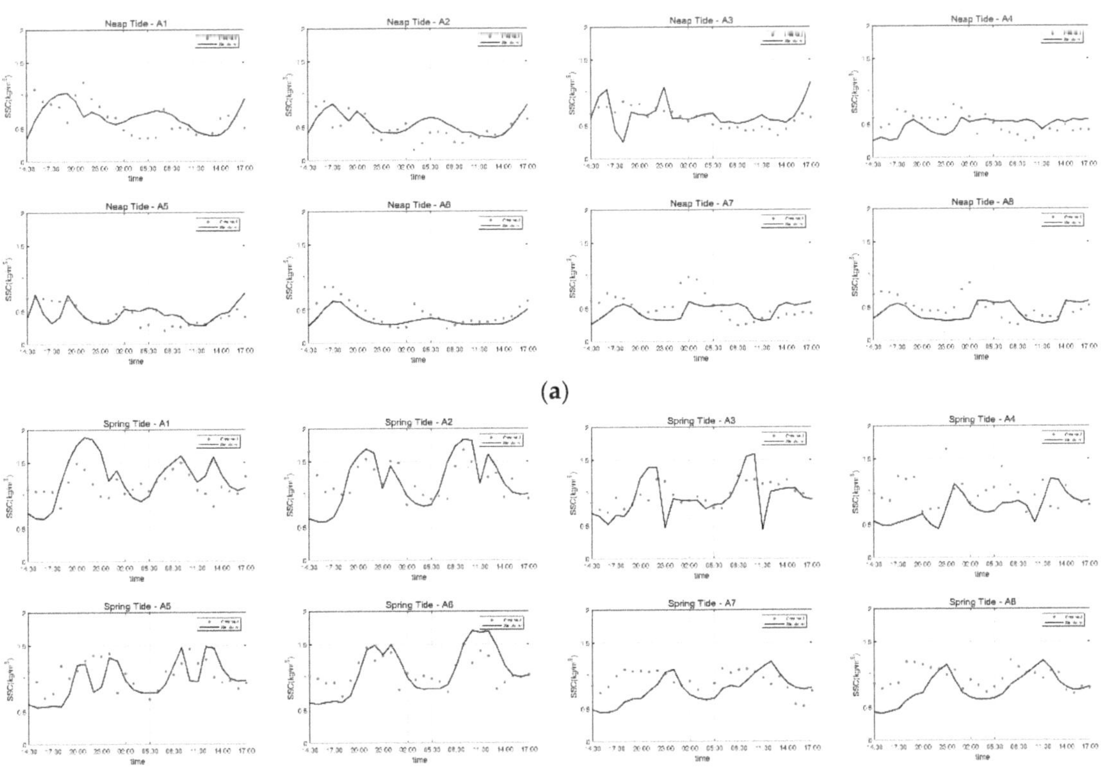

Figure 7. Validation of suspended sediment concentrations during tidal cycles. (**a**) Neap tide; (**b**) spring tide.

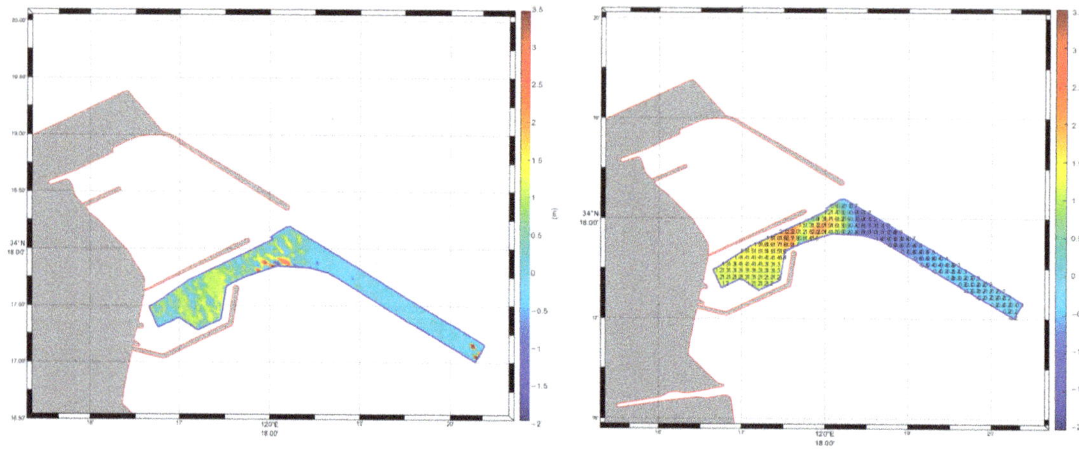

(a) (b)

Figure 8. Comparison of erosion and deposition in the study area. (**a**) Measured bathymetric changes; (**b**) simulated erosion and deposition distribution.

2.5. Validation of Erosion and Deposition

As demonstrated in Figure 8a, the bathymetric changes were measured over the period from 27 August to 9 November 2022. It is noteworthy that no dredging activities were conducted in the harbour basin or navigation channel during the observation period, thereby enabling the observations to serve as validation data for natural sedimentation. The results indicate that sedimentation in the navigation channel primarily occurs in the recirculation zone near the estuary, where bathymetric changes of 2–3 m were observed. Within the LNG harbour basin, the depth change was approximately 1 m, while the outer navigation channel exhibited an erosional trend.

The erosion and deposition distribution over a two-month period (September to October 2022) is presented in Figure 8b. The sediment distribution model results indicate that sedimentation is predominantly concentrated in the recirculation zone near the estuary. The sediment accumulation in the LNG navigation channel ranged from 2 to 3 m over the two-month period, while the harbour basin exhibited sedimentation of approximately 1–2 m. The intensity and spatial distribution of the modelled sedimentation align well with the measured results. The primary sedimentation zone in the navigation channel is concentrated in the recirculation zone, where weakened hydrodynamic forces result in sedimentation with a maximum thickness of 2.5–3.0 m. The accumulation extends in a tongue-shaped pattern into the harbour. In the LNG harbour basin, uniform sedimentation of 1.0 ± 0.2 m is observed, while the outer navigation channel exhibits an erosional trend due to increased tidal flux. These observations are indicative of the spatial heterogeneity of sediment transport pathways.

As demonstrated in Figure 8b, the modelled results under natural conditions, incorporating wave–current–sediment interactions for the study area from September to October 2022, are highlighted. The simulation successfully reproduces the observed sediment transport patterns, demonstrating that a sediment accumulation layer of 2.1–3.0 m forms monthly in the estuary recirculation zone, with a relative error of less than 15% compared to the measured values. Sediment accumulation along the axis of the LNG channel reaches 2.3–3.0 m. It is notable that the spatial gradient of sediment distribution and the isopleth morphology in the model closely match the measured data. In particular, in the estuary-to-harbour transition zone, the horizontal deviation between the simulated and observed sedimentation centres is less than 50 m. Since the construction of the northern breakwater of the coastal port area in June 2011, the jetting effect and localised wave reflection near the breakwater head have intensified flow velocities in this region, exacerbating seabed erosion near the breakwater head. The development of the scour pit near the breakwater head is outlined in Table 4. The scour pit has been observed to exhibit exponential growth, with a monthly average scour depth of 0.9–1.1 m, which is consistent with the modelled results of 0.95–1.05 m/month (relative error $\leq$ 8%).

Table 4. Monitoring results of punching pit depth development in Binhai Port Area.

Monitoring time	September 2015	9 April 2016	9 May 2016
Maximum blunt depth/m	−21.9	−35.7	−36.5

From the perspective of system validation, the model achieves engineering-level accuracy in several key metrics, including natural sedimentation intensity (annual sedimentation rate deviation < 15%), scour pit development rate (phase error < 5 days), and spatial differentiation of erosion and deposition (similarity coefficient > 0.78). Of particular note is the agreement in sedimentation thickness within the estuary zone, where measured values of 2.5–3.0 m closely correspond to the simulated result of 2.7 m, thereby demonstrating the model's capacity to capture the coupled processes of recirculation dynamics and flocculation-induced sedimentation.

3. Characteristics Analysis

3.1. Hydrodynamic Characteristics Analysis

As analysis nodes, the time of maximum ebb was set at 1:00 a.m. on 13 March 2021, while the time of maximum flood was chosen as 7:00 p.m. on 12 March 2021. The slack prior to flood occurrence was designated as 11:00 p.m. on 12 March 2021, and all slack preceding ebb times were set at 5:00 a.m. on 13 March 2021. The selection of these time nodes is based on tidal harmonic analysis and verified by the observed water levels (see Figure 2).

Tidal level analysis reveals that the study area is characterised by regular semidiurnal tidal currents [26]. The model utilises observational data from 2021 to perform a comparative analysis. As demonstrated in Figure 9, the overall flow pattern manifests smooth characteristics. During periods of flood tide, the tidal current flows predominantly in a southeastern direction, while during periods of ebb tide, the flow reverses to a northwestern direction. However, in the vicinity of the shoreline, a notable similarity is observed in the flow behaviour during both phases: the influence of boundaries and the underwater topography near the shore results in the flow direction aligning almost parallel with the shoreline and the isobaths. The nearly opposite directions of flood and ebb tides suggest that the study area exhibits a reciprocating flow pattern. Specifically, during flood tide, the flow direction ranges between 145° and 178°, while during ebb tide, it ranges between 319° and 353°.

During periods of spring tides, the flood tide moves from a northwest to a southeast direction. Upon reaching the northern breakwater, the current is obstructed by nearshore structures in the northwest section, which reduces the flow velocity. The current then moves towards the corner and southeastern sections of the northern breakwater, where these segments act as guiding structures, enhancing the flow velocity on their outer sides. The development of high-velocity zones is observed in the vicinity of the breakwater opening and the northern breakwater head, with maximum velocities reaching up to 1.2 m/s. As the current flows past the breakwater head, a clockwise eddy forms on the southern side of the breakwater. As this eddy develops, the water on its outer edge flows southeast along the southern breakwater, which also serves as a guiding structure, with velocities around 0.4 m/s. The current continues to evolve as it flows upstream along the southern breakwater and enters the harbour area. Upon encountering the southern side of the northern breakwater, the flow direction undergoes a change, with the northern breakwater acting as an additional guiding structure. This results in the formation of a counterclockwise eddy within the harbour. Within the semi-enclosed region of the harbour, flow velocities remain relatively slow, remaining below 0.2 m/s. The flow velocity continues to decrease as the flow enters the harbour's interior.

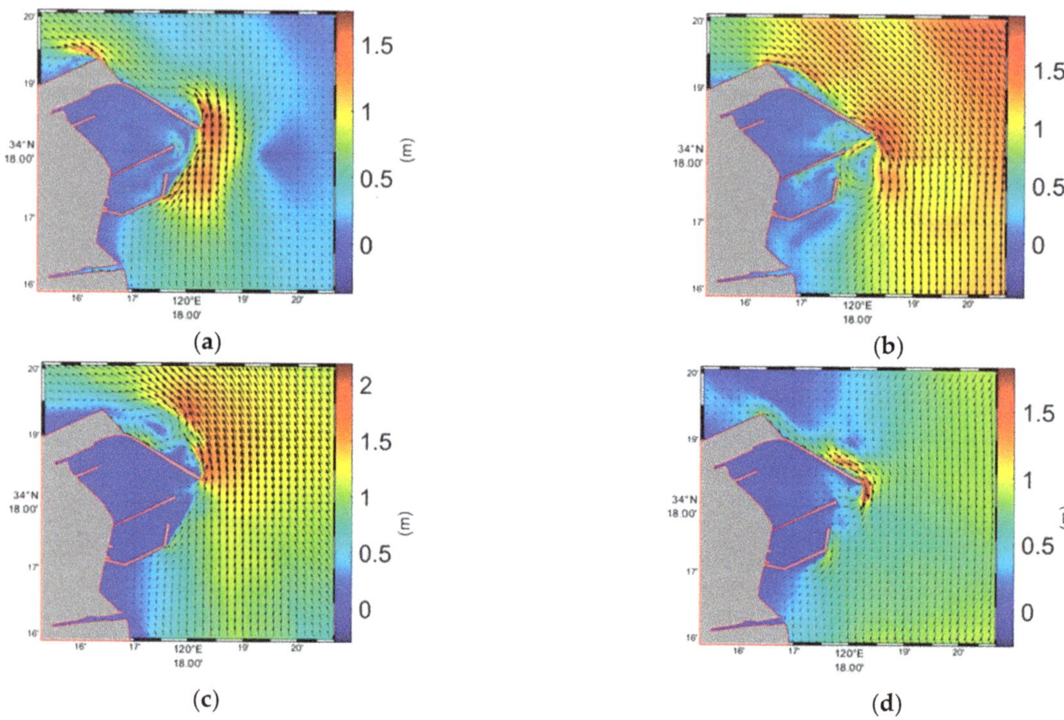

Figure 9. Flow field during maximum flood and ebb (spring tide). (**a**) Maximum flood; (**b**) slack before flood; (**c**) maximum ebb; (**d**) slack before ebb.

During the ebb tide, the tidal flow retreats in the opposite direction to that of the flood tide. The current is impeded by the southern breakwater, leading to a reduction in flow velocity. The tidal current is influenced by the guiding effect of the southern breakwater, flowing along its length and converging with the outflowing current near the breakwater opening, forming a transverse flow at the southern breakwater head. On the northern side of the northern breakwater, after the ebb current passes through the opening and flows past the northern breakwater head, a circular eddy is formed, with its elliptical major axis longer than the minor axis. As the tidal current strengthens, the elliptical eddy undergoes further development and extends northwestward, reaching as far as the turning point of the northern breakwater. At this turning point, the tidal flow bifurcates into two branches: one flows southwestward to merge with the main ebb current, while the other flows southeastward along the northern breakwater. The northern breakwater functions as a guiding structure, exhibiting a flow velocity of approximately 0.3 m/s.

Within the confines of the breakwater, a clockwise eddy also comes into being. Due to the pronounced sheltering effects within the harbour basin, the flow velocity is relatively low, remaining below 0.2 m/s. Seawater flowing outward from the breakwater opening eventually joins the main ebb current. While the configuration of the coastal harbour exerts a certain degree of influence on the directions of flood and ebb currents outside the breakwaters, it does not result in a substantial alteration to the overall flow pattern of the entire study area.

The velocity distribution demonstrates a tendency for higher flow velocities in offshore regions and lower velocities within the harbour. This suggests that the flow velocity within the coastal harbour is, on average, lower than that of the surrounding waters. At the harbour entrance, the influence of the breakwaters results in the observation of maximum velocities, with the high-velocity region gradually expanding towards the shallows. However, it is important to note that the distribution of flow velocity does not always align perfectly with the bathymetry. Specifically, velocities are comparatively low near the coastal harbour, while they are higher near the eastern and northern boundaries.

In the sea area proximate to the harbour entrance, the maximum flow velocity during peak flood and peak ebb reaches 1.5 m/s. It is noteworthy that in the vicinity of the former Yellow River estuary, the area exhibiting equivalent flow velocities during peak ebb is typically more extensive compared to that during peak flood. A comparison of the flow fields for peak flood and peak ebb indicates that apart from specific cases influenced by shallow areas, the flow paths during flood and ebb tides are essentially consistent.

Within and in the vicinity of the coastal harbour, the recirculation zone in proximity to the harbour entrance is the most prominent region. This phenomenon can be attributed to the contraction effect of the breakwaters, resulting in interactions between the main flow and the recirculation, which promote mixing. Furthermore, the velocity gradient, caused by the higher flow velocity outside the harbour basin and the relatively stagnant water head inside, induces shear stress, triggering secondary recirculations. This, in turn, serves to augment the overall turbulence of the flow pattern within the area.

As illustrated in Figure 10, the highest cross-flow velocity among the three calculated points during the spring tide is observed at Point V1, which is located in proximity to the entrance of the harbour basin. At this location, the jetting effect at the tip of the breakwater is intensified, and the region with the highest flow velocity is primarily concentrated near the navigation channel. The velocity component perpendicular to the tidal current direction is more pronounced due to the navigation channel's main axis being at a small angle to the current. This results in a more significant cross-flow phenomenon.

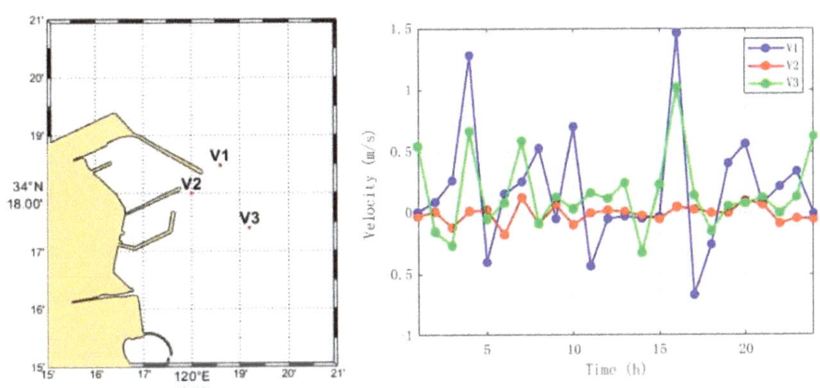

(a) (b)

Figure 10. Distribution of characteristic points and magnitudes of flow velocity in cross-flow calculations. (**a**) Characteristic points; (**b**) flow velocity.

3.2. *Analysis of Suspended Sediment Characteristics*

3.2.1. Characteristics of Sediment Concentration

As demonstrated in Figure 11 there is a clear correlation between the average suspended sediment concentration (SSC) and hydraulic parameters (flow velocity and water level) at four monitoring stations. During spring tides, the distribution of SSC demonstrates relatively stable fluctuation amplitudes, indicating hydrologically stable conditions in nearshore areas. Conversely, during the flood tide phase, the sustained rise in water level engenders an augmentation in sediment resuspension frequency through the intensification of hydrodynamic forcing, thereby precipitating a gradual rise in SSC. This phenomenon is hypothesised to originate from effective bed shear stress-driven sediment entrainment from channel beds and adjacent deposits.

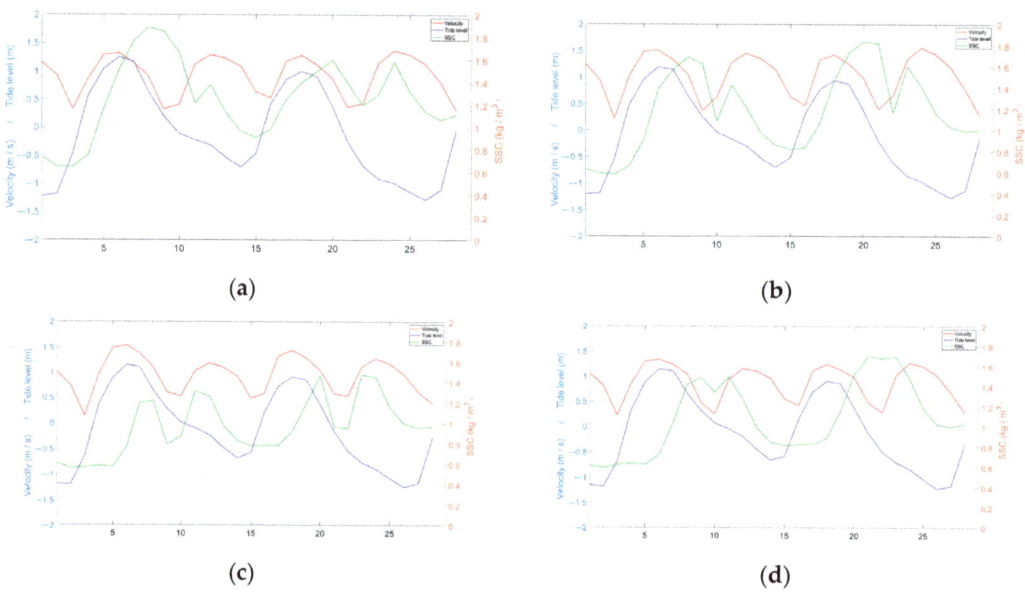

Figure 11. Relationship between average suspended sediment concentration, average flow velocity, and water level at each station. (**a**) Station A1; (**b**) Station A2; (**c**) Station A5; (**d**) Station A6.

Conversely, during ebb tide phases, diminished water levels reduce flow competence for sediment suspension, triggering particle settling and consequent SSC decrease. This inverse relationship between SSC and water level highlights the dynamic interplay between sediment transport and tidal hydrodynamics. Specifically, the synergistic effects of rising water levels and accelerated flows during flood tides significantly enhance sediment mobilisation and transport efficiency. Conversely, the ebb tide phase is characterised by a decline in SSC, which is associated with reduced hydraulic energy and enhanced sedimentation.

The synchronised fluctuations in water level, flow velocity, and SSC values across tidal cycles further substantiate the predominance of tides on sediment dynamics. Phase-locked increases in all three parameters during flood tide transition to synchronous decreases during ebb phases, thus demonstrating tidal pumping as the primary control mechanism for suspended sediment redistribution in this estuarine system.

3.2.2. Characteristics of Suspended Sediment Distribution Under Pure Tidal Currents

The distribution of suspended sediment in the harbour area during four tidal phases (flood surge, ebb surge, flood slack, and ebb slack) is illustrated in Figure 12. As is apparent from the figure, during the ebb phases, the spatial extent of the high-concentration suspended sediment regions is generally larger than during the flood phases. Specifically, during the flood surge phase, the maximum suspended sediment concentration occurs in the northwest area outside the harbour basin, while the suspended sediment concentration inside the basin also increases. During the flood slack phase, the high-concentration region shifts in a southeastward direction, likely due to the advection transport of suspended sediment.

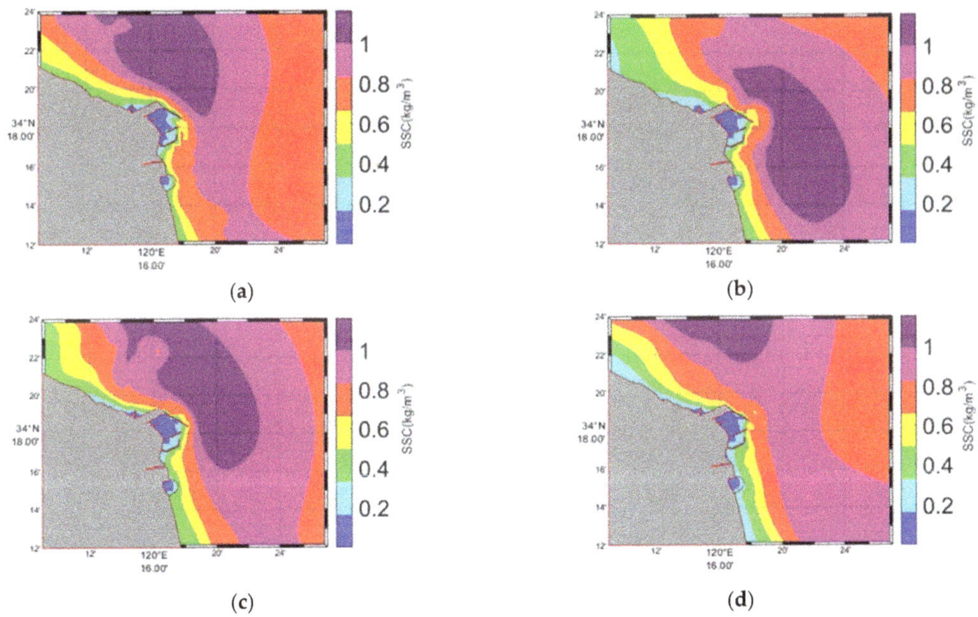

Figure 12. Distribution of suspended sediment concentration under tidal currents. (**a**) Flood surge and (**b**) flood slack; (**c**) ebb surge; (**d**) ebb slack.

During the ebb surge phase, the high-concentration region undergoes a further expansion, which may be indicative of the outward transport of suspended sediment. A similar phenomenon is observed during the ebb slack phase, wherein the high-concentration region shifts northwestward, indicating a redistribution of sediment influenced by tidal currents.

The presence of three distinct suspended sediment bands in the vicinity of the coast is observed under both flood and ebb tides. The formation of these bands is likely attributable to the transportation of sediment in both northward and southward directions, respectively, towards Haizhou Bay and the Subei Radial Sand Ridges [26]. Furthermore, at the harbour entrance, the sudden decrease in flow velocity leads to the accumulation of sediment, resulting in the formation of banded distributions along the transport pathways.

Within the confines of the harbour basin, the suspended sediment concentration is found to be relatively low. However, as one moves outward towards the sea, the concentration initially increases and then gradually decreases. The average suspended sediment concentration in the sea area exceeds 0.5 kg/m^3, which is generally higher than in the other coastal areas of China, thus classifying this region as one of the highest in terms of suspended sediment concentration along the Jiangsu coast.

This phenomenon can be attributed to two primary factors: (1) the presence of a substantial sediment supply resulting from coastal erosion and (2) the prevalence of stronger ebb currents near the coast in comparison to flood currents, which facilitate the transport of sediment towards the sea.

The isoconcentration contours of suspended sediment roughly follow the orientation of the coastline. This phenomenon can be explained by two factors: (1) the geomorphological features of the coastline dominate sediment transport patterns in the region, and (2) the coastal zone contains abundant terrigenous sediment inputs.

In summary, the spatial extent of suspended sediment distribution is greater in areas farther from the harbour basin, while the concentration inside the harbour basin is relatively low. During ebb phases, the maximum suspended sediment concentration and the area of high-concentration regions inside the harbour basin are significantly higher compared to flood phases.

3.2.3. Characteristics of Suspended Sediment Distribution Under Wave–Current Coupling

As demonstrated in Figure 13, during the flood surge phase, the intensification of wind may increase surface wave activity, leading to elevated suspended sediment concentrations on the northwest side outside the harbour basin. This phenomenon also promotes the transport of suspended sediment into the harbour basin. As the flood slack phase approaches, weakening wind speeds or shifts in wind direction enhance the southward movement of suspended sediment, significantly intensifying advective transport during this period.

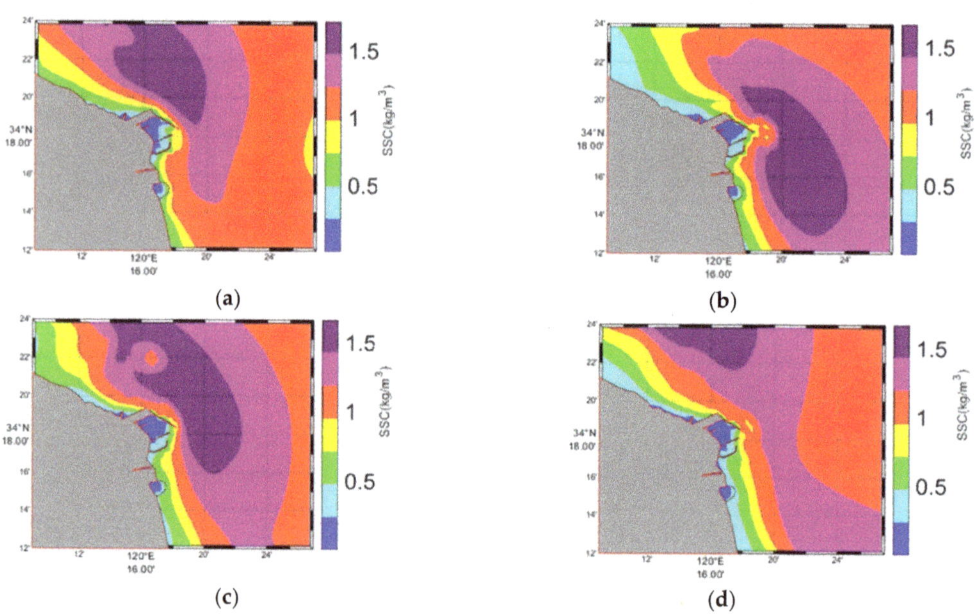

Figure 13. Suspended sediment concentration distribution under wave–current interactions. (**a**) Flood surge; (**b**) flood slack; (**c**) ebb surge; (**d**) ebb slack.

During the ebb surge phase, strong winds accelerate sediment transport outward, expanding the high-concentration region of suspended sediment. Even during the ebb slack phase, the continued influence of wind may cause the location of maximum suspended sediment concentration to shift upward. This suggests that wind not only influences water flow but also alters the deposition and redistribution patterns of suspended sediment.

Consequently, suspended sediment concentrations are observed to exhibit an upward trend, attributable to the modifying effects of wind strength and direction on wave characteristics, which in turn leads to alterations in wave energy.

In circumstances where wave activity is absent, variations in suspended sediment concentrations are predominantly influenced by tidal currents. However, under wave–current interactions, waves significantly impact suspended sediment distribution by altering hydrodynamic conditions. The presence of waves has been shown to facilitate higher suspended sediment concentrations, particularly in the northwest region outside the harbour basin and within the basin itself. The stirring effect of wind waves enhances the resuspension of sediments and facilitates the inward transport of suspended sediment into the harbour basin. The coupling of waves and currents leads to a more complex spatial variation in suspended sediment concentration. For instance, during the ebb slack phase, the persistent impact of wind waves causes the high-concentration zone to "shift upward," reflecting the influence of wind waves on sediment redistribution patterns. In conclusion, it is evident that waves significantly modify the distribution pattern of suspended sediment, which is otherwise predominantly influenced by tidal currents.

Tidal currents represent the predominant driving force in the distribution of suspended sediments. Changes in high-concentration areas and the extent of suspended sediment distribution are closely tied to the direction and strength of flood and ebb tidal currents. The stronger ebb currents in comparison to flood currents promote the outward transport of suspended sediment. The reduced flow velocity experienced in the vicinity of the harbour entrance gives rise to the deposition of sediment, resulting in a banded distribution pattern. This observation underscores the pivotal role of tidal currents in the transportation and deposition of sediments. However, the inclusion of wave activity significantly alters the dynamic characteristics of suspended sediment. Wind waves, by stirring water masses, enhance the capacity for the resuspension of sediment, particularly during high-energy events such as flood surges and ebb surges, where there is a significant increase in suspended sediment concentrations and rapid transportation. Furthermore, wind waves modify the direction and strength of tidal currents, thereby affecting the advective transport of suspended sediment and altering patterns of sediment deposition and redistribution.

In natural marine environments, tidal currents and waves frequently act in combination. A sediment transport model that considers only tidal currents may not accurately reflect real-world conditions. It is, therefore, vital to couple wave effects in order to achieve a more realistic simulation of suspended sediment concentration dynamics. The influence of waves on sediment movement patterns is significant, with alterations to hydrodynamic conditions, including bottom shear stress and wave–current interactions, being a primary factor. A more profound comprehension of the distribution of suspended sediment in the context of wave–current interactions is, therefore, essential to attain a comprehensive understanding of the intricate nature of sediment transport processes.

3.2.4. Characteristics of Suspended Sand Transport

As demonstrated in Figure 14, prior to the analysis of suspended sediment transport characteristics in the coastal area, it is acknowledged that the seabed sediment in this region predominantly comprises fine silt and clayey silt. In the context of high-velocity flows, these sediments are susceptible to erosion and transport. Within the port area, suspended sediment is primarily transported seaward. Moving outward from the port, the suspended sediment concentration is observed to be relatively high near the estuary, as illustrated in the figure. Furthermore, a clockwise circulation of suspended sediment transport is apparent near the breakwater, which likely promotes deposition in this region, contributing to sediment accumulation.

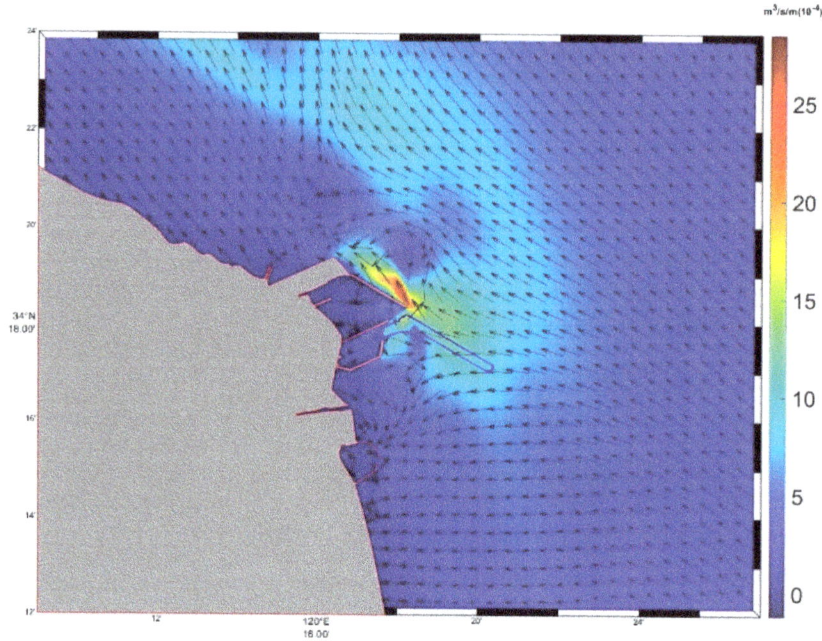

Figure 14. Suspended sediment concentration and distribution.

Exiting the port, the transport direction of the suspended sediment deflects, forming two primary transport pathways. One of these is directed northwestward toward Haizhou Bay, while the other heads in the opposite direction, southwestward toward the Subei Shoals. Consequently, the suspended sediment from the Huanghe River Estuary, which has been abandoned, is transported towards the southern tidal flats. Observations indicate that during periods of high-energy events, such as strong winds and waves, sediment transport becomes more pronounced in this region.

In summary, based on the numerical simulation results and previous studies, under normal conditions, suspended sediment transport in the Huanghe River Estuary occurs in two main directions: northwestward toward Haizhou Bay and southwestward toward the Subei Shoals [20]. The overall suspended sediment migration in this area is found to generally follow the northern trajectory along the Subei coastline. The predominant factors influencing sediment transport in this region are tidal currents and resuspension. It can be inferred that over extended periods, coastal areas are predominantly subject to erosion, with eroded sediments typically being transported southward, resulting in a pattern of erosion in the northern region and deposition in the southern region.

3.3. Analysis of Sediment Erosion and Deposition Characteristics

3.3.1. Erosion and Deposition Under Normal Weather Conditions

The analysis of field observation data and numerical simulation results indicates (as shown in Figure 15 that, prior to the implementation of the project, the sediment erosion and deposition processes in the study area exhibited significant spatiotemporal heterogeneity. In terms of spatial distribution, three distinct erosion–deposition units have been identified:

the estuary, the main harbour basin, and the northern breakwater head. The evolution of these units is jointly controlled by tidal dynamics and local geomorphological conditions.

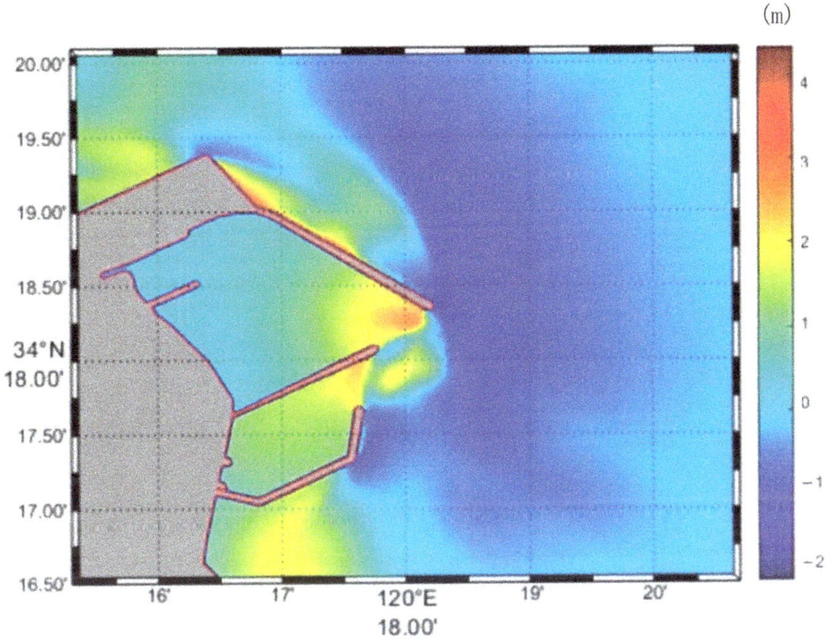

Figure 15. Erosion and deposition distribution in the study area.

During the flood tide stage, as the tidal current propagates from the open sea into the harbour, the combined effects of the estuary's narrowing and the abrupt change in harbour basin topography lead to the formation of a pronounced recirculation zone in the estuary region. The velocity field in this area exhibits a pronounced gradient, with maximum flow speeds of 3.5 m/s at the estuary rapidly decaying to below 0.8 m/s inside the harbour basin. This precipitous decline in flow velocity substantially mitigates fluid shear stress, consequently leading to the flocculation and settling of fine suspended sediments with a median grain size of d50 $\geq$ 0.03 mm. Over the course of 60 tidal cycles, a continuous deposition layer of a thickness ranging from 2.5 to 3.5 m forms in this region, exhibiting a vertical sorting pattern of finer sediments overlying coarser sediments.

During the ebb tide, the hydrodynamic system undergoes a reversal, resulting in the transport of eroded material from the main harbour basin downstream by the ebb current. The geometry of the basin, which is "funnel-shaped", serves to constrain flow velocities, which gradually increase near the estuary. This increase in velocity creates a "corridor effect" that facilitates secondary sediment transport. It is estimated that approximately 40% of the suspended sediment undergoes redeposition in the transition zone from the harbour basin to the estuary, forming a wedge-shaped deposition layer 2–3 m thick. It is noteworthy that this depositional belt exhibits an asymmetric cross-sectional distribution, with greater thickness on the eastern side. This asymmetry can be attributed to the deflection of flow paths caused by Coriolis forces.

The evolution of the scour trench at the northern breakwater head is of particular hydrodynamic significance. Observations made in situ demonstrate that during both flood and ebb tides, a jetting effect occurs at the breakwater head, resulting in a 35% increase in flow velocity. Furthermore, the analysis of field observations indicates that the velocities of tidal currents in proximity to the breakwater waterhead exceed the critical threshold for initiating the motion of muddy clay layers. This is identified as the primary factor driving scouring. The intense hydrodynamic conditions in this area have been shown to lead to severe seabed erosion. The data indicates a positive correlation between scour intensity and tidal range and a pulsating development pattern during spring and neap tidal cycles.

The mechanisms underlying the aforementioned erosion and deposition patterns can be attributed to the following: Firstly, net sediment transport asymmetry induced by tidal asymmetry. Secondly, local flow field distortion is caused by abrupt topographic changes. And thirdly, lag effects in the dynamic depositional response during sediment transport. It is noteworthy that the estuary region, as the critical chokepoint for water and sediment exchange, experiences the highest sedimentation rates and consequently becomes the key bottleneck limiting navigation channel capacity.

3.3.2. Sediment Change Under Extreme Weather Conditions

Abrupt deposition is a term used in the field of port and navigation engineering to describe the severe sedimentation caused by a single severe weather event, such as strong winds, which significantly impacts the normal operation and navigation of vessels. During storms or typhoons, storm surges and elevated offshore wave heights—often several times or even dozens of times greater than those under calm weather conditions—intensify sediment transport on the seabed and along the coast. This process leads to abrupt sedimentation in ports and navigation channels. In the context of engineering projects, abrupt deposition poses significant risks. In some cases, a single abrupt deposition event may exceed the cumulative sedimentation of several months. Irrespective of whether the port and navigation projects are located in silty or muddy coastal zones, abrupt deposition can occur under certain conditions.

The data sets for this study were obtained from two typhoons, "Lekima" in 2019 (the 9th typhoon of 2019, Figure 16b) and "Muifa" in 2022 (the 12th typhoon of 2022, Figure 16a). The weather processes during these events were selected for studying abrupt sedimentation under different terminal design scenarios in this project. The calculation period for each event is shown in Table 5. A 24 h timeframe was selected for the analysis of the impacts of each typhoon on different scenarios. Hourly outputs of significant wave height, peak wave period, and mean wave direction were extracted for the purposes of analysis.

As illustrated in Figure 17, the deposition distribution of each scenario is shown during the 24 h period of different typhoon events. During Typhoon "Lekima", the sediment deposition in the study area was primarily concentrated at the estuary, with a maximum deposition thickness of approximately 0.5 m. In contrast, during Typhoon "Muifa", the deposition at the estuary reached about 1 m. The characteristics of erosion and deposition distribution during Typhoon "Muifa" (Figure 17a) and Typhoon "Lekima" (Figure 17b) exhibit significant differences. As demonstrated in Figure 17b, during Typhoon "Lekima", the deposition was predominantly concentrated within the estuary region, exhibiting an augmented deposition area and a substantial escalation in intensity. The maximum deposition thickness attained a value of approximately 0.5 m. Concurrently, the channel region underwent significant erosion, a consequence of the potent hydrodynamic forces at play. In comparison with the standard conditions, the powerful storm surge and wave action during Typhoon "Lekima" significantly enhanced sediment transport and deposition.

In contrast, the erosion and deposition distribution during Typhoon "Muifa", as illustrated in Figure 17a, exhibited heightened intensity. In comparison with Typhoon "Lekima", both the intensity and extent of deposition in the estuary region were significantly greater, with the maximum deposition thickness reaching 1 m, which is twice that of Typhoon "Lekima". This suggests that the stronger storm surge during Typhoon "Muifa" accelerated rapid sediment accumulation. The channel region exhibited the most pronounced erosion, while the nearshore region experienced an expansion of the erosion area due to wave propagation and storm surge effects. Consequently, as typhoon intensity escalates, the disparities in erosion and deposition patterns become increasingly evident, manifesting predominantly as heightened deposition within the estuary region and erosion within the channel region.

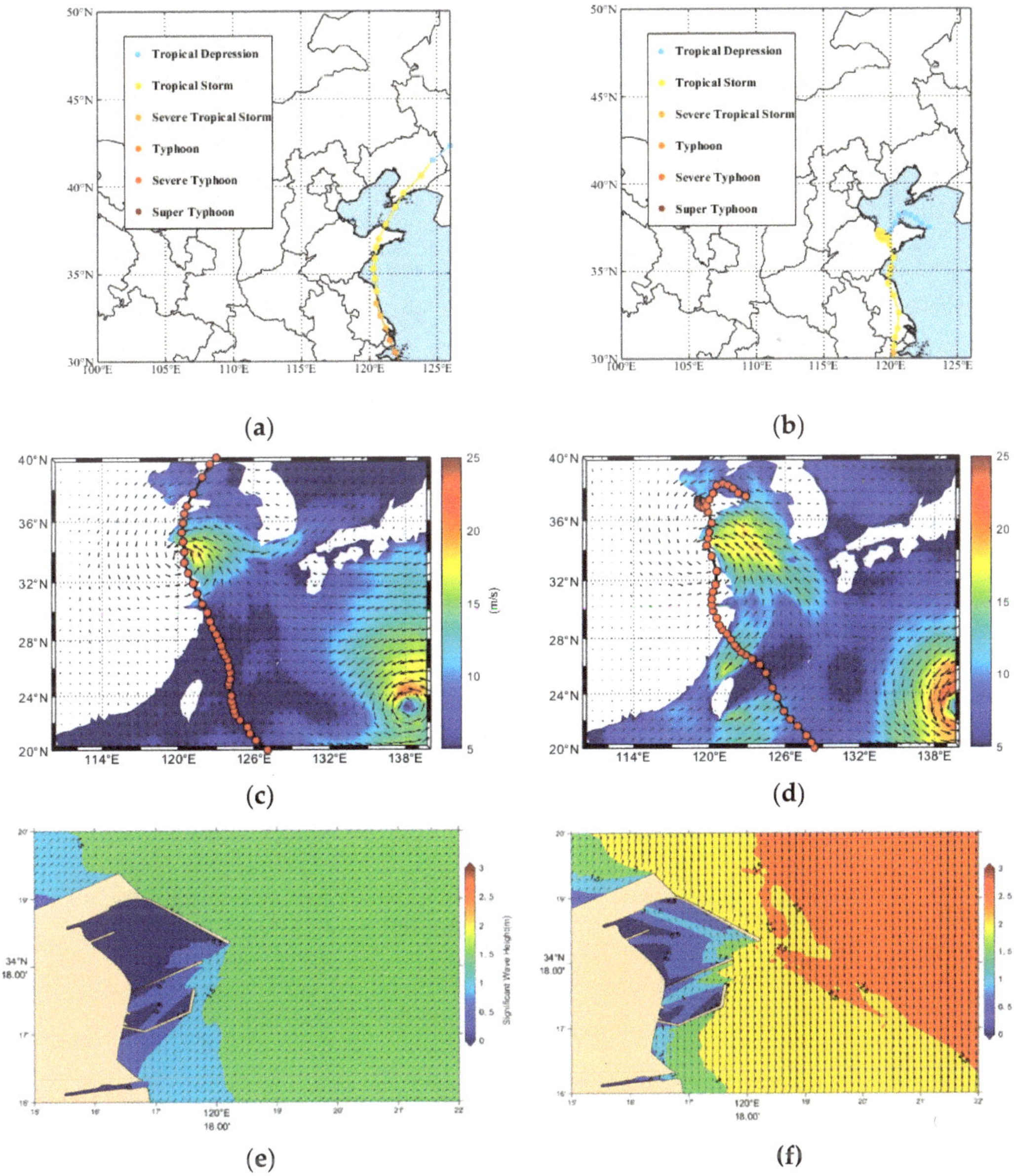

Figure 16. Wave field distribution under Typhoons "Muifa" and "Lekima". (**a**) Path of Typhoon "Muifa"; (**b**) path of Typhoon "Lekima"; (**c**) wind field during Typhoon "Muifa"; (**d**) wind field during Typhoon "Lekima"; (**e**) wave field during Typhoon "Muifa"; (**f**) wave field during Typhoon "Lekima".

Table 5. Calculation time of each typhoon process.

Typhoon Name	Calculation Time (Beijing Time)
Lekima (LEKIMA)	20:00 August 10–20:00 11 August 2019
Plum Blossom (Muifa)	0:00 September 15–0:00 16 September 2022

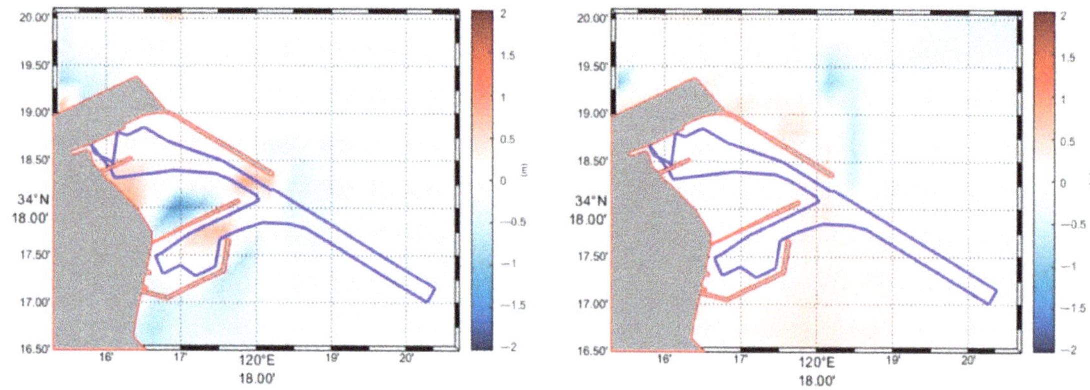

(**a**) (**b**)

Figure 17. Erosion and deposition distribution under different typhoons. (**a**) Typhoon "Muifa"; (**b**) Typhoon "Lekima".

4. Discussion

The study area exhibits a pronounced asymmetry in semi-diurnal tidal patterns [27], similar to the situation observed in the Yangtze River estuary: flood currents predominantly flow southeast, while ebb currents retreat northwest. The inflow of water within the channel is obstructed, resulting in a reduced ebb velocity. In addition, the effects of breakwaters and Coriolis forces further enhance the ebb and flow process in the channel. The breakwater structures create significant velocity gradients between the inner and outer harbour zones, resulting in flow rates in the harbour basin that are 40–60% lower than in adjacent coastal areas. This reduction in flow velocity favours sediment resuspension and localised deposition, a phenomenon also observed in the Dinh An estuary in Vietnam [28], which was mainly attributed to the reduction in flow velocity.

Suspended sediment concentration (SSC) shows a correlation with tidal patterns. Sediment transport rates vary between coastal ports depending on wave climate and other nearshore characteristics, while changes in hydrodynamic conditions are a major cause of severe siltation in curved channel sections [28]. As water levels and velocities increase, high tides increase SSC, whereas ebb tides decrease SSC as hydrodynamic energy decreases. Near the entrance, a sudden reduction in flow velocity triggers sediment accumulation, with SSC isopleths closely following depth contours.

Under regular tidal conditions, sediment transport is predominantly controlled by tidal currents and resuspension, similar to the situation in the Tay Estuary where sediment deposition is largely influenced by the morphology of the tidal currents [29]. This results in significant siltation within the harbour, while the outer channel experiences less erosion. Due to flow deflection, intense scour occurs near the breakwater heads, with scour depth positively correlated with tidal range (R = 0.78). Some alongshore drifting sediments are intercepted by the southern breakwater, making the entrance area a bottleneck for sediment transport and accumulating the highest levels of siltation, similar to conditions observed in the ports of Chennai and Visakhapatnam [30].

Extreme events such as typhoons Lekima and Muifa enhanced sediment dynamics through storm surges and wave action. The increased wave action caused sediments on the inner continental shelf to be resuspended by strong tidal currents and transported into the entrance channels [31], resulting in sedimentation within the harbour. Silt thickness at the entrance reached 0.5 m during Lekima and 1.0 m during Muifa, illustrating how climate-hydrodynamic interactions exacerbate sedimentation.

5. Conclusions

This study investigates sediment accumulation upstream of coastal harbour channels using field data. To further investigate the causes of sediment accumulation in deep water channels, a hydrodynamic numerical model was developed. The model was validated using measured tidal heights, flow velocities, directions and suspended sediment concentration (SSC). Two case studies were examined, one under normal weather conditions and the other under extreme weather conditions, to explain the mechanisms leading to entrance siltation. The results highlight the significant impact of the coastal harbour on the morphological dynamics of the coastal marine area, with a pronounced velocity gradient between the inner and outer harbour zones favouring sediment resuspension and localised deposition. The interaction between extreme climatic events and hydrodynamics amplified the sediment dynamics, increasing the sediment thickness at the entrance. Through a detailed analysis of the sedimentation challenges faced by coastal ports, the proposed modelling approach can be applied to other geographical regions with different hydrodynamic and morphological characteristics to assess their transferability. This will allow for comparative studies, making it applicable not only to the current port dredging management plans, but also to the planning of new port sites in coastal areas [30].

Author Contributions: Conceptualization, X.D. and Z.W.; methodology, X.D. and Z.W.; software, X.D.; validation, X.D.; investigation, X.D.; resources, Z.W. and X.M.; data curation, X.D.; writing—original draft preparation, X.D.; writing—review and editing, Z.W. and X.M.; visualization, X.D.; supervision, Z.W.; project administration, Z.W. and X.M.; funding acquisition, Z.W. and X.M. All authors have read and agreed to the published version of the manuscript.

Funding: This study was supported financially by the National Key Research and Development Program of China (2022YFC3104205) and the National Natural Science Foundation of China (42377457).

Institutional Review Board Statement: Not applicable.

Data Availability Statement: Due to privacy, the data supporting the results of this study are not publicly available. However, they can be accessed upon reasonable request from the corresponding author.

Acknowledgments: We would like to thank the Marine Big Data Center of the Institute for Advanced Ocean Study of the Ocean University of China for providing data conversion and storage, and for the professional and technical services provided by Yujie Dong.

Conflicts of Interest: The authors declare no conflict of interest.

References

1. Murray, N.J.; Phinn, S.R.; DeWitt, M.; Ferrari, R.; Johnston, R.; Lyons, M.B.; Clinton, N.; Thau, D.; Fuller, R.A. The global distribution and trajectory of tidal flats. *Nature* **2019**, *565*, 222–225. [CrossRef] [PubMed]
2. Tognin, D.; D'Alpaos, A.; Marani, M.; Carniello, L. Marsh resilience to sea-level rise reduced by storm-surge barriers in the Venice Lagoon. *Nat. Geosci.* **2021**, *14*, 906–911. [CrossRef]
3. Warrick, J.A.; Buscombe, D.; Vos, K.; Bryan, K.R.; Castelle, B.J.; Cooper, J.A.G.; Harley, M.D.; Jackson, D.W.T.; Ludka, B.C.; Masselink, G.; et al. Coastal shoreline change assessments at global scales. *Nat. Commun.* **2024**, *15*, 2316. [CrossRef] [PubMed]
4. Frazler, D.E. Recent deltaic deposits of the Mississippi River:their development and chronology. *Gulf Coast Assoc. Geol. Soc. Trans.* **1967**, *17*, 205–230.

5. Mikhailov, V.N.; Mikhailova, M.V. Delta formation processes at the Mississippi River mouth. *Water Resour.* **2010**, *37*, 515–530. [CrossRef]
6. Saha, R.C. Chattogram Port: A dedicated service institution to evolve the country boldly. *Marit. Technol. Res.* **2023**, *5*, 258294. [CrossRef]
7. Sakalayen, Q.; Chen, P.S.L.; Cahoon, S. The strategic role of ports in regional development: Conceptualising the experience from Australia. *Marit. Pol. Manag.* **2017**, *44*, 933–955. [CrossRef]
8. Santos, A.M.P.; Salvador, R.; Soares, C.G. A dynamic view of the socioeconomic significance of ports. *Marit. Econ. Logist.* **2018**, *20*, 169–189. [CrossRef]
9. Jia, H.Y.; Lampe, O.D.; Solteszova, V.; Strandenes, S.P. Norwegian port connectivity and its policy implications. *Marit. Pol. Manag.* **2017**, *44*, 956–966. [CrossRef]
10. Bottasso, A.; Conti, M.; Ferrari, C.; Tei, A. Ports and regional development: A spatial analysis on a panel of European regions. *Transport. Res. Pol. Pract.* **2014**, *65*, 44–55. [CrossRef]
11. Li, W.Y.; You, Z.J.; Cai, Z.Z.; Sui, Y. Club convergence and allometry in Chinse mainland coastal container ports. *Ocean Coast Manag.* **2022**, *230*, 106376. [CrossRef]
12. Zhu, G.R.; Xie, Z.L.; Xu, H.; Liang, M.; Cheng, J.; Gao, Y.; Zhang, L. Land reclamation pattern and environmental regulation guidelines for port clusters in Bohai Sea, China. *PLoS ONE* **2021**, *16*, e0259516. [CrossRef]
13. Chen, X.Y.; Liu, D.H.; Lu, J. The sedimentary environment of current estuary and abandoned estuary at the modern Yellow River delta. *Adv. Mater. Res.* **2012**, *356*, 914–919. [CrossRef]
14. Su, M.; Yao, P.; Wang, Z.B.; Zhang, C.K.; Stive, M.J. Exploratory morphodynamic hindcast of the evolution of the abandoned Yellow River delta, 1578–1855 CE. *Mar. Geol.* **2017**, *383*, 99–119. [CrossRef]
15. Zhang, L.; Cheng, J.; Jiang, H.; Xie, H. Impact of port planning on hydrodynamics and water environment: Case study of Dongjiakou Port area of Qingdao Port. In Proceedings of the International Conference on Smart Transportation and City Engineering (STCE 2022), Chongqing, China, 12–14 August 2022; p. 1246002.
16. Li, X.; Xie, M.X.; Li, M.G. Numerical Modeling of the Hydrodynamics for Zhuanghe Fishing Port. *Adv. Mater. Res.* **2013**, *765–767*, 2952–2956. [CrossRef]
17. Saengsupavanich, C.; Yun, L.S.; Lee, L.H.; Sanitwong-Na-Ayutthaya, S. Intertidal intercepted sediment at jetties along the Gulf of Thailand. *Front. Mar. Sci.* **2022**, *9*, 970592. [CrossRef]
18. Tan, Z. Risk Management Research in Port Depth Maintenance Program. Master's Thesis, Dalian Maritime University, Dalian, China, 2012. (In Chinese)
19. Pei, Y.; Lu, P. Study on the Coastal Stability of Binhai Harbour and It's Impact from Deep Waterway Project. *Mod. Transp. Technol.* **2011**, *8*, 87–91. (In Chinese)
20. He, R.; Xu, B.; Yao, L.; Qing, Y.; Hu, Y. Construction technology optimization of maintenance dredging project in Binhai port area, Yancheng. *Port Waterw. Eng.* **2019**, *7*, 217–225. (In Chinese)
21. Jin, H. Coastal Evolution Process and Nearshore Suspended Sediment Research of the Abandoned Yellow River Delta. Master's Thesis, East China Normal University, Shanghai, China, 2014. (In Chinese).
22. Zhang, R.; Lu, L.; Wang, Y. The mechanism and trend of coastal erosion of Jiangsu Province in China. *Geogr. Res.* **2002**, *21*, 469–478. (In Chinese)
23. Cheng, H.; Miao, R.; Yu, H.; Xing, G.; Gao, P.; Zhao, Y. *Hydrological Observation Report for the Optimization of Overall Layout of Binhai Port*; Technical Report; CCCC First Harbor Consultants Co., Ltd.: Tianjin, China, 2021. (In Chinese)
24. Fu, Y.; Zhou, X.; Zhou, D. Accuracy analysis of ocean tidal model over China seas based on the gauge data. *Sci. Surv. Mapp.* **2017**, *42*, 28–32. (In Chinese)
25. Jiangsu 908 Special Project Office. *Comprehensive Survey and Assessment Report of Jiangsu Offshore Marine Areas*; Science Press: Beijing, China, 2012. (In Chinese)
26. Zhang, L.; Chen, S.; Yi, L. The sediment source and transport trends around the abandoned Yellow River Delta, China. *Mar. Georesources Geotechnol.* **2016**, *34*, 440–449. [CrossRef]
27. Kuang, C.P.; Chen, W.; Gu, J.; He, L.L. Comprehensive analysis on the sediment siltation in the upper reach of the deepwater navigation channel in the Yangtze Estuary. *J. Hydrodyn.* **2014**, *26*(2), 299–308. [CrossRef]
28. Nguyen, V.T.; Zheng, J.H.; Zhang, J.S. Mechanism of back siltation in navigation channel in Dinh An Estuary, Vietnam. *Water Sci. Eng.* **2013**, *6*, 178–188.
29. Dobereiner, C.; McManus, J. Turbidity maximum migration and harbor siltation in the Tay Estuary. *Can. J. Fish. Aquat. Sci.* **1983**, *40*, s117–s141. [CrossRef]

30. Sarma, K.G.S. Siltation and coastal erosion at shoreline harbours. *Procedia Eng.* **2015**, *116*, 12–19. [CrossRef]
31. Chuanteng, L.; Yufang, H.; Xiaofeng, L.; Xianbo, Z. Numerical Simulation of Sudden Siltation in Yangtze Estuary under Extreme Weather. In Proceedings of the 15th International Conference on Hydroinformatics 2024, Beijing, China, 27–31 May 2024; p. 142.

Journal of
Marine Science and Engineering

MDPI

Article

Joint Probability Distribution of Wind–Wave Actions Based on Vine Copula Function

Yongtuo Wu [1], Yudong Feng [1], Yuliang Zhao [2,*] and Saiyu Yu [1]

[1] Shandong Electric Power Engineering Consulting Institute Corp., Ltd., Jinan 250013, China; appder@126.com (Y.W.); fengyudong2021@163.com (Y.F.); hkdysy@163.com (S.Y.)
[2] College of Engineering, Ocean University of China, Qingdao 266100, China
* Correspondence: zhaoyuliang@ouc.edu.cn

Abstract: During its service life, a deep-sea floating structure is likely to encounter extreme marine disasters. The combined action of wind and wave loads poses a threat to its structural safety. In this study, elliptical copula, Archimedean copula, and vine copula models are employed to depict the intricate dependence structure between wind and waves in a specific sea area of the Shandong Peninsula. Moreover, hourly significant wave height, spectral peak period, and 10 m average wind speed hindcast data from 2004 to 2023 are utilized to explore the joint distribution of multidimensional parameters and environmental design values. The results indicate the following: (1) There exists a significant correlation between wind speed and wave parameters. Among them, the C-vine copula model represents the optimal trivariate joint distribution, followed by the Gaussian copula, while the Frank copula exhibits the poorest fit. (2) Compared with the high-dimensional symmetric copula models, the vine copula model has distinct advantages in describing the dependence structure among several variables. The wave height and period demonstrate upper tail dependence characteristics and follow the Gumbel copula distribution. The optimal joint distribution of wave height and wind speed is the t copula distribution. (3) The identification of extreme environmental parameters based on the joint probability distribution derived from environmental contour lines is more in line with the actual sea conditions. Compared with the design values of independent variables with target return periods, it can significantly reduce engineering costs. In conclusion, the vine copula model can accurately identify the complex dependency characteristics among marine variables, offering scientific support for the reliability-based design of floating structures.

Keywords: wind–wave parameter; joint probability distribution; copula theory; vine copula; marine structure design

1. Introduction

Deep-sea floating offshore new-energy projects are faced with formidable challenges. These challenges stem from the intricate wind and wave loads, the substantial movement of floating foundations, and the development of large-scale units [1,2]. In the marine environment, the combined impact of multiple loads is the primary determinant of structural safety and engineering costs [3,4]. Traditionally, in nearshore and coastal engineering, the annual extreme value method has been prevalently employed [5]. This method independently determines the extreme value distribution models of various environmental variables, calculates the design parameters with a specific recurrence interval (e.g., occurring once every period of years), and thereby defines the design loads that engineering structures must endure from environmental forces [6]. However, in reality, there exists a strong correlation among metocean data. The failure or even collapse of most deep-sea floating structures usually does not result from a single environmental load surpassing the critical value. Instead, it is often due to multiple combined loads reaching or exceeding the limit-state function. To guarantee the safety of marine engineering and prevent waste, it is essential to take into account the dependence structure between multiple variables and their extreme values. Only by doing so can we reasonably determine the design parameters of the marine environment [7–9].

To better simulate the impacts of intricate multidimensional environmental loads, scholars have dedicated themselves to researching joint distribution models of diverse environmental variables [10–13]. The conditional probability model is a prevalent joint probability model for characterizing marine environmental factors. In particular, the Weibull–Lognormal conditional joint model for wave height and period has gained wide acceptance [14]. This model first fits the marginal distribution of the primary controlling element. Subsequently, it conducts nested fitting on the conditional distribution of the remaining variables, based on the previously determined element. Its construction is relatively straightforward. Nevertheless, when depicting the complex correlations among multivariate environmental factors, this model often simplifies by assuming independence, thereby reducing its accuracy. Simão et al. [15] introduced a multidimensional long-term joint probability model for environmental parameters grounded in conditional functions. The Nataf distribution, based on the Gaussian assumption, can be utilized to construct joint models for any high-dimensional variables [16]. However, the actual dependence structure of marine environmental variables is nonlinear and non-Gaussian. As a result, significant deviations occur in the fitting results [17]. The copula theory, by integrating the marginal distribution of marine environmental variables with copula functions, enables the construction of a joint distribution model. This approach has found increasing application in sea-state assessment [18,19]. Bivariate copulas have been extensively employed to simulate two-dimensional correlated variables in the reliability analysis of marine engineering [20]. Regarding multidimensional random variables, the vine copula model offers a more flexible means of describing their multivariate dependency structures [21]. For instance, Yang et al. [22] proposed an optimized Archimedean copula to simulate the multivariate joint distribution of five-dimensional wind and wave parameters in the sea-crossing bridge region. Additionally, Dong et al. [23] put forward a three-dimensional maximum entropy distribution of effective wave height, wind speed, and load direction, providing a theoretical reference for determining the design parameters of marine environments.

The vine copula model effectively decomposes multidimensional distribution functions into a combination of multiple nested bivariate copulas and the product of univariate marginal distributions. This unique decomposition enables the model to describe the correlations among multiple variables with great flexibility. As a result, it has found extensive application in joint probability analysis within the fields of hydrology and meteorology [24,25]. For instance, Montes-Iturrizaga and Heredia-Zavoni [26] utilized C-vine copulas to construct multivariate environmental contours. They based their work on significant wave height, peak spectral period, and wind velocity data obtained from storm hindcasts. Similarly, Lin and Dong [27] conducted a wave energy assessment. They employed the trivariate vine copula distribution of significant wave height, mean period, and direction. In their study, the marginal distributions were fitted using the maximum entropy distribution and a mixture of von Mises distributions. In hydrology and ocean engineering, when dealing with problems involving several correlated variables that demand highly flexible dependence modeling, the pair-copula decomposition approach, as offered by the vine copula model, is highly recommended. This approach allows for a more accurate and adaptable representation of the complex relationships among variables, thereby enhancing the reliability and effectiveness of relevant analyses.

Moreover, the environmental contour (EC) approach is widely utilized to determine the design values of multivariate marine variables [28–31]. This is crucial for accurately assessing the structural response influenced by correlated environmental loads. The EC approach offers the combined extreme environmental conditions within a specific return period, thus enabling the prediction of the maximum structural response at the target return period level. It is recommended in numerous international standards, such as those of DNV GL [32]. Haver [33] was the first to introduce the concept of EC to depict the joint probability distribution of significant wave height and wave period. Subsequently, Winterstein et al. [34] proposed the EC method in the transformed standard normal space of expected exceedance probability, relying on the Inverse First Order Reliability Method (IFORM). The IFORM algorithm has since found extensive application in ocean engineering. Montes-Iturrizaga and Heredia-Zavoni [35] applied a derived formulation to construct IFORM-based environmental contours using bivariate copulas. Their study revealed that the choice of copula models significantly impacts the resulting ECs. Specifically, the IFORM-type contours assume a convex form for the structural failure boundary, while the ISORM- and highest density regions-based methods assume a concave form [36,37]. Furthermore, Huseby et al. [38] employed Monte Carlo simulations to establish ECs without the necessity of transformations. Vanem [39] conducted a comparative study on the estimation of extreme structural responses from different EC methods. The results indicated that in certain cases, the differences can be substantial and consequential. Most applications demonstrating ECs derived from different algorithms have been carried out using bivariate variables. For instance, Clarindo and Guedes Soares [40] compared the contours constructed based on the Burr–Lognormal distribution with those considering the Weibull–Lognormal distribution. Their findings suggested that the maximum values obtained from the Burr–Lognormal distribution were more favorable due to the superior predictive power of its contour lines. In addition, Vanem et al. [41] conducted a simulation study on the uncertainty of ECs caused by sampling variability across different estimation methods. The establishment of the multivariate joint distribution of environmental variables is a pivotal step in constructing multidimensional ECs. Heredia-Zavoni and Montes-Iturrizaga [42] utilized three-dimensional vine copulas to model directional ECs, discovering that directionality can have a significant impact on ECs. Bai et al. [43] established three-dimensional direct sampling-based ECs using a semi-parametric joint probability model. In this model, a log-transformed KDE–Paretotails approach was proposed to fit marginal distributions, and vine copulas were used to estimate the joint models. Fang et al. [44] optimized the C-vine copula and constructed ECs for the joint wind–wave environment of sea-crossing bridges. They proposed a one-step optimization method to identify an optimized canonical vine copula. Meng and Li [45] used the R-vine copula and a direct sampling approach to calculate three-dimensional ECs of wind and wave, taking into account different sampling methods and seasonal effects. Wu et al. [46] employed pair copulas to construct the multivariate joint probability distribution and generated ECs for data-driven applications. While the errors in extreme response evaluation resulting from the contour approximation itself may be relatively low compared to response-based analysis, the fitted models for joint distributions contribute significantly to the overall errors [47].

In conclusion, for the risk assessment and management of floating structures, an accurate statistical portrayal of relevant extreme environmental conditions is essential [48]. The environmental contour approach plays a pivotal role as an input in the design of marine structures, which must endure the loads exerted by environmental forces. This research utilizes the vine copula function to formulate a joint probability model of wind and wave parameters. It then contrasts this model with high-dimensional symmetric copula models. The overarching goal is to precisely depict the multidimensional dependence structure of environmental parameters, thereby offering a scientific foundation for the safety and reliability evaluation of offshore structures. Drawing upon the hourly significant wave height, spectral peak period, and 10 m average wind speed hindcast data in a specific sea area of the Shandong Peninsula from 2004 to 2023, various copula models were employed to compute their joint probability distribution and establish an environmental isosurface. This project not only supplies reasonable environmental design parameters for Shandong's offshore new-energy structures but also provides a theoretical basis for marine disaster risk assessment and the formulation of disaster prevention and reduction strategies.

The marine environments in different regions have unique characteristics, and the applicability of existing research results in specific areas, such as a specific sea area of the Shandong Peninsula, remains to be further verified. There is a lack of effective models and methods that can accurately describe the joint probability distribution of multiple parameters under the complex marine environmental conditions in this region. This study aims to fill these gaps. By using the vine copula function to construct a joint probability model of wind and wave parameters, it fully considers the correlations among marine environmental parameters, provides reasonable design environmental parameters for offshore new-energy structures in the Shandong Peninsula, and offers a theoretical basis for marine disaster risk assessment and the formulation of disaster prevention and reduction strategies.

The objective of this study is to establish reasonable design environmental parameters for calculating wind–wave actions on floating structures. This is achieved by taking into account the joint correlated characteristics of sea states. In Section 2, we delve into fundamental multivariate methodologies and the design approach grounded in trivariate joint probability distribution models. Here, we comprehensively review the theoretical underpinnings and practical applications of these methods, providing a solid foundation for subsequent analysis. Section 3 focuses on the construction of the environmental surface for the evaluation of design values. We detail the procedures and considerations involved in creating this surface, which are crucial for accurately assessing the loads that floating structures are likely to encounter. Section 4 elaborates on the original data and the data-processing steps carried out in accordance with the multidimensional analysis framework. Additionally, the results obtained from this analysis are presented and discussed in this section, offering insights into the implications of the data. Finally, Section 5 summarizes the conclusions of this work.

2. Multivariate Distribution Theory

This section expounds on the construction theory of copula-based trivariate joint distributions for wind–wave parameters, along with the derivation of environmental surfaces. The methodological framework for multivariate analysis is presented in Figure 1. Initially, the fundamental copula theory and diverse trivariate models are introduced. Additionally, a concise description of the parameter estimation method and goodness of fit is provided. Subsequently, the kernel estimation utilized for marginal distribution fitting are presented. The IFORM-based environmental surfaces are deduced to ascertain the maximal design loads. This process is essential for accurately determining the loads that wind–wave parameters may impose, thereby contributing to more informed engineering and research decisions in the context of marine-related studies.

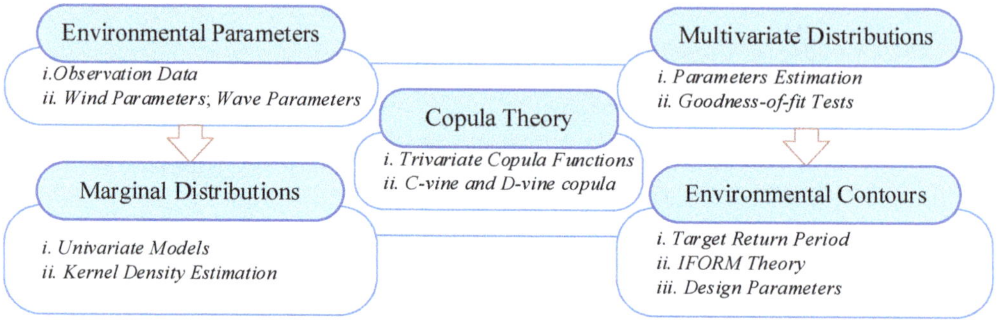

Figure 1. Flowchart of multi-load design concepts.

2.1. Basic Copula Theory

The copula theory provides a means to describe multidimensional related environmental variables. Based on Sklar's theorem [49], multidimensional joint probability models can be formulated by integrating copula functions with the marginal distributions of random variables. Suppose a k-dimensional random variable is $\boldsymbol{X} = (x_1, x_2, \cdots, x_k)$ and its marginal cumulative distribution functions are $F_1(x_1), F_2(x_2), \cdots, F_k(x_k)$, respectively. If these marginal distributions are continuous and strictly increasing, then there exists a copula function $C(u_1, u_2, \cdots, u_k)$ $(u_i = F_i(x_i), i = 1, 2, \cdots, k)$ that can be used to establish the joint distribution as follows:

$$F(x_1, x_2, \cdots, x_k) = C(u_1, u_2, \cdots, u_k) = C(F_1(x_1), F_2(x_2), \cdots, F_k(x_k)) \tag{1}$$

where $F(x_1, x_2, \cdots, x_k)$ represents the joint cumulative distribution of random variables. The corresponding probability density function can be derived through differentiation as follows:

$$\begin{aligned} f(x_1, x_2, \cdots, x_k) &= \frac{\partial^k F(x_1, x_2, \cdots, x_k)}{\partial x_1 \partial x_2 \cdots \partial x_k} = \frac{\partial^k F(x_1, x_2, \cdots, x_k)}{\partial F_1(x_1) \partial F_1(x_2) \cdots \partial F_k(x_k)} \cdot \frac{\partial F_1(x_1) \partial F_1(x_2) \cdots \partial F_k(x_k)}{\partial x_1 \partial x_2 \cdots \partial x_k} \\ &= c_{12\cdots k}(F_1(x_1), F_2(x_2), \cdots, F_k(x_k)) \cdot f_1(x_1) f_2(x_2) \cdots f_k(x_k) \end{aligned} \tag{2}$$

where $f(\cdot)$ and $c_{12\cdots k}(\cdot)$ represent the marginal probability density functions of environmental variables and the copula density, respectively. The typical procedures for constructing a multivariate joint distribution by means of copula theory entail two main steps. Firstly, it is necessary to fit the marginal distribution of each variable. This step involves identifying the appropriate probability distribution that best describes the behavior of each individual environmental variable. Subsequently, an appropriate copula function needs to be selected. This function is crucial as it serves to accurately depict the correlation structure among the random variables, thereby enabling a comprehensive and accurate representation of the multivariate relationships within the dataset.

2.2. Marginal Distribution Model

By leveraging a nonparametric kernel density function, the marginal distribution of hourly environmental data can be estimated [50]. This approach effectively circumvents the impact of errors associated with parametric distribution fitting. Subsequently, the maximum likelihood method can be adopted to estimate the copula parameters. The kernel density estimation of the environmental dataset $\boldsymbol{X} = (x_1, x_2, \cdots, x_k)$ can be expressed as follows:

$$\hat{f}_{X_i}(\hat{x}_i) = \frac{1}{N h_{X_i}} \sum_{j=1}^{N} K_{X_i}\left(\frac{\hat{x}_i - x_{ij}}{h_{X_i}}\right) \tag{3}$$

where K_X represents the non-negative kernel density function, h_X is the bandwidth parameter, and N denotes the total number of data points. The marginal distribution derived by means of the Gaussian kernel density function can be expressed as follows:

$$\hat{u}_1 = \int_{-\infty}^{\hat{x}_1} \hat{f}_{X_1}(x)dx = \frac{1}{N}\sum_{i=1}^{N} f\left(\frac{\hat{x}_1 - x_{1i}}{h_{X_1}}\right) \tag{4}$$

2.3. Trivariate Copula Model

The trivariate copula models most frequently employed encompass elliptical copulas (such as Gaussian and t copulas) and Archimedean copulas (including Clayton, Frank, and Gumbel copulas). The Gaussian copula offers a convenient means to simulate multidimensional variables. Specifically, the cumulative distribution function of the trivariate Gaussian copula can be represented as follows [51]:

$$\begin{aligned} C_G(u_1, u_2, u_3; \boldsymbol{\rho}) &= \Phi_\rho\left[\Phi^{-1}(u_1), \Phi^{-1}(u_2), \Phi^{-1}(u_3)\right] \\ &= \int_{-\infty}^{\Phi^{-1}(u_1)} \int_{-\infty}^{\Phi^{-1}(u_2)} \int_{-\infty}^{\Phi^{-1}(u_3)} \frac{1}{\sqrt{(2\pi)^3|\rho|}} \exp -\left(\frac{1}{2}\mathbf{x}^T\rho^{-1}\mathbf{x}\right)d\mathbf{x} \end{aligned} \tag{5}$$

where $\Phi^{-1}(\cdot)$ is the inverse function of the standard normal distribution; u_i represents the univariate cumulative distribution $F(x_i)$; $\mathbf{x} = [x_1, x_2, x_3]^{\mathrm{T}}$ denotes the three-dimensional environmental variable vector; and $\boldsymbol{\rho} = \begin{bmatrix} 1 & \rho_{12} & \rho_{13} \\ \rho_{21} & 1 & \rho_{23} \\ \rho_{31} & \rho_{32} & 1 \end{bmatrix}$ represents the linear correlation coefficient matrix.

The cumulative distribution function of the trivariate Student t copula can be expressed in the following manner [51]:

$$\begin{aligned} C_t(u_1, u_2, u_3; \boldsymbol{\rho}, v) &= T_{\rho,v}\left[T_v^{-1}(u_1), T_v^{-1}(u_2), T_v^{-1}(u_3)\right] \\ &= \int_{-\infty}^{T_v^{-1}(u_1)} \int_{-\infty}^{T_v^{-1}(u_2)} \int_{-\infty}^{T_v^{-1}(u_3)} \frac{\Gamma[(v+3)/2]}{\Gamma(v/2)\sqrt{(\pi v)^3|\rho|}} \left(1 + \frac{\mathbf{x}^T\rho^{-1}\mathbf{x}}{v}\right)^{-((v+3)/2)} d\mathbf{x} \end{aligned} \tag{6}$$

where $T_v^{-1}(\cdot)$ and $\Gamma(\cdot)$ are the inverse function of the Student t distribution with degree of freedom v and the gamma function, respectively.

Archimedean copulas have found extensive applications in hydrology and marine-related scenarios [52]. The distribution functions of the trivariate Clayton, Frank, and Gumbel copulas are presented below, respectively:

$$C_C(u_1, u_2, u_3; \theta) = \left(u_1^{-\theta} + u_2^{-\theta} + u_3^{-\theta} - 2\right)^{-1/\theta}, \theta \in [-1, \infty) \tag{7}$$

$$C_F(u_1, u_2, u_3) = -\frac{1}{\theta}\ln[1 + \frac{(e^{-\theta u_1} - 1)(e^{-\theta u_2} - 1)(e^{-\theta u_3} - 1)}{(e^{-\theta} - 1)^2}], \theta \neq 0 \tag{8}$$

$$C_G(u_1, u_2, u_3) = \exp\left\{-[(-\ln u_1)^\theta + (-\ln u_2)^\theta + (-\ln u_3)^\theta]^{1/\theta}\right\}, [1, \infty) \tag{9}$$

where θ represents the parameter of the copula function, which serves to characterize the correlation structure among random variables.

2.4. Vine Copula Model

The vine copula theory uses bivariate copula functions as building blocks to depict the correlation structure among multiple variables. Compared with some multivariate copula functions (such as elliptical and Archimedean copula functions), its construction method is more flexible, as it can take into account the different correlations between pairs of variables within the multiple variables. Additionally, the types of multivariate copula functions are relatively limited and often come with many restrictive conditions. In contrast, there is a wide variety of bivariate copula types. This provides the foundation for the extensive application of the vine copula theory [53].

Consider the three-dimensional random vector $\boldsymbol{X} = (x_1, x_2, \cdots, x_k)$ with a joint probability density of $f_{123}(x_1, x_2, x_3)$ and marginal densities of $f_1(x_1), f_2(x_2), f_3(x_3)$ respectively. Then, $f_{123}(x_1, x_2, x_3)$ can be decomposed using conditional densities as follows:

$$f_{123}(x_1, x_2, x_3) = f_3(x_3) f_{2|3}(x_2 \big| x_3) f_{1|2,3}(x_1 \big| x_2, x_3) \tag{10}$$

According to Sklar's theorem, we have the following:

$$f_{2|3}(x_2|x_3) = \frac{f_{23}(x_2, x_3)}{f_3(x_3)} = c_{23}[F_2(x_2), F_3(x_3)] f_2(x_2) \tag{11}$$

$$f_{1|2,3}(x_1|x_2, x_3) = \frac{f_{13|2}(x_1, x_3|x_2)}{f_{3|2}(x_3|x_2)} = c_{13|2}[F_{1|2}(x_1|x_2), F_{3|2}(x_3|x_2)] f_{1|2}(x_1|x_2) \tag{12}$$

If $f_{1|2}(x_1|x_2)$ is decomposed in a manner similar to Equation (11), then,

$$f_{1|2,3}(x_1|x_2, x_3) = c_{13|2}[F_{1|2}(x_1|x_2), F_{3|2}(x_3|x_2)] c_{12}[F_1(x_1), F_2(x_2)] f_1(x_1) \tag{13}$$

Combining Equations (11)–(13), we can obtain,

$$f_{123}(x_1, x_2, x_3) = c_{23}[F_2(x_2), F_3(x_3)] c_{12}[F_1(x_1), F_2(x_2)] c_{13|2}[F_{1|2}(x_1|x_2), F_{3|2}(x_3|x_2)] f_1(x_1) f_2(x_2) f_3(x_3) \tag{14}$$

Therefore, the following can be obtained:

$$c_{123}[F_1(x_1), F_2(x_2), F_3(x_3)) = c_{23}[F_2(x_2), F_3(x_3)] c_{12}[F_1(x_1), F_2(x_2)] c_{13|2}[F_{1|2}(x_1|x_2), F_{3|2}(x_3|x_2)] \tag{15}$$

The trivariate copula can be decomposed into the product of bivariate copulas and conditional bivariate copulas. However, the decomposition methods are not unique. Generally, the joint probability density of an n-dimensional random variable $\boldsymbol{X}$ can be decomposed in the following form:

$$f_{12\cdots n}(x_1, x_2, \cdots, x_n) = f_n(x_n) f_{n-1|n}(x_{n-1}|x_n) \cdots f_{1|2\cdots n}(x_1|x_2, \cdots, x_n) \tag{16}$$

The conditional density can be further decomposed using the pair-copula method:

$$f(x|v) = c_{xv_j|v_{-j}}[F(x|v_{-j}), F(v_j|v_{-j})] f(x|v_{-j}) \tag{17}$$

Among them, v is an m-dimensional vector and v_{-j} is the vector obtained by removing the variable v_j from v.

The marginal conditional distribution function is as follows:

$$F(x|v) = \frac{\partial C_{xv_j|v_{-j}}[F(x|v_{-j}), F(v_j|v_{-j})]}{\partial F(v_j|v_{-j})} = C_{x|v}(u_1, u_2) = h(u_1, u_2; \boldsymbol{\theta}) \tag{18}$$

The decomposition of multidimensional random variables is not unique. As the number of dimensions increases, the number of decomposition methods will increase extremely rapidly.

A regular vine is a set of trees. The edges of tree i serve as the nodes of tree $I + 1$, and two edges of tree i are connected in tree $I + 1$ if and only if they share a common node in tree i. C-vine and D-vine are decomposed according to a specific rule. If there is only a single node with degree $n - i$ in each tree T_i, that is, the root node, then such a vine structure is called a C-vine. The joint density of n-dimensional random variables is decomposed into the corresponding C-vine as follows:

$$f_{1\cdots n}(x_1, \cdots, x_n) = \prod_{k=1}^{n} f_k(x_k) \cdot \prod_{j=1}^{n-1} \prod_{i=1}^{n-j} c_{j,j+i|1,\cdots,j-1}[F(x_j|x_1, \cdots, x_{j-1}), F(x_{j+i}|x_1, \cdots, x_{j-1})] \tag{19}$$

In the C-vine structure, each tree has a root node, which is particularly effective in the case of a random vector with key variables.

If the tree T_i is a path and the degree of all nodes does not exceed 2, such a vine structure is called a D-vine. The joint density of n-dimensional random variables is decomposed into the corresponding D-vine as follows:

$$f_{1\cdots n}(x_1, \cdots, x_n) = \prod_{j=1}^{n-1} \prod_{i=1}^{n-j} c_{i,i+j|i+1,\cdots,i+j-1}[F(x_i|x_{i+1}, \cdots, x_{i+j-1}), F(x_{i+j}|x_{i+1}, \cdots, x_{i+j-1})] \cdot \prod_{k=1}^{n} f_k(x_k) \tag{20}$$

The tree structures of C-vine and D-vine for 5-dimensional variables are shown in Figure 2. The nodes of each tree represent variables, and the edges represent copula functions or conditional copula functions. The nodes of the subsequent tree are the edges of the previous tree. From the figures, the star-shaped structure feature of C-vine and the chain-shaped structure feature of D-vine can be seen. The conditional multidimensional distribution functions can be constructed by utilizing h-functions in combination with bivariate copulas and marginal distributions. Each bivariate copula is capable of describing the unique dependence structure of different datasets. In this work, seven types of bivariate copula probability functions are selected to depict the dependence structure of wind and wave parameters. The candidate distribution functions are presented in Table 1.

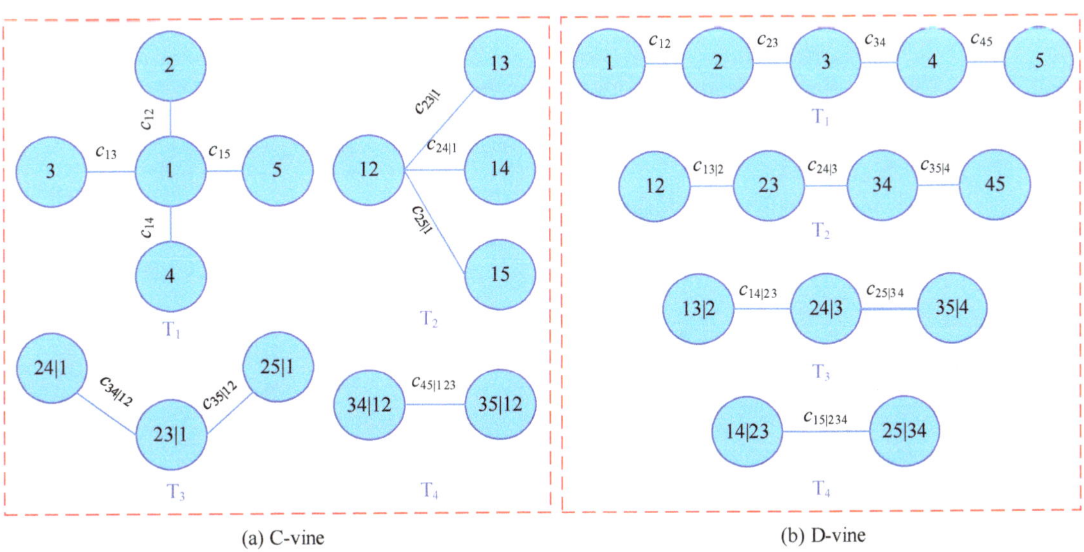

Figure 2. The tree structures of C-vine and D-vine.

Table 1. Candidate bivariate copula models and parameters.

Copula	Function	Parameter
Gaussian	$\Phi_2(\Phi^{-1}(u_1), \Phi^{-1}(u_2); \rho)$	$(-1, 1)$
t	$T_{2,v}(T_v^{-1}(u_1), T_v^{-1}(u_2); \rho, v)$	$(-1, 1), [1, 1000000]$
Clayton	$(u_1^{-\theta} + u_2^{-\theta} - 1)^{-1/\theta}$	$[0.00001, 150]$
Gumbel	$\exp\{-[(-\ln u_1)^{\theta} + (-\ln u_2)^{\theta}]^{1/\theta}\}$	$[1, 120]$
Frank	$-1/\theta \ln\{[1 - e^{-\theta} - (1 - e^{-\theta u_1})(1 - e^{-\theta u_2})]/(1 - e^{-\theta})\}$	$[-700, 700] \backslash \{0\}$
Plackett	$1/(2\theta - 2)\{1 + (\theta - 1)(u_1 + u_2) - [(1 + (\theta - 1)(u_1 + u_2))^2 - 4\theta(\theta - 1)u_1 u_2]^{1/2}\}$	$[0, 10000000]$
Clayton	$u_1 + u_2 - 1 + [(1 - u_1)^{-\theta} + (1 - u_2)^{-\theta} - 1]^{-1/\theta}$	$[0, 10000000]$

2.5. Parameter Estimation

The maximum likelihood estimation (MLE) method is simple and widely employed for copula parameter estimation. The logarithmic likelihood function of candidate copulas can be given as follows:

$$L(\theta) = \sum_{i=1}^{N} \ln c(u_{1i}, u_{2i}, u_{3i}; \theta) \quad (21)$$

The unknown parameter θ can be calculated by maximizing $L(\theta)$:

$$\widetilde{\theta} = \text{argmax} L(\theta) \quad (22)$$

Before referring to the maximum likelihood estimation of vine copula, it is necessary to illustrate the sequential estimation method. Its estimation steps are as follows: (1) Estimate the parameters of the bivariate copulas in the first tree using the original data. (2) Calculate the variable values (conditional distribution function values) of the second tree using the copulas estimated in the previous step. (3) Estimate the parameters of the bivariate copulas between the corresponding variables using the variable values of the second tree calculated above. (4) Estimate the parameters of the bivariate copulas of the remaining trees using steps similar to (2) and (3). The copula parameters of all trees can be estimated by following the above steps. In sequential estimation, the selection of the bivariate copula function uses the AIC criterion.

The maximum likelihood function of C-vine is as follows:

$$L_C(\theta) = \sum_{j=1}^{n-1} \sum_{i=1}^{n-j} \sum_{t=1}^{T} \ln\left\{c_{j,j+i|1,\cdots,j-1}[F(x_{j,t}|x_{1,t}, \cdots, x_{j-1,t}), F(x_{j+i,t}|x_{1,t}, \cdots, x_{j-1,t})]\right\} \quad (23)$$

The likelihood function of D-vine is as follows:

$$L_D(\theta) = \sum_{j=1}^{n-1} \sum_{i=1}^{n-j} \sum_{t=1}^{T} \ln\left\{c_{i,i+j|i+1,\cdots,i+j-1}[F(x_{i,t}|x_{i+1,t}, \cdots, x_{i+j-1,t}), F(x_{i+j,t}|x_{i+1,t}, \cdots, x_{i+j-1,t})]\right\} \quad (24)$$

The fitting accuracy and validation can be attained by comparing the empirical distributions with the statistical parameters of various candidate copula models, such as the root mean square error (*RMSE*) and the Akaike Information Criterion (AIC) values.

$$RMSE = \sqrt{\frac{1}{N} \sum_{i=1}^{N} [F_c(x_{1i}, x_{2i}, x_{3i}) - F_e(x_{1i}, x_{2i}, x_{3i})]^2} \quad (25)$$

$$AIC = -2\ln(L) + 2k \quad (26)$$

where F_e and F_c are the empirical values defined by the original variables and the estimated probabilities using the trivariate distribution function, respectively. N is the total number of original data. The empirical distribution of the trivariate variables can be calculated as follows:

$$F_e(x_{1i}, x_{2i}, x_{3i}) = \frac{1}{N}\sum_{j=1}^{N} I(x_{1j} \le x_{1i}, x_{2j} \le x_{2i}, x_{3j} \le x_{3i}) \tag{27}$$

where (x_1, x_2, x_3) represents the original variables, N is the length of each dataset, and $I = 1$ when x_{1j}, x_{2j}, and x_{3j} satisfy $x_{1j} \le x_{1i}$, $x_{2j} \le x_{2i}$ and $x_{3j} \le x_{3i}$; otherwise, $I = 0$.

3. Environmental Surfaces Using Copulas

Environmental contours are defined in the original physical space $\boldsymbol{X}$ of random variables, and they can be generated by mapping the random variables into the standard space $\boldsymbol{U}$. In the standard space, a probability density contour configuration with specific properties is constructed by determining a circle or a sphere centered at the origin with a specific radius. Then, the probability density contour configuration is transformed back into the original physical space, so that the corresponding environmental contour configuration in the original physical space can be obtained. The obtained environmental contour configuration can withstand the action of environmental loads corresponding to a specific exceedance probability (or a specific return period). This transformation can be performed by the inverse Rosenblatt transformation or the inverse Nataf transformation.

The probability edge based on the FORM theory in the standard normal space can be obtained as follows:

$$\beta_F = 1 - \Phi(P_f) \tag{28}$$

The target failure probability, P_f, which is associated with extreme sea conditions having a T_r-year return period and the duration of observed wave data, T_s (h), can be defined as follows:

$$P_f = \frac{T_s}{365.25 \times 24 \times T_r} \tag{29}$$

The design sea states along environmental contours are defined by the values of the original variables in the physical space $\boldsymbol{X}$, which are transformed from the values of the vector $\boldsymbol{z}$ in the standard normal space $\boldsymbol{Z}$ with an equivalent reliability edge, $\|\boldsymbol{z}\|^2 = \beta_F^2$. Any correlated random variables can be transformed into independent random variables in the standard normal space. This process is called the Rosenblatt transformation, as shown in the following formula:

$$\begin{cases} \Phi(z_1) = F_1(x_1) \\ \Phi(z_2) = F_{2|1}(x_2|x_1) \\ \cdots \\ \Phi(z_n) = F_{n|1,2,\cdots,n-1}(x_n|x_1,\cdots,x_{n-1}) \end{cases} \Leftrightarrow \begin{cases} x_1 = F_1^{-1}(\Phi(z_1)) \\ x_2 = F_{2|1}^{-1}(\Phi(z_2)|x_1) \\ \cdots \\ x_n = F_{n|1,2,\cdots,n-1}^{-1}(\Phi(z_n)|x_1,\cdots,x_{n-1}) \end{cases} \tag{30}$$

where $\Phi(\cdot)$ is the cumulative distribution function in the standard normal distribution; $F_{i|1,2,\cdots,i-1}(\cdot)$ is the conditional cumulative distribution function of variable x_i given x_1, x_2, $\cdots$,x_{i-1}; $F_{i|1,2,\cdots,i-1}^{-1}(\cdot)$ is its inverse form; $\boldsymbol{x} = (x_1, x_2, \cdots, x_n)$ and $\boldsymbol{z} = (z_1, z_2, \cdots, z_n)$ represent the random variables in the original physical and standard normal space, respectively.

The conditional distributions can be calculated by deriving the joint distribution in the following manner:

$$F_{2|1}(x_2|x_1) = \frac{\partial C_{21}(F_2(x_2), F_1(x_1))}{\partial F_1(x_1)} = h_{21}(F_2(x_2), F_1(x_1)) \tag{31}$$

$$F_{3|1,2}(x_3|x_1,x_2) = \frac{\partial C_{32|1}(F_{3|1}(x_3|x_1), F_{2|1}(x_2|x_1)}{\partial F_{2|1}(x_2|x_1)} = \frac{\partial C_{32|1}[h_{31}(F_3(x_3), F_1(x_1)), h_{21}(F_2(x_2), F_1(x_1))]}{\partial[h_{21}(F_2(x_2), F_1(x_1))]} \tag{32}$$

Equation (32) can be simplified to the following form:

$$F_{3|1,2}(x_3|x_1,x_2) = h[h_{31}(F_3(x_3), F_1(x_1)), h_{21}(F_2(x_2), F_1(x_1))] \tag{33}$$

4. Environmental Information

4.1. Marginal Probabilistic Distributions

Taking the EAR5 wind field as the background wind field, high-precision wind field data are obtained through hindcast simulation using the WRF meteorological model. Then, numerical wave simulation is carried out based on the SWAN wave model. A total of 175,320 hourly data of wind speed V_s, wave height H_s, and wave period T_p in a certain sea area of Shandong from 1 January 2004, to 31 December 2023, are simulated. The sample scatter plot is shown in Figure 3a. Figure 3 presents the empirical distributions and kernel density estimation results of various environmental parameters in the target sea area. It can be seen that the cumulative probabilities of the sample data of V_s, H_s, and T_p are in good agreement with the kernel density estimation curves, which can accurately describe the marginal distributions of the variables. Based on the copula theory, different models are adopted to describe the joint characteristics of multiple variables for the evaluation of design loads. In the present simulation, we utilized a standard desktop computer (1 core 3.3 GHz, 16 GB RAM), and employed MATLAB R2016b with the multivariate analysis framework to build and run the simulation models. The simulation and estimation without the plotting takes about 10 min for one case, and the duration of parameter estimation is related to the length of the dataset.

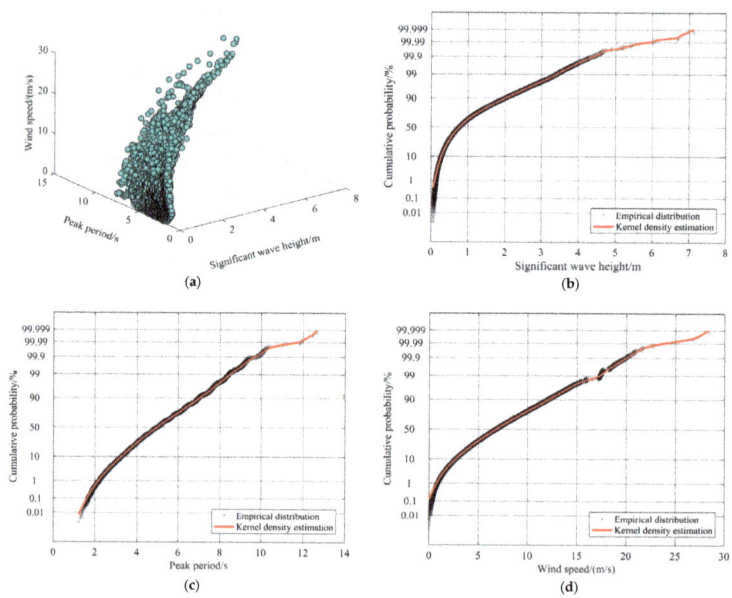

Figure 3. Scatter plot of wind–wave parameters and fitting of marginal distributions. (**a**) Scatter plot of (Hs, Tp, Vs). (**b**) Marginal distribution fitting of Hs. (**c**) Marginal distribution fitting of Tp. (**d**) Marginal distribution fitting of V_s.

4.2. Joint Distribution of Wind–Wave Parameters

After determining the optimal marginal distributions for each variable, the joint probability of the three-dimensional variables was calculated based on the copula theory. The MLE method was employed to estimate the parameters of three-dimensional elliptical and Archimedean copulas, which are presented in Table 2. The corresponding joint probability densities of Gaussian and Clayton copulas are depicted in Figure 4. As can be seen from statistical test values such as the AIC and RMSE, the fitting performance of the elliptical copula is superior to that of the Archimedean copula.

Table 2. The estimated parameters of trivariate copula functions.

Model		Parameter	*RMSE*	AIC
Metaelliptical	Gaussian	$\rho_{12} = 0.86$; $\rho_{13} = 0.90$; $\rho_{23} = 0.62$	0.0168	3.97×10^5
	t	$\rho_{12} = 0.85$; $\rho_{13} = 0.93$; $\rho_{23} = 0.64$; $v = 5.79$	0.0155	3.84×10^5
Archimedean	Clayton	$\theta = 1.46$	0.0652	4.25×10^5
	Frank	$\theta = 0.19$	0.1490	6.05×10^5
	Gumbel	$\theta = 1.12$	0.1300	5.66×10^5

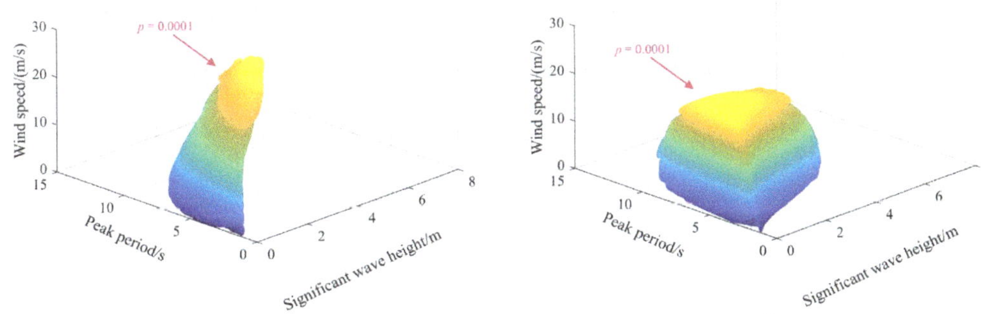

Figure 4. Joint density probability diagram based on trivariate symmetric copula. (**a**) Gaussian copula. (**b**) Clayton copula.

The joint probability distributions of three-dimensional variables are constructed using the C-vine and D-vine copula models. According to Equations (19) and (20), it can be seen that different combinatorial forms of the three-dimensional joint distribution can be obtained based on binary copulas. The main variable is determined based on the correlation between the variables. The Pearson correlation coefficient requires the variables to be continuously normally distributed and have a linear relationship, while the Kendall correlation coefficient does not have strict requirements for sample data. The Kendall values of (H_s, T_p), (H_s, V_s), and (T_p, V_s) are 0.67, 0.75, and 0.44, respectively. Since the correlation coefficients between H_s and T_p and between H_s and V_s are both higher than that between T_p and V_s, H_s is selected as the main control variable. In this study, the binary combinations of (H_s, T_p) and (H_s, V_s) are chosen, and the vine copula model is used to calculate the three-dimensional joint distribution.

Table 3 lists the best-fitting binary copulas and their parameter estimates. The results show that in the C-vine and D-vine models, the binary *t* copula is the optimal model for fitting the bivariate (H_s, T_p). The binary Gumbel copula is selected to describe (H_s, V_s) and (T_p, V_s), and the best-fitting distributions of the conditional probabilities (T_p, V_s; H_s) and

(H_s, V_s; T_p) are the Plackett and Frank copulas, respectively. The bivariate joint density probabilities of (H_s, T_p) and (T_p, V_s) are shown in Figure 5a,b, respectively. It is worth noting that different copula models can be selected to fit the correlation structures of bivariate variables with different tail characteristics. The contours of $c_{23|1}$ and $c_{13|2}$ corresponding to different return periods are shown in Figure 5c,d, respectively. The trivariate joint densities with p = 0.0001 obtained by the C-vine and D-vine copulas are depicted in Figure 5e,f, respectively. Meanwhile, the AIC and *RMSE* values are calculated to evaluate the degree of fitting, as shown in Table 3. Statistical tests indicate that the C-vine and D-vine models fit better than the three-dimensional elliptical copula function because the best-fitting binary copulas are selected to describe the correlations between different variables during the decomposition process of the joint probability model.

Table 3. The best-fitted pair copulas and estimated parameters in vine copula models.

Model	Variable	Pair Copula	Parameter	*RMSE*	AIC
C-vine	H_s, T_p	t	ρ = 0.86; v = 52.85	0.0121	3.63×10^5
	H_s, V_s	Gumbel	θ = 4.09	0.0121	5.41×10^5
	T_p, V_s; H_s	Plackett	θ = 0.05	0.0121	4.68×10^5
D-vine	H_s, T_p	t	ρ = 0.86; v = 52.85	0.0148	3.95×10^5
	T_p, V_s	Gumbel	θ = 1.77	0.0148	6.72×10^5
	H_s, V_s; T_p	Frank	θ = 19.15	0.0148	5.98×10^5

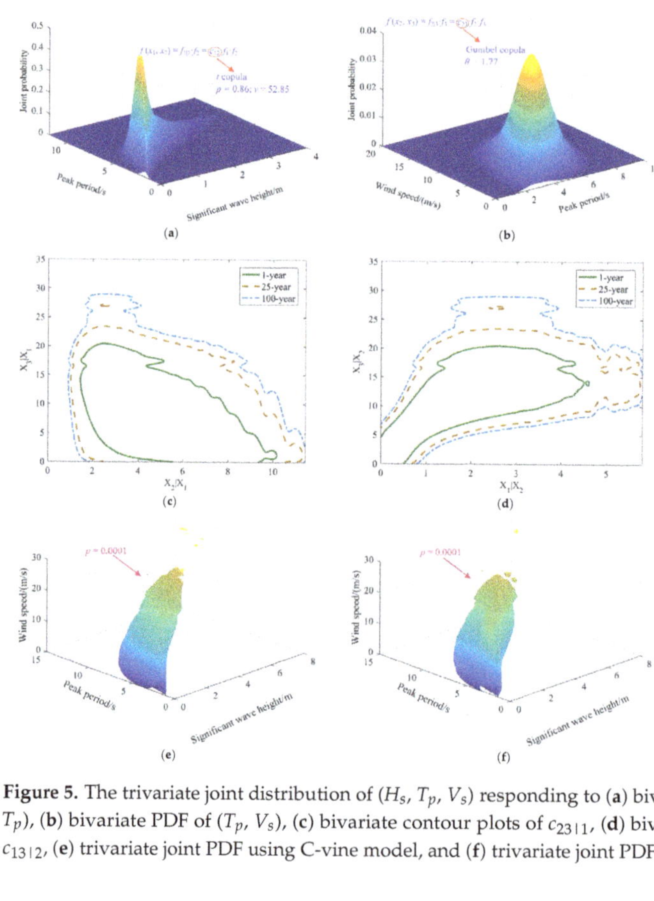

Figure 5. The trivariate joint distribution of (H_s, T_p, V_s) responding to (**a**) bivariate joint PDF of (H_s, T_p), (**b**) bivariate PDF of (T_p, V_s), (**c**) bivariate contour plots of $c_{23|1}$, (**d**) bivariate contour plots of $c_{13|2}$, (**e**) trivariate joint PDF using C-vine model, and (**f**) trivariate joint PDF using D-vine model.

4.3. Environmental Surfaces and Load Assessment

For the design and risk assessment of offshore structures, it is necessary to estimate the extreme sea conditions corresponding to multi-year return periods. The environmental surface is constructed based on the environmental contour method. Environmental surfaces obtained from all candidate trivariate models are compared. Figure 6 shows the three-dimensional measured data and the environmental surfaces with a 20-year return period obtained by using different copula models combined with the IFORM method. The proximity to the return period of the sample data indicates the degree of fitting of the trivariate joint distribution. It can also be seen from Figure 6 that there are significant differences in the predictions of extreme sea conditions corresponding to different joint distribution models. The Archimedean copula model performs worse in fitting environmental parameters compared to other models, while the C-vine copula shows a better performance than others.

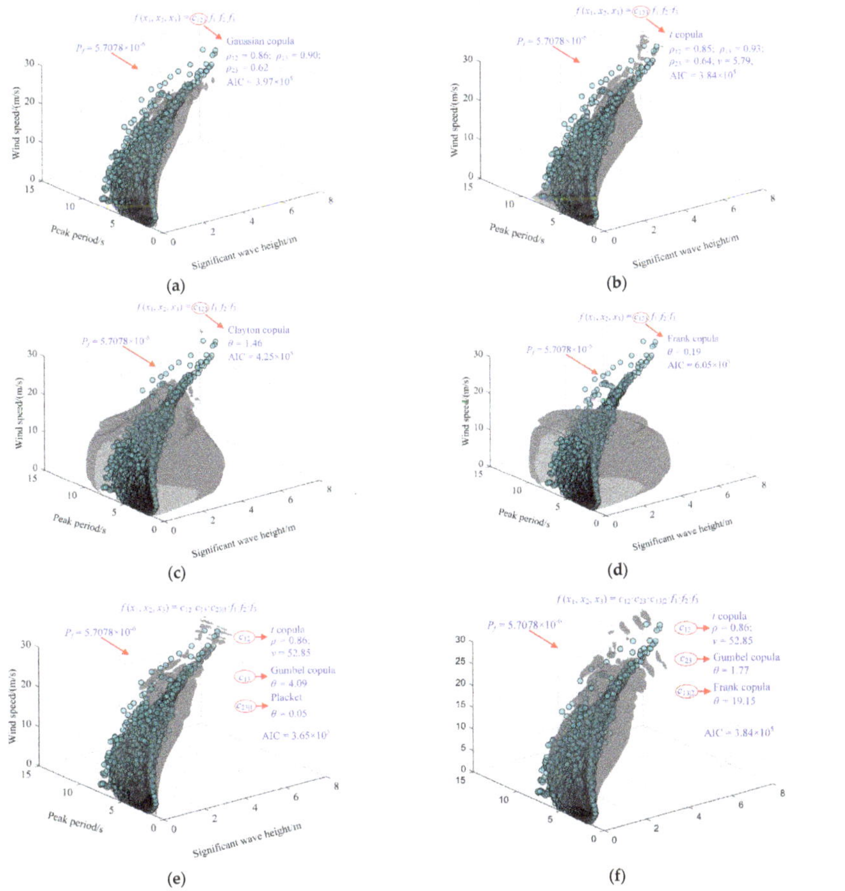

Figure 6. Original metocean variables and 10-year extreme environmental surfaces responding to **(a)** Gaussian, **(b)** t, **(c)** Clayton, **(d)** Frank, **(e)** C-vine, and **(f)** D-vine models.

The contour lines of the two-dimensional variables are drawn under the condition of the given variables. Figure 7 illustrates the conditional joint distribution contour plots of (H_s, T_p) for V_s values of 4.0 m/s, 8.0 m/s, 12.0 m/s, and 16.0 m/s. Meanwhile, Figure 8 presents the conditional joint distribution contour plots of (H_s, V_s) when T_p takes on values of 5.0 s, 6.0 s, 7.0 s, and 8.0 s. Moreover, Figure 9 shows the conditional joint distribution contour plots of (T_p, V_s) when H_s is given as 1.0 m, 2.0 m, 3.0 m, and 4.0 m. Based on the 20-year hindcast data, the two-dimensional conditional contour lines under the given variables derived from the 20-year return period vine copula model encompass nearly all the sample data, and their shapes show a good match. As can be observed from the figures, the t copula function aligns well with H_s and T_p under the given V_s condition. The distribution of H_s and V_s under the given T_p condition complies with the Frank copula function. The results indicate that the C-vine copula model is capable of effectively describing the correlation of two-dimensional conditional variables. This is due to the fact that the multidimensional variables are decomposed into multiple binary variable combinations, and the best-fitting copula function for each binary variable dataset is determined individually, thereby achieving an ideal match between the data characteristics and the copula model. Different binary copula models are suitable for characterizing the diverse tail correlations of the original data. Generally speaking, a three-dimensional symmetric copula function with specific characteristics is unable to perfectly fit multiple variables that exhibit complex correlation structures.

The annual extreme wave data in the target sea area are selected for single-element distribution fitting. The results show that the Gumbel model fits the wave height and wind speed well, as shown in Figure 10. Table 4 lists the independent 100-year return values of wave height and wind speed obtained by the annual extreme value method. The environmental surface with a 100-year return period is constructed based on the C-vine copula model. The determination of design parameters should be based on the response characteristics of the offshore structure, and Table 4 only provides the maximum wind speed–wave height design values on the isosurface and accompanying element values. It can be observed that the design values determined by the environmental contour method, which takes into account the correlation of environmental parameters, are smaller than those of the independent annual extreme value method.

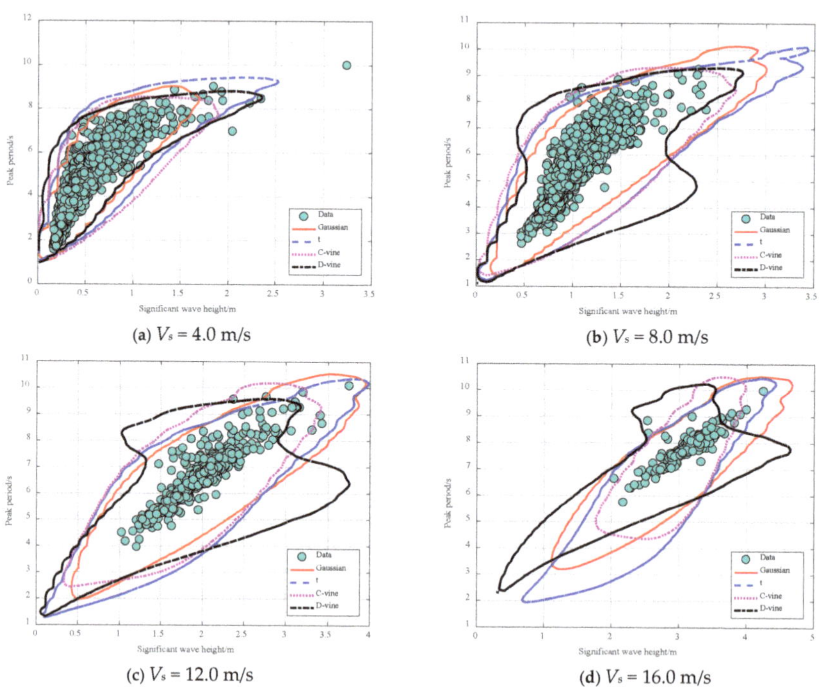

(a) V_s = 4.0 m/s (b) V_s = 8.0 m/s (c) V_s = 12.0 m/s (d) V_s = 16.0 m/s

Figure 7. Environmental contours of (H_s, T_p) given V_s based on various copula models.

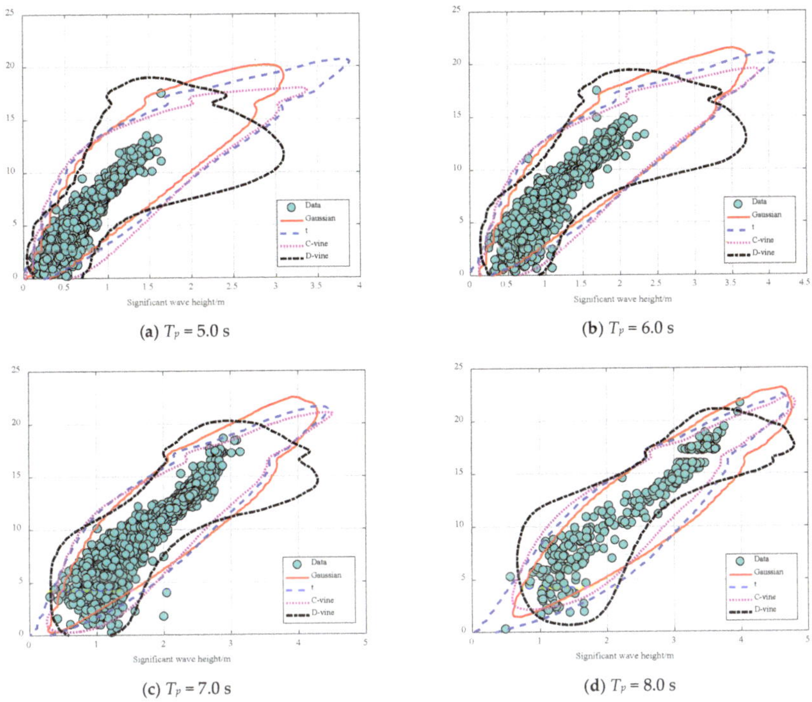

(**a**) $T_p = 5.0$ s (**b**) $T_p = 6.0$ s (**c**) $T_p = 7.0$ s (**d**) $T_p = 8.0$ s

Figure 8. Environmental contours of (H_s, V_s) given T_p based on various copula models.

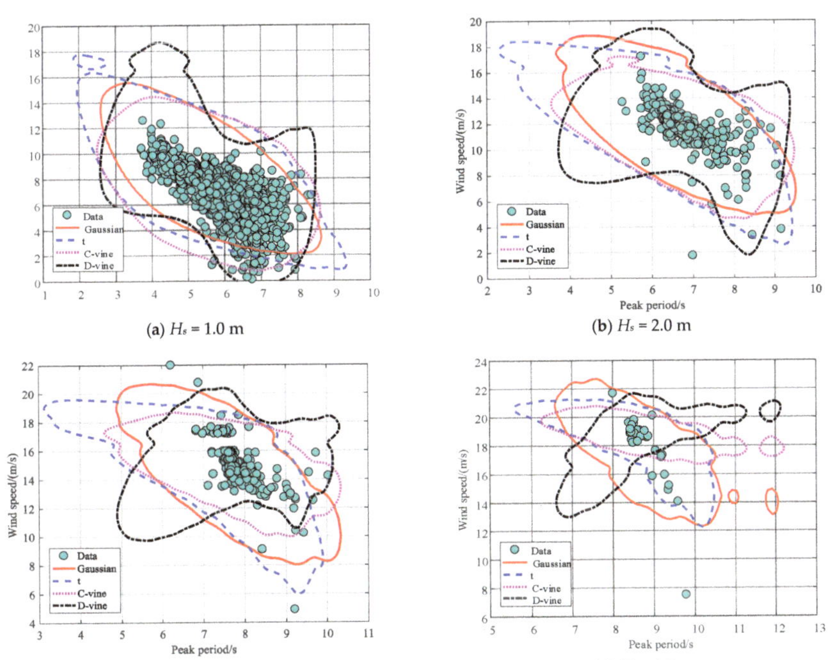

(**a**) $H_s = 1.0$ m (**b**) $H_s = 2.0$ m (**c**) $H_s = 3.0$ m (**d**) $H_s = 4.0$ m

Figure 9. Contour plots of (Hs, T_p) conditional on V_s using various copulas.

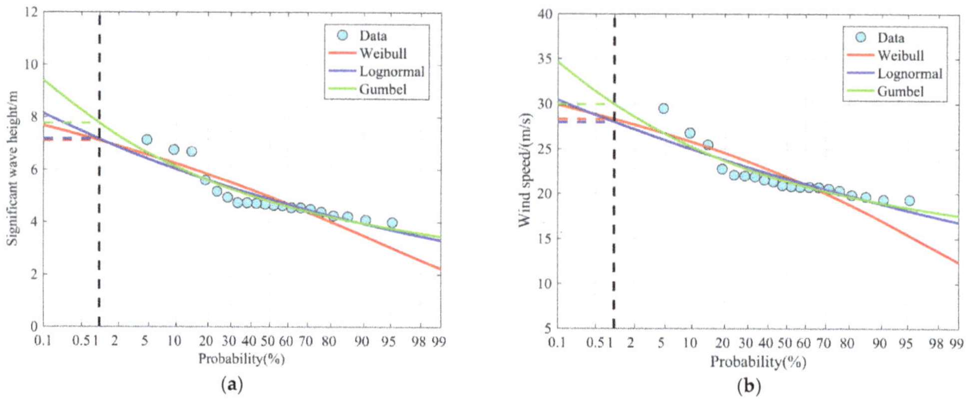

Figure 10. Distribution fitting of annual extreme wind and wave parameters. (**a**) Significant wave height. (**b**) Wind speed.

Table 4. One-hundred-year environmental design parameters determined by various models.

Model	H_s (m)	V_s (m/s)
Annual extreme value method	7.78	29.95
Environmental contour method	7.40 7.31	29.61 29.71

5. Conclusions

This study explored an analytical method for evaluating multidimensional design loads of offshore structures based on the three-dimensional joint probability of wind–wave parameters. The determination was carried out through the copula theory combined with the environmental contour method. The analysis process of three-dimensional wind–wave actions mainly consists of three steps: First, determine the marginal distributions of wind speed, wave height, and wave period; then, construct the joint distribution of variables based on the copula theory; finally, calculate the environmental isosurfaces and bivariate conditional contour lines to evaluate the design loads of marine structures.

The joint probability density and cumulative probability of wind–wave parameters in a certain sea area of the Shandong Peninsula were calculated using kernel density estimation and copula models. The results showed that the C-vine copula model could effectively construct the joint distribution of (H_s, T_p, V_s). In this case, the trivariate elliptical copula fitted the hindcast environmental variables better than the Archimedean copula. The conditional contour plots indicated that, within the three-dimensional model, the vine copula could flexibly describe the correlations of bivariate variables. A goodness-of-fit analysis demonstrated that the vine copula model was more suitable than the trivariate symmetric copula function for expressing the statistical distribution characteristics of marine environmental parameters.

The environmental contour method and the annual extreme value method based on the IFORM theory were employed to estimate the design parameters of this sea area. The vine copula model, combined with the environmental contour method, can enhance the accuracy and efficiency of predicting extreme structural responses under the target return period. The annual extreme value method assumes that marine environmental factors are independent of each other, which does not conform to the actual sea conditions. The overly high environmental design parameter values lead to conservative design. Establishing an accurate joint distribution model and reasonably evaluating environmental loads can provide technical support for the reliability design, disaster prevention, and mitigation of deep-sea floating offshore new-energy projects.

For future research, several directions can be explored. First, the current study focused on a certain sea area of the Shandong Peninsula. Expanding the research scope to other sea areas with different geographical and oceanographic characteristics can further verify the universality and adaptability of the proposed method. This will help to determine whether the C-vine copula model and the environmental contour method can still maintain high accuracy and efficiency in different marine environments. Secondly, the application of the model can be further deepened. For example, more complex offshore structures, such as large-scale floating wind farms and deep-sea oil and gas platforms, can be considered. By applying the established joint distribution model and environmental load evaluation method to these structures, more accurate design parameters can be obtained, which is conducive to the safety and stability of these structures. Moreover, emerging technologies like artificial intelligence and big data can be integrated. Machine-learning algorithms can be used to optimize the parameter determination process in the copula model, improving the accuracy of the model. Big data can provide more abundant data sources for model calibration and verification, enhancing the reliability of the research results. In this way, continuous improvement and innovation can be achieved in the field of evaluating multidimensional design loads of offshore structures, providing more solid technical support for the development of the marine engineering industry.

Author Contributions: Conceptualization, Y.W., Y.F. and Y.Z.; Methodology, Y.W., Y.Z. and S.Y.; Software, Y.Z. and S.Y.; Validation, Y.Z. and S.Y.; Formal analysis, Y.Z. and S.Y.; Investigation, Y.F. and S.Y.; Resources, Y.W., Y.F. and S.Y.; Data curation, Y.W., Y.F. and S.Y.; Writing – original draft, Y.F., Y.Z. and S.Y.; Visualization, Y.Z. and S.Y.; Project administration, Y.W.; Funding acquisition, Y.W., Y.Z. and S.Y. All authors have read and agreed to the published version of the manuscript.

Funding: This study was funded by the Natural Science Foundation of Shandong Province (ZR2023QE016), National Natural Science Foundation of China (W2411039), and Qingdao Postdoctoral Applied Research Grant (QDBSH20220202093).

Institutional Review Board Statement: Not applicable.

Informed Consent Statement: Not applicable.

Data Availability Statement: The original contributions presented in this study are included in the article. Further inquiries can be directed to the corresponding author.

Conflicts of Interest: Yongtuo Wu, Yudong Feng, and Saiyu Yu were employed by Shandong Electric Power Engineering Consulting Institute Corp., Ltd. The remaining author declares that the research was conducted in the absence of any commercial or financial relationships that could be construed as a potential conflict of interest.

References

1. Xie, S.; Li, Y.; Li, Y.; Kan, Y.Z. Preliminary study of the dynamics and environmental response of multiple floating offshore wind turbines on a single large-scale platform. *Ocean Eng.* **2025**, *316*, 120008. [CrossRef]
2. Ju, S.H.; Huang, Y.C. Study on multiple wind turbines in a platform under extreme waves and wind loads. *Energy Convers. Manag. X* **2025**, *25*, 100877. [CrossRef]
3. Safari, M.; Ghasemi, S.H.; Nowak, A.S. Performance-based target reliability analysis of offshore wind turbine mooring lines subjected to the wind and wave. *Probabilistic Eng. Mech.* **2024**, *77*, 103673. [CrossRef]
4. Ahmed, S.; Li, T.; Yi, P.; Chen, R. Environmental impact assessment of green ammoinapowered very large tanker ship for decarbonized future shipping operations. *Renew. Sustain. Energy Rev.* **2023**, *188*, 113774. [CrossRef]
5. Dong, S.; Fan, D.; Tao, S. Joint occurrence period of wind speed and wave height based on both service term and risk probability. *J. Ocean Univ. China* **2012**, *11*, 488–494. [CrossRef]

6. Vanem, E. Analysing multivariate extreme conditions using environmental contours and accounting for serial dependence. *Renew. Energy* **2023**, *202*, 470–482. [CrossRef]
7. Eckert-Gallup, A.C.; Sallaberry, C.J.; Dallman, A.R.; Neary, V.S. Application of principal component analysis (PCA) and improved joint probability distributions to the inverse first-order reliability method (I-FORM) for predicting extreme sea states. *Ocean Eng.* **2016**, *112*, 307–319. [CrossRef]
8. Cheng, Z.; Svangstu, E.; Moan, T.; Gao, Z. Long-term joint distribution of environmental conditions in a Norwegian fjord for design of floating bridges. *Ocean Eng.* **2019**, *191*, 106472. [CrossRef]
9. Zhao, Y.; Dong, S. A multi-load joint distribution model to estimate environmental design parameters for floating structures. *Ocean Eng.* **2020**, *217*, 107818. [CrossRef]
10. Vanem, E.; Zhu, T.; Babanin, A. Statistical modelling of the ocean environment—A review of recent developments in theory and applications. *Mar. Struct.* **2022**, *86*, 103297. [CrossRef]
11. Song, Y.; Chen, J.; Sørensen, J.D.; Li, J. Multi-parameter full probabilistic modeling of long-term joint wind-wave actions using multi-source data and applications to fatigue analysis of floating offshore wind turbines. *Ocean Eng.* **2022**, *247*, 110676. [CrossRef]
12. James, J.P.; Panchang, V. Assessment of joint distributions of wave heights and periods. *Ocean Eng.* **2024**, *313*, 119501. [CrossRef]
13. Bruserud, K.; Haver, S.; Myrhaug, D. Joint description of waves and currents applied in a simplified load case. *Mar. Struct.* **2018**, *58*, 416–433. [CrossRef]
14. Haver, S. Wave climate off northern Norway. *Appl. Ocean Res.* **1985**, *7*, 85–92. [CrossRef]
15. Simão, M.L.; Sagrilo, L.V.S.; Videiro, P.M. A multi-dimensional long-term joint probability model for environmental parameters. *Ocean Eng.* **2022**, *255*, 111470. [CrossRef]
16. Silva-González, F.; Heredia-Zavoni, E.; Montes-Iturrizaga, R. Development of environmental contours using Nataf distribution model. *Ocean Eng.* **2013**, *58*, 27–34. [CrossRef]
17. Salvadori, G.; Tomasicchio, G.R.; D'alessandro, F. Practical guidelines for multivariate analysis and design in coastal and offshore engineering. *Coast. Eng.* **2014**, *88*, 1–14. [CrossRef]
18. Dong, S.; Wang, N.; Lu, H.; Tang, L. Bivariate distributions of group height and length for ocean waves using Copula methods. *Coast. Eng.* **2015**, *96*, 49–61. [CrossRef]
19. Marinella, M.; Lamberti, A.; Archetti, R. Coastal flooding: A copula based approach for estimating the joint probability of water levels and waves. *Coast. Eng.* **2015**, *97*, 37–52.
20. Fu, J.; Khan, F. Monitoring and modeling of environmental load considering dependence and its impact on the failure probability. *Ocean Eng.* **2020**, *199*, 107008. [CrossRef]
21. Aghatise, O.; Khan, F.; Ahmed, S. Reliability assessment of marine structures considering multidimensional dependency of the variables. *Ocean Eng.* **2021**, *230*, 109021. [CrossRef]
22. Yang, Y.; Fang, C.; Li, Y.; Xu, C.; Liu, Z. Multivariate joint distribution of five-dimensional wind and wave parameters in the sea-crossing bridge region using Hierarchical Archimedean Copulas. *J. Wind Eng. Ind. Aerodyn.* **2024**, *247*, 105684. [CrossRef]
23. Dong, S.; Tao, S.; Li, X.; Soares, C.G. Trivariate maximum entropy distribution of significant wave height, wind speed and relative direction. *Renew. Energy* **2015**, *78*, 538–549. [CrossRef]
24. Shen, Y.; Wang, S.; Zhang, B.; Zhu, J. Development of a stochastic hydrological modeling system for improving ensemble streamflow prediction. *J. Hydrol.* **2022**, *608*, 127683. [CrossRef]
25. She, Z.; Huang, L.; Cai, H.; Fan, M.; Yu, L.; Li, B.; Lan, X.; Chen, X.; Liu, Z. Detecting human interventions by spatial dependence of extreme water levels using a high dimensional conditional probability approach over the Pearl River Delta. *J. Hydrol.* **2023**, *622*, 129681. [CrossRef]
26. Montes-Iturrizaga, R.; Heredia-Zavoni, E. Multivariate environmental contours using C-vine copulas. *Ocean Eng.* **2016**, *118*, 68–82. [CrossRef]
27. Lin, Y.; Dong, S. Wave energy assessment based on trivariate distribution of significant wave height, mean period and direction. *Appl. Ocean Res.* **2019**, *87*, 47–63. [CrossRef]
28. Speers, M.; Randell, D.; Tawn, J.; Jonathan, P. Estimating metocean environments associated with extreme structural response to demonstrate the dangers of environmental contour methods. *Ocean Eng.* **2024**, *311*, 118574. [CrossRef]
29. Mackay, E.; Hauteclocque, G. Model-free environmental contours in higher dimensions. *Ocean Eng.* **2023**, *273*, 113959. [CrossRef]
30. Coe, R.G.; Manuel, L.; Haselsteiner, A.F. On limiting the influence of serial correlation in metocean data for prediction of extreme return levels and environmental contours. *Ocean Eng.* **2022**, *266*, 113032. [CrossRef]
31. Beshbichi, O.E.; Rødstøl, H.; Xing, Y.; Ong, M.C. Prediction of long-term extreme response of two-rotor floating wind turbine concept using the modified environmental contour method. *Renew. Energy* **2022**, *189*, 1133–1144. [CrossRef]
32. *DNVGL-RP-C205*; Environmental Conditions and Environmental Loads. Det Norske Veritas (DNV): Oslo, Norway, 2017.
33. Haver, S. On the joint distribution of heights and periods of sea waves. *Ocean Eng.* **1987**, *14*, 359–376. [CrossRef]
34. Winterstein, S.R.; Jha, A.K.; Kumar, S. Reliability of floating structures: Extreme response and load factor design. *J. Waterw. Port Coast. Ocean Eng.* **1999**, *125*, 163–169. [CrossRef]

35. Montes-Iturrizaga, R.; Heredia-Zavoni, E. Environmental contours using copulas. *Appl. Ocean Res.* **2015**, *52*, 125–139. [CrossRef]
36. Chai, W.; Leira, B.J. Environmental contours based on inverse SORM. *Mar. Struct.* **2018**, *60*, 34–51. [CrossRef]
37. Haselsteiner, A.F.; Ohlendorf, J.-H.; Wosniok, W.; Thoben, K.-D. Deriving environmental contours from highest density regions. *Coast. Eng.* **2017**, *123*, 42–51. [CrossRef]
38. Huseby, A.B.; Vanem, E.; Natvig, B. A new approach to environmental contours for ocean engineering applications based on direct Monte Carlo simulations. *Ocean Eng.* **2013**, *60*, 124–135. [CrossRef]
39. Vanem, E. A comparison study on the estimation of extreme structural response from different environmental contour methods. *Mar. Struct.* **2017**, *56*, 137–162. [CrossRef]
40. Clarindo, G.; Guedes Soares, C. Environmental contours of sea states by the I-FORM approach derived with the Burr-Lognormal statistical model. *Ocean Eng.* **2024**, *291*, 116315. [CrossRef]
41. Vanem, E.; Gramstad, O.; Bitner-Gregersen, E.M. A simulation study on the uncertainty of environmental contours due to sampling variability for different estimation methods. *Appl. Ocean Res.* **2019**, *91*, 101870. [CrossRef]
42. Heredia-Zavoni, E.; Montes-Iturrizaga, R. Modeling directional environmental contours using three dimensional vine copulas. *Ocean Eng.* **2019**, *187*, 106102. [CrossRef]
43. Bai, X.; Jiang, H.; Huang, X.; Song, G.; Ma, X. 3-Dimensional direct sampling-based environmental contours using a semi-parametric joint probability model. *Appl. Ocean Res.* **2021**, *112*, 102710. [CrossRef]
44. Fang, C.; Xu, Y.; Li, Y. Optimized C-vine copula and environmental contour of joint wind-wave environment for sea-crossing bridges. *J. Wind Eng. Ind. Aerodyn.* **2022**, *225*, 104989. [CrossRef]
45. Meng, X.; Li, Z. 3-Dimensional environmental contours of winds and waves accounting for different sampling methods and seasonal effects. *Ocean Eng.* **2024**, *304*, 117724. [CrossRef]
46. Wu, X.; Ma, C.; Zhang, J. Multivariate reliability method using the environment contour model based on C-vine copulas. *Ocean Eng.* **2024**, *299*, 117282. [CrossRef]
47. de Hauteclocque, G.; Mackay, E.; Vanem, E. Quantitative comparison of environmental contour approaches. *Ocean Eng.* **2022**, *245*, 110374. [CrossRef]
48. Mohapatra, S.C.; Amouzadrad, P.; Bisop, I.B.d.S.; Guedes Soares, C. Hydrodynamic Response to Current and Wind on a Large Floating Interconnected Structure. *J. Mar. Sci. Eng.* **2025**, *13*, 63. [CrossRef]
49. Sklar, A. *Fonctions de Répartition à N Dimensions Et Leurs Marges*; L'Institut de Statistique de l'Université: Paris, France, 1959.
50. Djaloud, T.S.; Seck, C.T. Nonparametric kernel estimation of conditional copula density. *Stat. Probab. Lett.* **2024**, *212*, 110154. [CrossRef]
51. Wang, W.; Yang, H.; Huang, S.; Wang, Z.; Liang, Q.; Chen, S. Trivariate copula functions for constructing a comprehensive atmosphere-land surface-hydrology drought index: A case study in the Yellow River basin. *J. Hydrol.* **2024**, *642*, 131784. [CrossRef]
52. Li, J.; Pan, S.; Chen, Y.; Gan, M. The performance of the copulas in estimating the joint probability of extreme waves and surges along east coasts of the mainland China. *Ocean Eng.* **2021**, *237*, 109581. [CrossRef]
53. Yuan, X.; Huang, Q.; Song, D.; Xia, E.; Xiao, Z.; Yang, J.; Dong, M.; Wei, R.; Evgeny, S.; Joo, Y.-H. Fatigue Load Modeling of Floating Wind Turbines Based on Vine Copula Theory and Machine Learning. *J. Mar. Sci. Eng.* **2024**, *12*, 1275. [CrossRef]

Journal of
Marine Science and Engineering

MDPI

Article

Interannual Characteristics of Tropical Cyclones in Northwestern Pacific Region in Context of Warm Pool and Monsoon Troughs

Junru Guo [1,2,3], Shichao Wang [1,2,3], Xin He [1,2,3], Jun Song [1,2,3,4,*], Yanzhao Fu [1,2,3,4] and Yu Cai [1,2,3]

1 School of Marine Science and Environment Engineering, Dalian Ocean University, Dalian 116023, China; guojunru@dlou.edu.cn (J.G.); wangsc9905@163.com (S.W.); hexin@dlou.edu.cn (X.H.); yzfu@dlou.edu.cn (Y.F.); caiyu@dlou.edu.cn (Y.C.)
2 Operational Oceanographic Institution, Dalian Ocean University, Dalian 116023, China
3 Liaoning Key Laboratory of Marine Real-Time Forecast and Risk Warning, Dalian 116023, China
4 Dalian Engineering Research Center for Applied Oceanography, Dalian 116023, China
* Correspondence: songjun2017@dlou.edu.cn

Abstract: This study utilizes the typhoon path ensemble dataset from the Marine Science Big Data Center, surface temperature data from NOAA's COBE Sea Surface Temperature dataset, and wind field data from the NCEP/DOE Reanalysis II dataset. It employs analytical techniques such as wavelet analysis, correlation analysis, and the Mann–Kendall test to investigate the temporal and spatial variations in tropical cyclones in the Northwest Pacific, focusing on aspects such as frequency, genesis regions, and Accumulated Cyclone Energy. The research examines the influence of environmental factors, including warm pool thermal conditions and monsoon troughs, on the behavior of tropical cyclones. Findings indicate that over the past 39 years, there has been an upward trend in the frequency of tropical cyclones, whereas ACE has exhibited a slight downward trend. The results of the M-K test suggest that following a period of rapid increase, cyclone frequency experienced a significant shift in 1996, subsequently displaying a downward trend. Notably, the frequency of cyclones aligns with mutation points corresponding to warm pool thermal conditions and the Monsoon Trough Index. Wavelet analysis reveals that cyclone frequency, ACE, warm pool thermal conditions, and MTI exhibit similar small scale periodic variations. The observed differences in the genesis regions of tropical cyclones are attributed to fluctuations in warm pool thermal conditions. Specifically, years characterized by cooler warm pool conditions correspond with a stronger MTI, while warmer conditions are associated with a weaker MTI. The genesis regions of cyclones predominantly lie within the monsoon trough, where environmental conditions favorable for cyclone development are intensified during years of cooler warm pool conditions, resulting in heightened convective activity.

Keywords: warm pool; monsoon troughs; tropical cyclones; interannual variations

1. Introduction

As one of the world's high sea surface temperature (SST) areas, the Northwest Pacific is one of the main areas for the formation of tropical cyclones (TCs) and typhoons in the world, and according to statistics, about one-third of the world's TCs are generated over this sea area. The study of the TC formation process, intensity, structural change, and activity law has always been a frontier scientific issue in the fields of marine hydrology, meteorology, and geographical environment science. The occurrence and development of TC is affected by the atmospheric environment field over the upper ocean and ocean surface, and the formation and development of TC is the result of complex interactions between the atmosphere, the ocean, and the structure of the TC itself [1–3]. At the same time, the generation and development trends of TCs are influenced by different physical factors on various time scales. Consequently, research on TC activities ranges from studying the characteristics of individual cyclones to examining the statistical characteristics of changes on seasonal, interannual, and decadal scales [4–6]. Nevertheless, the study of TC activities faces significant challenges due to the scarcity of detailed field observations and the limited accumulation of long-term data pertaining to TCs in the expansive oceanic environment. Additionally, there exists a deficiency in comprehensive and systematic understanding of the physical mechanisms that govern the formation and evolution of TCs.

Approximately 90% of the world's warm sea water is concentrated in the tropical Western Pacific Ocean, making this region the warmest sea area globally. This area, commonly referred to as the warm pool, is characterized by the most intense air–sea interactions in the world and serves as a significant genesis of latent heat release in the global atmosphere [7–9]. Research indicates that over the past few decades, global temperatures have consistently risen, leading to an increase in the temperature of the upper layers of the oceans. As one of the warmest ocean regions on Earth, the sea surface temperature in the warm pool of the Western Pacific Ocean has risen markedly. This area not only represents the upper branch of the Walker circulation but also acts as a strong convergence zone for airflow and water vapor, facilitating the development of intense convective activity and precipitation [8,10,11]. Numerous studies have demonstrated that the Western Pacific warm pool directly influences the thermal and dynamic aspects of tropical cyclone activity in the Northwest Pacific. Furthermore, tropical cyclone activity exhibits significant intraseasonal, interannual, and interdecadal variations due to the thermal changes occurring in the Western Pacific warm pool [12–14]. There is also a discernible trend of increasing tropical cyclone activity (in terms of frequency, intensity, etc.) in regions where sea surface temperatures are rising [15,16]. Nevertheless, certain research findings suggest that, within the framework of global warming, the intensity of regional TCs has escalated, whereas their frequency may be subject to a potential decline [13,17,18]. There exists a relationship between TC activity and ocean thermodynamic changes; however, regarding interannual variations, the impact of SST changes on TCs may not be the primary influencing factor [16]. Recent research has demonstrated that the thermal regime of the warm pool in the Northwest Pacific affects the monsoon trough, a crucial component of the atmospheric circulation system in the tropics and subtropics, particularly in the Asian monsoon region [18,19]. This phenomenon is closely linked to the monsoon and has significant implications for regional climate, precipitation, and weather systems [20]. Prior research has indicated that the intensity of the Northwest Pacific Monsoon Trough's low-level zonal winds plays a crucial role in determining both the location and path of TCs. Additionally, the Northwest Pacific Monsoon Trough creates vital large-scale environmental conditions that are conducive to the development of TCs [3,15,21]. These conditions include the convergence of airflow, cyclonic relative vorticity, divergence of upper-level airflow, minimal vertical wind shear, and a sufficient supply of water vapor, all of which collectively create a conducive environment for the formation and development of TCs. Given the severe disasters caused by TCs in the Northwest Pacific, many researchers have systematically investigated the generation, structure, development, and movement patterns of TCs in this region [22,23]. As a result, the investigation into the characteristics and mechanisms underlying interannual and interdecadal variations in TC activity in the Northwest Pacific, along with the large-scale alterations in environmental factors stemming from thermal changes in the warm pool, has attracted the interest of Chinese researchers and pertinent scientists.

In this study, we utilized the NOAA extended reconstruction of sea surface temperature data from 1984 to 2022, the NCEP/DOE Reanalysis II dataset (https://psl.noaa.gov/data/gridded/data.ncep.reanalysis2.html, accessed on 1 March 2024), and the typhoon track dataset offshore of China to analyze the spatiotemporal variation characteristics of TCs in the Northwest Pacific over the past 40 years. Additionally, we explored the relationship between TC activity and monsoon trough dynamics in the Western Pacific Ocean, considering the context of changes in the warm pool.

2. Data and Methods

2.1. Data Sources

The tropical cyclone data used in this study is derived from the China Offshore Typhoon Track Ensemble Dataset of the Ocean Science Big Data Center (https://msdc.qdio.ac.cn/data/metadata-special-detail?id=1422759994058625025, accessed on 1 March 2024), which includes the dataset of typhoon tracks in China's offshore waters from 1945 to 2023, including the true track information of each typhoon, typhoon intensity, air pressure, central wind speed, moving speed, moving direction.

The sea temperature data comes from NOAA's COBE Sea Surface Temperature dataset (https://psl.noaa.gov/data/gridded/data.cobe.html, accessed on 1 March 2024), with a spatial resolution of 1.0° × 1.0°. This dataset includes monthly sea surface temperature data from JMA from 1891 to 2023.

Wind field data and relative humidity data are from the monthly average data of NOAA's NCEP/DOE Reanalysis II dataset (https://psl.noaa.gov/data/gridded/data.ncep.reanalysis2.html, accessed on 1 March 2024) with a spatial resolution of 2.5° × 2.5° and a monthly average for the time range from 1979 to 2024.

Longwave radiation data (https://psl.noaa.gov/data/gridded/data.olrcdr.interp.html, accessed on 1 March 2024) are derived from NOAA interpolated Outgoing Longwave Radiation (OLR) with a spatial resolution of 2.5° × 2.5°.

2.2. Method Introduction

2.2.1. Wavelet Transform Method

Wavelet transform is a method that simultaneously analyzes signals in both the time domain and frequency domain. Wavelet transform can reflect the local variation characteristics of signal frequency in a time series, and it is effective in analyzing the changes in sequences over time. In Matlab (2022a), we can conveniently perform wavelet analysis using the wavelet toolbox or related functions such as "wavelet". This article uses the continuous wavelet transform method to conduct periodic analysis on data such as cyclone occurrence frequency, ACE, Western Pacific warm pool SSTA, and MTI. The main function for wavelet transform is as follows [24,25].

$$WT(a,b) = \frac{1}{\sqrt{a}} \int_{-\infty}^{+\infty} f(t)\psi\left(\frac{t-b}{a}\right)dt \quad (1)$$

In the formula, a is the scale, b is the translation parameter, $f(t)$ corresponds to the time series, and ψ represents the wavelet basis function. Here, the Morlet wavelet is selected as the wavelet basis function using the wavelet analysis method.

$$\psi(t) = \pi^{-\frac{1}{4}}\exp(-\frac{t^2}{2})\exp(iw_0 t) \quad (2)$$

In the formula, w_0 represents the dimensionless frequency, and the wavelet variance is employed to better characterize the strength of periodic signals in the time series. The calculation formula for wavelet variance is as follows:

$$\mathrm{Var}(a) = \frac{1}{N}\sum_{t=0}^{N-1}(WT(a,t))^2 \tag{3}$$

In the formula, N represents the length of the time series.

2.2.2. Mann–Kendall Method

The Mann–Kendall (M-K) trend test is a non-parametric statistical testing method that is widely used to analyze the long-term trends of various meteorological element time series data. It has advantages such as a simple calculation process, a wide detection range, low interference, and a high degree of quantification. The principle of the M-K test is as follows [26,27]:

$$S = \sum_{i=1}^{n-1}\sum_{j=i+1}^{n} sgn(X_j - X_i) \tag{4}$$

In the formula, S represents the M-K statistic; sgn is the sign function; n is the number of sample; X_i and X_j are the i-th and j-th data values of the time series, respectively; its variance is calculated as shown in formula (not considering tie cases, that is, $X_i = X_j$) (5).

$$\mathrm{Var}(S) = \frac{n(n-1)(2n+5)}{18} \tag{5}$$

The Z values corresponding to different S intervals in the M-K statistic formula are as follows:

$$Z = \left\{ \frac{(S-1)}{\sqrt{\mathrm{Var}(S)}}, \quad S > 0 \right. \tag{6}$$

In a two-sided trend test, a positive Z indicates that the sequence has an increasing trend, while a negative Z indicates that the sequence has a decreasing trend. At a given confidence level, if $Z \geq Z_{\alpha/2}$, the trend of the sequence is significant; if $Z \leq Z_{\alpha/2}$, the trend of the sequence is not significant. The critical values of the statistic Z for significance levels of 0.1, 0.05, and 0.01 are 1.64, 1.96, and 2.58, respectively.

Next, use the M-K method for mutation analysis, structured as follows:

$$S_k = \sum_{i=1}^{k}\sum_{j}^{i-1} a_{ij}, k = 2,3,4,\ldots,n \tag{7}$$

In the formula, $a_{ij} = \begin{cases} 1, X_i > X_j \\ 0, X_i < X_j \end{cases} 1 \leq j \leq i$.

Assuming the time series is random, the statistics are as follows:

$$UF_K = \frac{|S_k - E(S_k)|}{\sqrt{\mathrm{Var}(S_k)}}, k = 1,2,3,4,\ldots,n \tag{8}$$

In the formula, $E(S_k)$ represents the mean, and $\mathrm{Var}(S_k)$ represents the variance, calculated according to Formulas (9) and (10).

$$E(S_k) = \frac{k(k+1)}{4} \tag{9}$$

$$\mathrm{Var}(S_k) = \frac{k(k-1)(2k+5)}{72} \tag{10}$$

According to the time series in reverse order $X_n, X_{n-1}, \ldots, X_1$, calculate UB_k using the process described above. Within the confidence interval $|U| \leq 1.96$, if $UF_K = UB_k$, the intersection point is identified as the mutation point of the time series. When $UF_K > 0$, the sequence shows an upward trend; when $UF_K < 0$, the sequence shows a downward trend.

2.2.3. Monsoon Trough Intensity Index

In order to better study the interannual variations in the monsoon trough, this paper adopts the definition of the Monsoon Trough Intensity Index (MTI) proposed by Zhang Xiang et al., which uses MTI to quantify the intensity of monsoon trough activity. The formula is as follows [28]:

$$\mathrm{MTI} = \frac{\xi' - \overline{\xi}}{\sqrt{\sigma^2}} \tag{11}$$

In the formula, ξ' represents the average convective low-level 850 hPa relative vorticity over the region of 215–165° E and 5–20° N in the Western Pacific during the autumn (September to November) of a certain year. $\overline{\xi}$ is the climatic average of the convective low-level 850 hPa relative vorticity for this region from 1984 to 2022 during autumn. $\sqrt{\sigma^2}$ is the mean square deviation of the convective low-level 850 hPa relative vorticity for this region. According to the formula, if MTI > 0, the Western Pacific Monsoon Trough is stronger; conversely, if MTI < 0, the Northwestern Pacific Monsoon Trough is weaker.

3. Results

3.1. Temporal and Spatial Variation Characteristics of Tropical Cyclones in Northwest Pacific

By analyzing the occurrence frequency of tropical cyclones in the Western Pacific over the past 39 years, we identified that their activity is primarily concentrated during the summer and autumn months, spanning from June to November. Consequently, this study focuses on cyclones that attain tropical storm intensity (wind speed $\geq$ 17.3 m/s) within these seasons. Figure 1 illustrates the interannual distribution of tropical cyclone frequency derived from statistical methods. Specifically, Figure 1a depicts the total frequency of cyclones over the 39-year period, totaling 789 occurrences. The year 1994 recorded the highest frequency with 30 cyclones, while 1998 had the lowest with only 10. From a long-term perspective, the frequency of cyclones exhibits a slight upward trend, increasing at a rate of 0.0623 occurrences per year, though this trend is not statistically significant. Figure 1b shows the trend significance and abrupt changes in the frequency of Western Pacific TCs from 1984 to 2022 based on the M-K test. The results reveal an abrupt change in cyclone frequency in 1986, followed by a significant upward trend between 1993 and 1995, suggesting that the period from 1991 to 1996 was characterized by heightened cyclone activity. However, after the abrupt change in 1996, the frequency of cyclones began to decline, although this decrease is not statistically significant. This lack of significance in the declining trend may be attributed to the substantial short-term fluctuations in cyclone occurrence, which obscure a clear long-term downward trend.

In order to explore the frequency variation cycles of tropical cyclones in summer and autumn in the Northwest Pacific Ocean from 1984 to 2022, the wavelet analysis method was used to obtain the wavelet power spectrum variation diagram (Figure 2a) and the wavelet full spectrum diagram (Figure 2b) of tropical cyclone frequency. The area enclosed by the thick black line in Figure 2a represents the wavelet power that passes the 95% confidence test. The black dashed line indicates the critical line of the cone of influence (COI), and the area below the dashed line represents the region significantly affected by the edge effects of the continuous wavelet transform. In Figure 2b, the red dashed line represents the critical line for the 95% significance level.

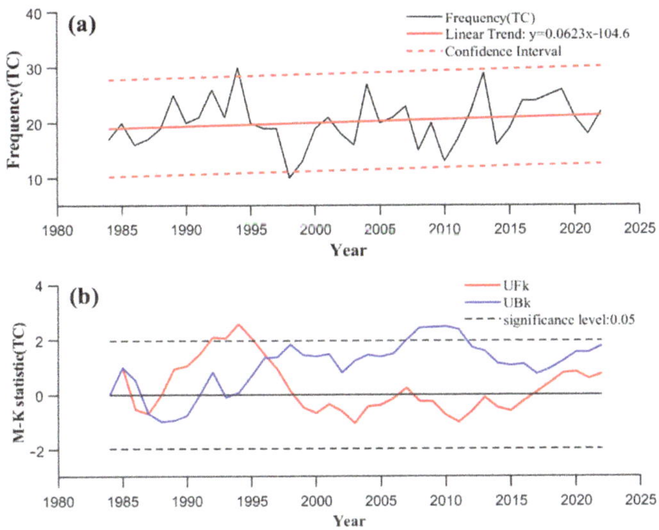

Figure 1. Interannual variation and M-K trend test of frequency of tropical cyclone bottles from 1984 to 2022. (**a**) is Interannual variation and (**b**) is M-K trend test.

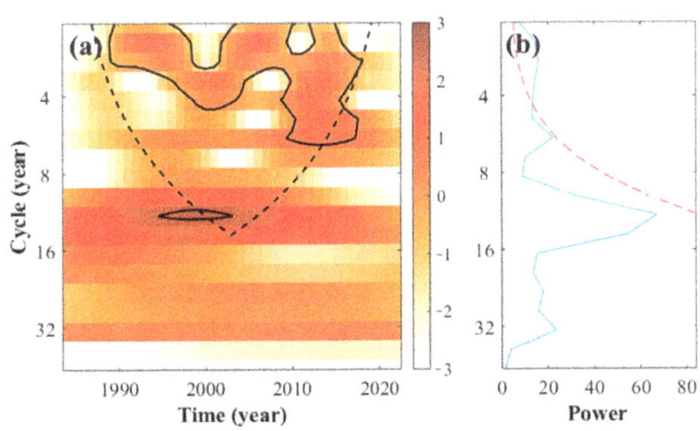

Figure 2. Wavelet power spectrum variation chart and global wavelet power map of tropical cyclone frequency in Northwest Pacific from 1984 to 2022. (**a**) is wavelet power spectrum variation diagram and (**b**) is wavelet full spectrum diagram.

You can see from Figure 2a that the tropical cyclone frequency wavelet power change cycle is the most prominent at small size scales, and the TC frequency by the 95% confidence level inspection cycle is roughly 1–3 years and 1 to 6 years, of which 1989–1995 for 1–3 years scale change cycles, and 2011–2018 is characterized by 1–6 years scale change cycles. It can also be seen from the global spectrum of wavelet in Figure 2b that the period of passing the 95% confidence test is mainly within 6 years. It can be concluded that the frequency of tropical cyclones in the Northwest Pacific Ocean in recent 39 years has a small scale periodic change of 1–3 years and 1–6 years [29].

Accumulated Cyclone Energy (ACE) is a quantitative metric for assessing the intensity of tropical cyclones, and was first proposed by Bell in 2000. This measure encompasses all tropical cyclones that exceed the intensity threshold for tropical storms, and it is computed by summing the squares of their maximum wind speeds at six-hour intervals over a designated period [30]. Consequently, compared to studies focusing on single indicators such as tropical cyclone frequency or intensity, the analysis of ACE provides a more comprehensive understanding of the overall trends in tropical cyclone activity. Therefore, we calculated the ACE values for TCs with an intensity greater than or equal to tropical storm intensity (wind speed of at least 17.2 m/s) over the past 39 years.

The ACE trend and long-term trend of TCs from June to November for the years 1984 to 2022 are illustrated in Figure 3a. This figure reflects the ACE trend of TCs over the past 39 years. The analysis indicates that ACE exhibits a generally negligible downward trend, characterized by a decline rate of -0.145×10^3 (m^2/s^2). The maximum ACE value was recorded in 1994, while the minimum occurred in 1999. According to the M-K test, the average annual ACE of TCs over the 39-year period indicated no significant increase or decrease. There was no notable upward trend in ACE prior to 1998, but there was a non-significant downward trend from 1998 to 2004, from 2007 to 2015, and after 2020. The first mutation point was identified in 1998, after which the ACE displayed a non-significant alternating trend of increases and decreases, with two additional mutation points in 2007 and 2013.

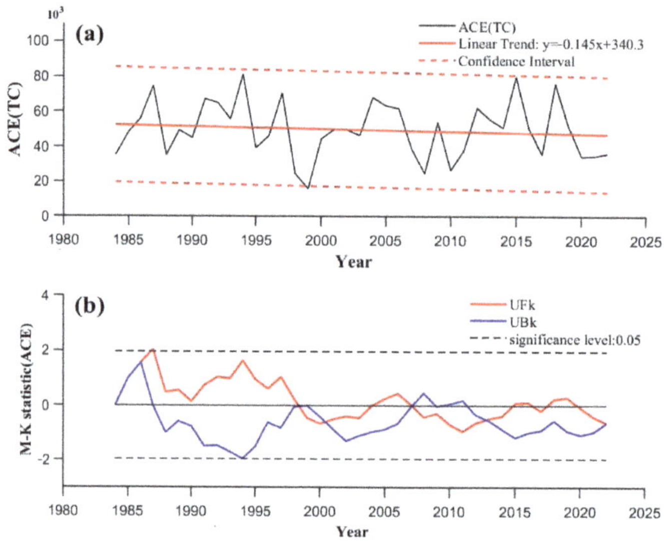

Figure 3. Interannual variation and M-K trend test of ACE from 1984 to 2022. (**a**) is Interannual variation and (**b**) is M-K trend test.

Figure 4 illustrates the periodic fluctuations of ACE. As depicted in Figure 4a, there is a notable periodic variation in ACE associated with tropical cyclones on a relatively small scale. The periodicity of ACE for these cyclones, assessed at a 95% confidence level, is identified as occurring within the ranges of 1–5 years and 2–4 years. Specifically, the period from 1987 to 2004 is characterized by a 1–5 year oscillation, whereas the period from 2007 to 2020 is characterized by a 2–4 year oscillation. Furthermore, the global wavelet power spectrum presented in Figure 4b corroborates that the periodicity of ACE, also evaluated at a 95% confidence level, predominantly resides within the 1–5 year range, with a wavelet power peak observed at 2–3 years. This suggests that variations in ACE are particularly pronounced during this timeframe. In conclusion, the ACE of tropical cyclones in the Northwest Pacific over the past 39 years demonstrates low-frequency oscillatory periodic changes of 1–5 years and 2–4 years.

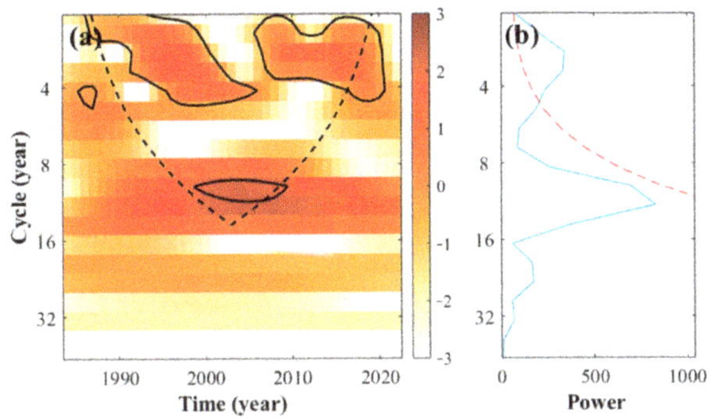

Figure 4. Wavelet power spectrum variation chart and global wavelet power map of ACE in Northwest Pacific from 1984 to 2022. (**a**) is wavelet power spectrum variation diagram and (**b**) is wavelet full spectrum diagram.

In order to gain a deeper insight into the spatial distribution patterns of tropical cyclone origins within the Northwest Pacific region, we divided the sea areas between 0 and 40° N and 100 and 160° E into 1° × 1° grids. We then calculated the frequency of TCs in each grid to create a spatial distribution map of tropical cyclone genesis of various categories in the Northwest Pacific from June to November over the past 39 years (Figure 5). The analysis reveals that tropical cyclone geneses are primarily concentrated between 5 and 24° N and 110 and 155° E. Specifically, there are three areas of high tropical cyclone frequency: 10–20° N, 112–120° E; 8–20° N, 125–135° E; and 8–20° N, 140–152° E. The frequencies of TCs in these three areas account for approximately 16%, 24%, and 21% of the total number of TCs, respectively, while the remaining TCs are distributed around these three concentration zones.

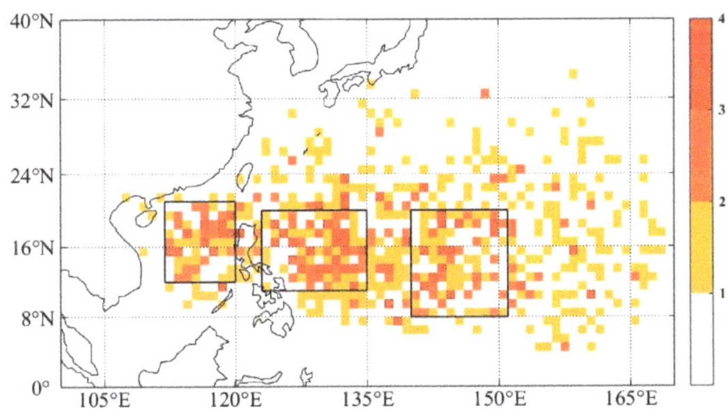

Figure 5. Spatial distribution of tropical cyclone genesis in Western Pacific from 1984 to 2022.

We have classified tropical cyclones that have occurred over the past 39 years in accordance with the classification criteria established by the China Meteorological Administration. The classification of tropical cyclones is predicated on their maximum sustained wind speeds, which are categorized into five distinct classifications: Tropical Storm (TS, with a maximum sustained wind speed exceeding 17.2 m/s), Severe Tropical Storm (STS, with a maximum sustained wind speed exceeding 24.5 m/s), Typhoon (TY, with a maximum sustained wind speed exceeding 32.7 m/s), Severe Typhoon (STY, with a maximum sustained wind speed exceeding 41.5 m/s), and Super Typhoon (Super TY, with a maximum sustained wind speed exceeding 51.0 m/s). Figure 6 depicts the spatial distribution of the genesis regions for these various categories of tropical cyclones in the Northwest Pacific.

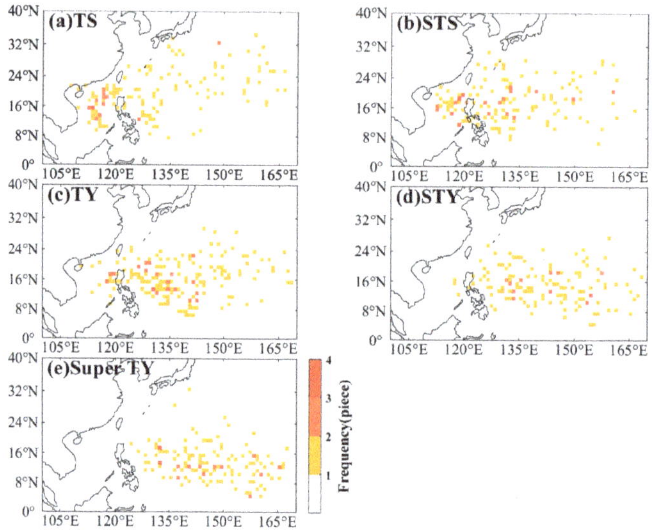

Figure 6. Spatial distribution of tropical cyclone geneses of different grades in Western Pacific from 1984 to 2022.

The analysis presented in Figure 6a indicates that the primary genesis region for TS is located in the South China Sea, with a secondary area of high frequency situated to the east of the Philippine Islands in the Western Pacific Ocean. Figure 6b illustrates the spatial distribution of STS, which predominantly originate from the southern part of the South China Sea, west of the Philippines, and from the Northwestern Pacific Ocean, east of the Philippine Islands. The genesis region for TY is primarily concentrated in the Central Northwestern Pacific Ocean, specifically to the east of the Philippine Islands, with a significant focus in the area defined by the coordinates 8–20° N and 118–150° E (Figure 6c). In the case of STY and Super TY, the genesis regions shift further eastward, with the principal areas of origin located in the eastern part of the Western Pacific, predominantly within the longitudinal range of 120–160° E. As the intensity of tropical cyclones increases, the genesis regions in the South China Sea appear to diminish (Figure 6d,e), while the high-frequency centers progressively migrate eastward. This observation suggests that the South China Sea is less favorable for the formation of high-intensity tropical cyclones.

The following is a statistical analysis of the latitude and longitude distribution of cyclone genesis regions in the Northwest Pacific from 1984 to 2022 (Figures 7 and 8). From Figure 7, it can be clearly observed that the main latitude position of TCs has not changed significantly over the past 39 years, remaining consistently concentrated in the range of 8–20° N. However, from the perspective of longitude distribution (Figure 8), there is a slight westward trend in the generation locations of TCs. Specifically, before 2000, TCs were mainly generated near 125–150° E; during the period from 2000 to 2010, this range slightly shifted eastward to 130–145° E; and from 2010 to 2022, the main generation locations of TCs further shifted westward, concentrating between 120 and 140° E.

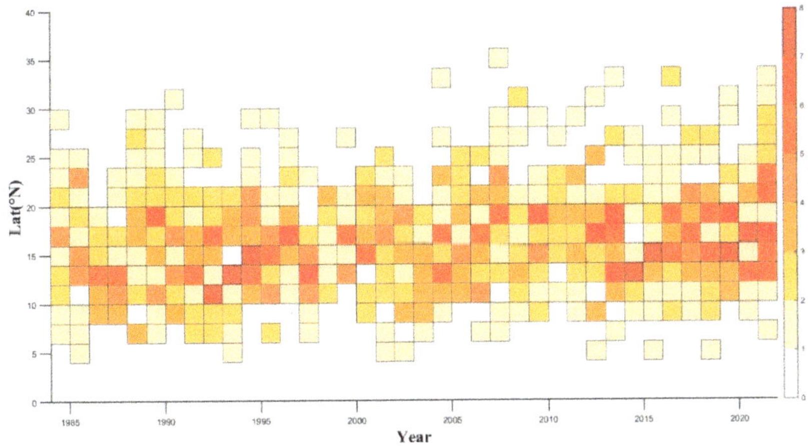

Figure 7. Interannual distribution of latitude of TC origins in Northwest Pacific from 1984 to 2022.

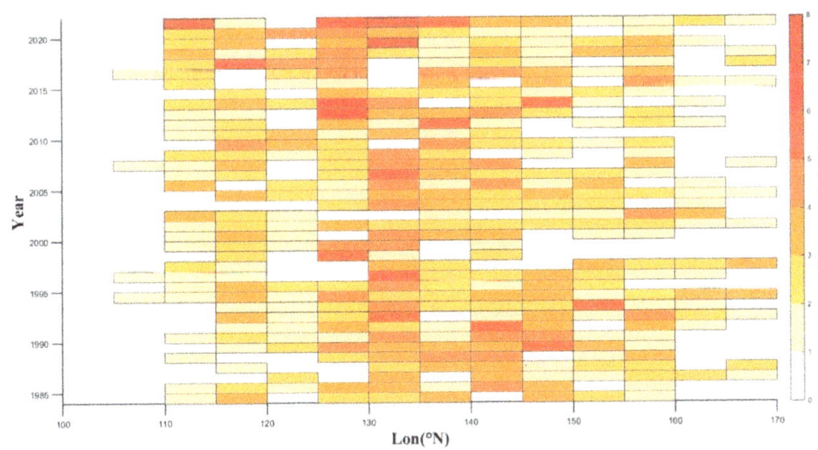

Figure 8. Interannual distribution of longitude of TC origins in Northwest Pacific from 1984 to 2022.

3.2. The Relationship Between Tropical Cyclone Activity in the Northwest Pacific and Temperature Changes in the Warm Pool in the Western Pacific

TC activity demonstrates a substantial correlation with alterations in ocean thermodynamics. Research suggests that the warming anomaly of SST may be a critical factor contributing to the rise in annual tropical cyclone frequency. However, the relationship between tropical cyclone activity and sea surface temperature continues to be a topic of ongoing debate. While warm SSTs create a more favorable environment for TC development, some studies suggest that SST changes are not the primary driver influencing the number of typhoons [16].

The Northwest Pacific Ocean is recognized as one of the principal regions for tropical cyclone genesis globally, primarily attributable to the existence of the Western Pacific warm pool, which is the largest and warmest body of warm water on the planet. The thermal conditions of this warm pool exert a significant influence on the formation and development of TCs in the region. To investigate the relationship between the interannual variability of the warm pool and TC activity in the Western Pacific, a study region within the bounds of 0–16° N and 125–165° E in the Western Pacific Ocean was designated as the warm pool [31,32]. The mean sea surface temperature anomaly (SSTA) from June to

November in this area was calculated and used as an indicator of the thermal state of the warm pool (Figure 9). It is not difficult to see that the warm pool heat conditions in the past 20 years have been warmer compared to the previous 20 years. To better reflect the changes in warm pool heat conditions, this article defines years with warm pool SSTA > 0.2 °C (SSTA < −0.2 °C) as significantly warm (cold) years. Statistics show that in the past 39 years, there have been 7 significantly warm years and 9 significantly cold years for the warm pool.

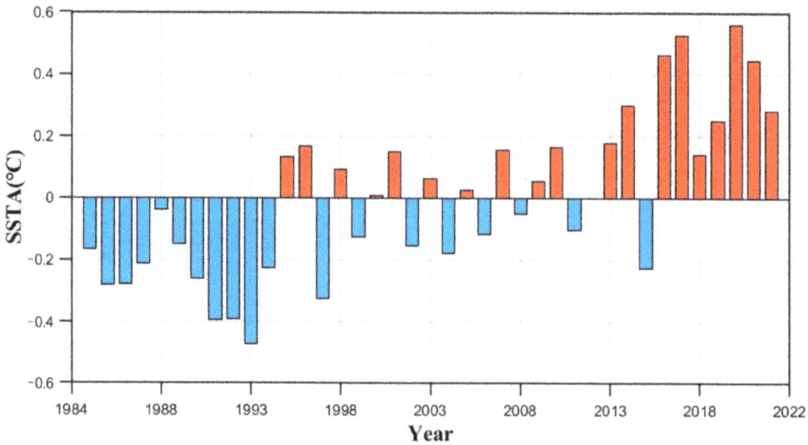

Figure 9. Interannual variability of SSTA in Northwest Pacific warm pool from 1984 to 2022.

The results indicate a significant increase in the thermal environment of the Western Pacific warm pool over the past 39 years, as demonstrated by the interannual variations illustrated in Figure 10a. The sea surface temperature within the warm pool displays a consistent upward trend, rising at an annual rate of 0.0169 °C. To gain deeper insights into the thermal state of the warm pool, the long-term trends and potential climate shifts in the SSTA of the warm pool in the Northwest Pacific were analyzed using the M-K test method, as depicted in Figure 10b. The analysis indicates a significant mutation point in the SSTA of the warm pool in 1990, while no notable thermal changes were detected in 1986. From 1986 to 1991, the thermal conditions of the warm pool remained relatively stable, with no significant upward trend observed. Similarly, between 1991 and 1995, there was no substantial decline in the thermal conditions. However, following 1996, the thermal environment of the warm pool has shown a consistent upward trajectory, with a particularly pronounced increase observed after 2006.

Through the application of wavelet analysis, we identified the periodic variations in the SSTA in the Western Pacific warm pool (see Figure 11). The findings suggest that the SSTA in this region demonstrates notable periodic variations on a small scale with cycles of 1–3 years and 2–5 years. From 1996 to 2004, it showed a 1–3 year periodic variation, while from 2011 to 2021, it exhibited a 2–5 year periodic variation (Figure 11a). By observing the wavelet global spectrum (Figure 11b), it can be found that the wavelet power, through confidence testing, has a periodicity of 1–4 years, with a peak at 3 years, indicating that the periodic variation in the warm pool is most significant around 3 years.

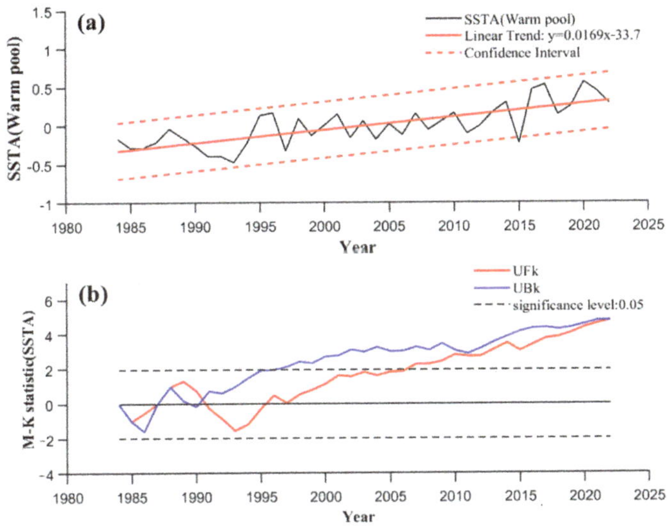

Figure 10. Interannual variation and M-K trend test of the SSTA in Northwest Pacific warm pool from 1984 to 2022. (**a**) is Interannual variation and (**b**) is M-K trend test.

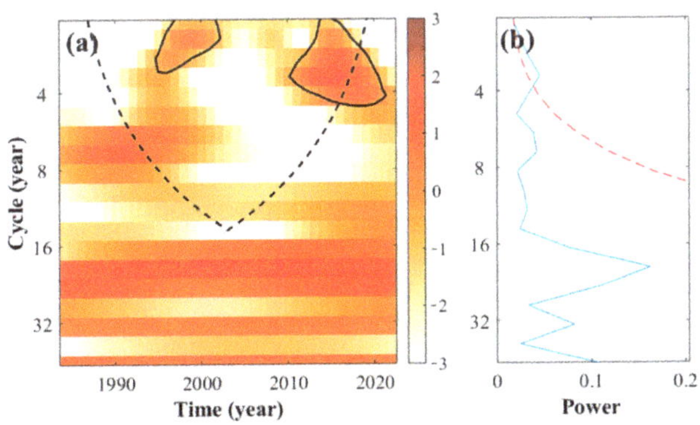

Figure 11. Wavelet power spectrum variation chart and global wavelet power map of SSTA in Northwest Pacific warm pool from 1984 to 2022. (**a**) is wavelet power spectrum variation diagram and (**b**) is wavelet full spectrum diagram.

To examine how tropical cyclone activity responds to variations in SSTA within the Northwest Pacific warm pool, we conducted a statistical analysis of TC distribution patterns over the past 39 years under both cold and warm thermal conditions of the warm pool, as illustrated in Figure 12. During years with a warm thermal state of the warm pool, the 29 °C isotherm is positioned approximately at 22° N (Figure 12a), extending across much of the sea areas south of the South China Sea and northwest of the Philippine Islands. The primary genesis regions for TCs are located in the central and western parts of the Northwest Pacific, south of the 29 °C isotherm, which can be roughly categorized into three distinct areas: the South China Sea region (8–20° N, 112–120° E), the sea area west of the Philippine Islands (8–20° N, 125–134° E), and the Northwest Pacific region (10–20° N, 140–145° E).

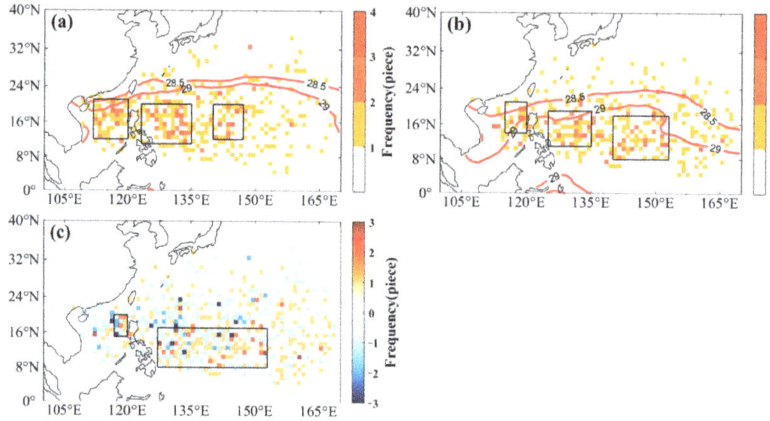

Figure 12. Distribution of cyclone genesis regions and abnormal distribution of cyclone genesis regions in Western Pacific warm pool from 1984 to 2022 (red solid line represents isotherm of warm pool, and black box indicates high-frequency cyclone genesis regions). (**a**) is the warm pool thermal state in warm years, (**b**) is the warm pool thermal state in cold years, and (**c**) is the distribution of cyclone frequency anomalies.

In contrast, during years with a cold thermal state of the warm pool (Figure 12b), the 29 °C isotherm shifts to around 20° N, covering most of the central and western parts of the West Pacific. Under these conditions, the genesis regions for TCs are primarily concentrated south of the 28.5 °C isotherm, with high-frequency areas also located south of the 29 °C isotherm. These regions can be divided into three main areas: the South China Sea region (8–20° N, 112–120° E), the sea area west of the Philippine Islands (5–20° N, 125–135° E), and the Northwest Pacific region (8–20° N, 140–155° E).

Figure 12c highlights the differences in cyclone frequency during anomalous years of warm pool thermal conditions, revealing positive anomalies in a small area of the South China Sea east of the Philippine Islands and in the central–southern part of the Northwest Pacific (8–16° N, 128–155° E). Conversely, negative anomalies are observed in the northern part of the Northwest Pacific region. These findings suggest that during years with a cold thermal state of the warm pool, the genesis regions for cyclones shift further south and east.

3.3. Relationship Between Tropical Cyclone Activity and Monsoon Trough Activity in Northwest Pacific

Given that warm pools are not the primary factor influencing the frequency of TCs, what factors determine the spatial distribution of these phenomena? The prevalence of TCs in the Western Pacific Ocean can be attributed not only to the relatively high surface sea temperatures in the region, which fulfill the thermal requirements for TC formation, but also to the frequent presence of active monsoon troughs during the summer and autumn months. The monsoon trough is an important feature of atmospheric circulation, typically referring to a low-pressure trough that forms in tropical and subtropical regions during the monsoon season (usually summer). It is usually located near the equator and extends northward to the Asian continent. When the monsoon trough develops significantly, it can extend eastward to the Indochinese Peninsula and connect with the near-equatorial trough around the Philippines [28,33]. Relevant studies indicate that approximately 75% of the world's TCs originate within monsoon troughs, with around 80% of the TCs formed in the Northwest Pacific occurring in these troughs [33,34]. Figure 13 shows the interannual variation in the MTI in the Northwest Pacific from 1984 to 2022, calculated using the above definitions and wind field data. To clearly reflect the characteristics of monsoon trough activity, this paper defines years with MTI > 1 (MTI < −1) as years with significantly strong (weak) monsoon troughs. It can be seen that in the past 39 years, there have been 6 years with significantly strong monsoon troughs and 5 years with significantly weak monsoon troughs.

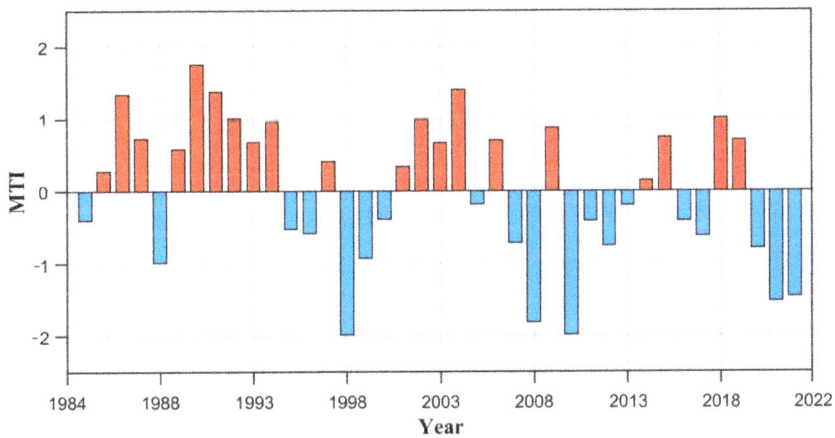

Figure 13. Interannual variation in MTI in Pacific Northwest from 1984 to 2022.

Through a long-term trend study of MTI over the past 39 years, we found that MTI shows a non-significant downward trend; the M-K test indicated a mutation in 1996. Before 1996, the intensity of MTI showed an increasing trend, while after 1996, it mainly exhibited a weakening trend. By employing wavelet analysis, we obtained the wavelet power spectrum change diagram of MTI (Figure 14a) and the global wavelet power diagram (Figure 14b). The results show that the periodicity of MTI, verified at a 95% confidence level, is 1–5 years and 1–6 years. From 1985 to 1993, it exhibited a variation cycle of 3–5 years, from 1993 to 2002, a variation cycle of 1–4 years, from 2007 to 2016, a variation cycle of 1–3 years, and from 2016 to 2021, primarily a variation cycle of 3–6 years. Additionally, there is a relatively significant periodic variation around 2–3 years in MTI.

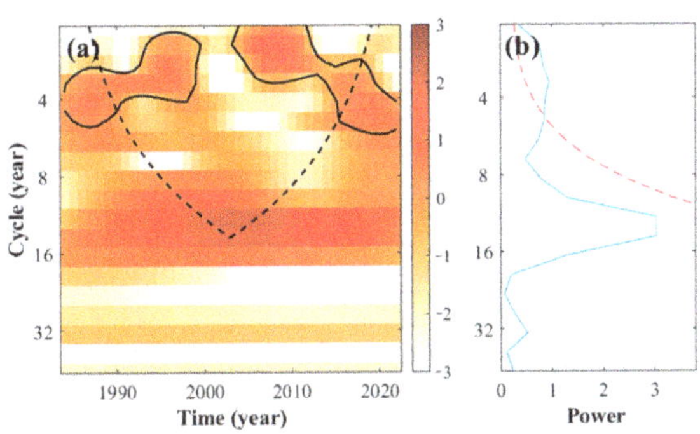

Figure 14. Wavelet power spectrum variation chart and global wavelet power map of MTI in Northwest Pacific from 1984 to 2022. (**a**) is wavelet power spectrum variation diagram and (**b**) is wavelet full spectrum diagram.

The results of the aforementioned research demonstrate a discernible weakening trend in the intensity of the monsoon trough in the Northwest Pacific over the past 39 years. Previous research has indicated that the interannual variability of the monsoon trough in this region is predominantly affected by the thermal conditions of the Northwest Pacific warm pool. Furthermore, the activity of TCs in the Western Pacific is closely linked to the interannual fluctuations in the monsoon trough. To investigate these relationships, we conducted an analysis of the correlations among TC frequency, ACE, SSTA of the warm pool, and MTI [21,24,26]. Figure 15 presents the Pearson correlation coefficients among TC frequency, ACE, warm pool SSTA, and MTI. The analysis reveals that TC frequency is significantly positively correlated with both ACE (r = 0.4675) and MTI (r = 0.4252) at the 95% confidence level ($p < 0.05$). Additionally, ACE exhibits a strong positive correlation with MTI (r = 0.7219, $p < 0.001$), suggesting that MTI has a more substantial influence on both cyclone frequency and ACE. In contrast, warm pool SSTA is significantly negatively correlated with both ACE (r = −0.4474, $p < 0.01$) and MTI (r = −0.5495, $p < 0.001$), while showing only a weak and insignificant positive correlation with TC frequency. This pattern suggests that the thermal state of the warm pool significantly influences MTI. As the thermal state of the warm pool varies, a potential weakening of MTI may ensue, which could result in a decrease in TC frequency and a reduction in ACE.

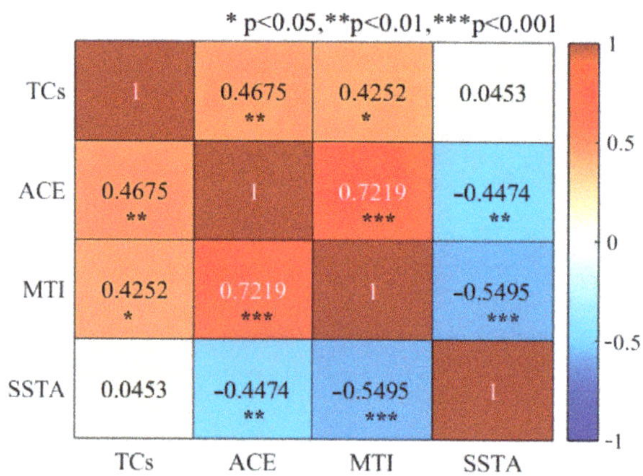

Figure 15. Heatmap of correlation coefficients for tropical cyclone frequency, ACE, SST, and MTI.

In order to better study the activity trends of the monsoon trough under varying conditions of warm pool thermal status, we analyzed the distribution of the monsoon trough under warm and cold thermal conditions of the warm pool. The figure below shows the distribution of the monsoon trough at 850 hPa in the Northwestern Pacific under warm (Figure 16a) and cold (Figure 16b) thermal conditions of the warm pool. Under warm thermal conditions, the 29 °C isotherm is relatively positioned to the north, located around 24° N, covering a vast area of the Northwestern Pacific south of 24° N. At this time, the activity area of the monsoon trough is shifted westward, mainly concentrated west of 130° E, and can extend over the Philippine Islands. In contrast, when the warm pool thermal status is cold, the 29 °C isotherm is positioned further south, around 16° N. In this context, the monsoon trough exhibits a southeastward orientation, with its activity range located between 5° N and 16° N, extending southeastward to 155° E. This distribution characteristic reveals that under cold conditions of the warm pool, the activity area of the monsoon trough is broader, covering most of the southeastern region of the Northwestern Pacific.

In the preceding analysis, it was determined that the thermal conditions of the warm pool exert a significant influence on the activity trends of the monsoon trough. Consequently, it is imperative to investigate whether the activity of TCs exhibits a discernible pattern in response to fluctuations in MTI in the Northwest Pacific. This study aims to examine the variations in the 850 hPa flow field associated with the monsoon trough, as well as the distribution of TC genesis regions from June to September during years characterized by strong and weak MTI. Since the monsoon trough typically weakens and dissipates in October–November each year, the analysis did not include cyclone activity during this period.

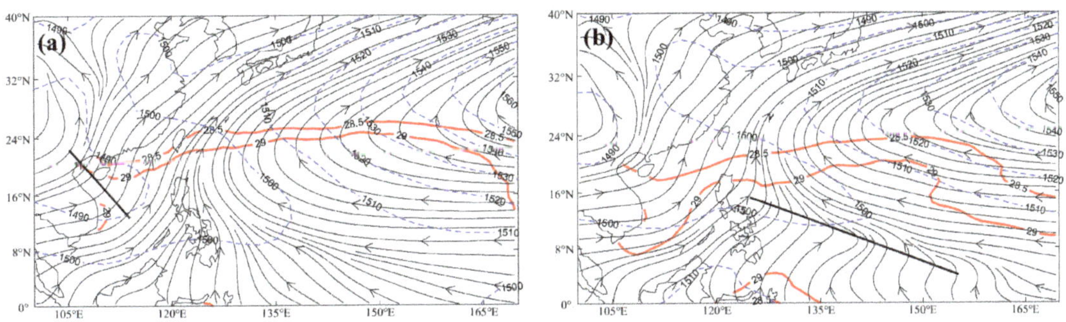

Figure 16. Average distribution of 850 hPa monsoon troughs in warm pool thermal conditions from 1984 to 2022 warm (**a**) and cold (**b**) (red solid line is warm pool SST contour, blue dashed line is 850 hPa contour, black streamline is 850 hPa wind field, and heavy black line indicates approximate location of monsoon trough).

In years characterized by a stronger MTI, the monsoon trough extends from the South China Sea to the Central–Western Pacific Ocean at 150° E during the months of June to July, encompassing latitudes between 5° and 16° N, as illustrated in Figure 17a. During this timeframe, the primary regions for TC genesis are situated in the South China Sea and the Western Pacific, specifically to the east of the Philippines, with cyclone formation predominantly occurring within the confines of the monsoon trough. Subsequently, from August to September, the monsoon trough migrates northward, covering the northern portion of the South China Sea and the Northwest Pacific at latitudes ranging from 8° to 24° N, as depicted in Figure 17b. The trough further extends eastward into the Western Pacific at 165° E.

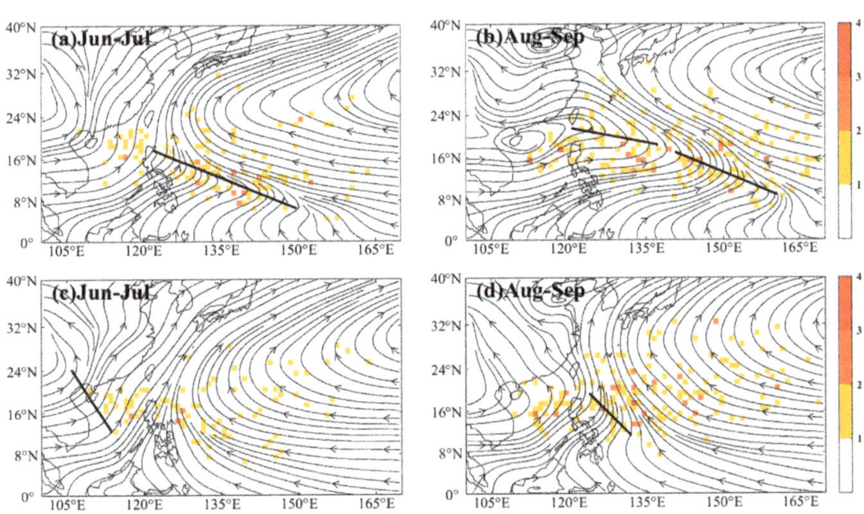

Figure 17. Average distribution of monsoon trough at 850 hPa and distribution of TC genesis from autumn 1984 to 2022.

In contrast, during June and July in years of weak MTI (Figure 17c), the monsoon trough is primarily located over the South China Sea at latitudes of 5° to 16° N, with the main TC genesis regions concentrated in the South China Sea and the Northwest Pacific east of the Philippine Islands. From August to September, the monsoon trough shifts eastward over the Northwest Pacific near longitudes of 120–130° E, with the primary TC genesis regions located to the east–southeast of the South China Sea and in the Northwest Pacific near the Philippine Islands.

In summary, during years of strong MTI, the TC genesis regions are predominantly within the monsoon trough. In contrast, during years of weak MTI, TC genesis is more likely to occur in the northward airflow belt, which results from the convergence of the northwesterly equatorial airflow and the airflow from the Eastern Pacific. This observation highlights the strong relationship between tropical cyclone activity and the dynamics of the monsoon trough.

3.4. Characteristics of Environmental Factors Above Northwest Pacific Under Abnormal Warm Pool Heat Conditions

Through a comparative analysis, it is evident that the thermal conditions of the warm pool significantly influence the spatial distribution and activity trends of the monsoon trough (Figure 16). Specifically, during warmer warm pool conditions, the monsoon trough's activity is primarily confined to the western and northern parts of the South China Sea. In contrast, during colder warm pool conditions, the monsoon trough shifts to the central–eastern regions of the Northwest Pacific, expanding in a southeastern direction. These findings provide essential insights into the relationship between the thermal state of the warm pool and the dynamics of the monsoon trough. Prior research has emphasized a substantial connection between the monsoon trough and cyclone activity. One key question remains: why do cyclones frequently form within the monsoon trough? Outgoing Longwave Radiation (OLR) is a crucial metric for assessing convective activity, with lower OLR values typically indicating intense convection and high cloud tops, while higher OLR values suggest weaker convection. By examining the distribution and variability of OLR within the monsoon trough, we can gain deeper insights into the underlying mechanisms connecting monsoon trough activity to cyclone formation and development [28,34].

Figure 18 provides the 850 hPa streamlines and the composite distribution of OLR (W m^{-2}) over the Northwest Pacific during June–July, August–September, and October–November in years characterized by strong and weak monsoon troughs. Figure 18a illustrates that in strong MTI years, the monsoon trough extends eastward from the South China Sea to the Central Northwest Pacific during the months of June and July, spanning approximately 5–24° N, with heightened convective activity in this region. By August–September (Figure 18b), the monsoon trough shifts significantly northward, covering latitudes of 10–20° N over the South China Sea and the Northwest Pacific, and extends eastward over the Eastern Northwest Pacific near 165° E. Convective activity also migrates northward, reaching the South China Sea and the Northwest Pacific at latitudes of 10–18° N and longitudes of 110–160° E. Figure 18c depicts the monsoon trough's variation from October to November, showing a gradual weakening and southward movement due to the westward expansion of the easterly airflow. During this period, the monsoon trough transitions into a cyclonic vortex centered over the Northwest Pacific near 10° N, 135° E.

Figure 18d–f illustrates the average 850 hPa streamlines and OLR over the Northwest Pacific during the periods of June–July, August–September, and October–November in years characterized by weak MTI. Specifically, Figure 18d indicates that during weak MTI years, the monsoon trough is predominantly situated over the western part of the South China Sea and the Northwest Pacific, at latitudes ranging from 8° to 20° N, during the months of June and July, with notable convective activity in this area. From August to

September (Figure 18e), the monsoon trough shifts southward, covering the South China Sea and the Northwest Pacific at latitudes of 8–18° N, and extends eastward over the Central Northwest Pacific near 140° E. Convective activity also migrates northward, affecting the South China Sea and the Northwest Pacific at latitudes of 8–18° N and longitudes of 110–140° E. Figure 18f reveals that from October to November, the monsoon trough experiences a weakening along the southern boundary of the Western Pacific subtropical high and subsequently shifts southward in conjunction with the easterly airflow, evolving into a cyclonic vortex centered over the western part of the Northwest Pacific near 8° N, 115° E.

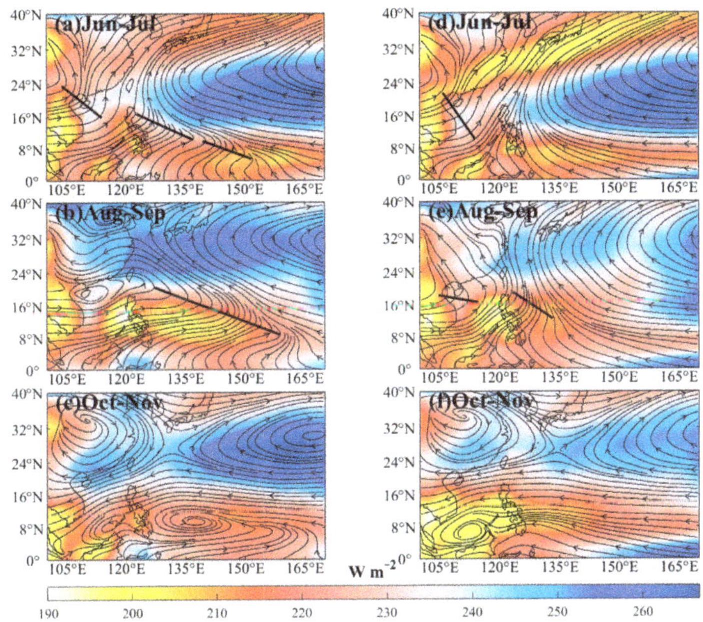

Figure 18. Average distribution of 850 hPa monsoon trough and OLR distribution during strong (**a–c**) and weak (**d–f**) MTI years in summer and autumn from 1984 to 2022 (shaded areas, unit: W m^{-2}).

In summary, during strong MTI years, the monsoon trough exhibits greater intensity, with its activity range expanding northward and eastward. The trough line tends to be positioned more northerly or easterly, resulting in broader coverage over the eastern part of the Western Pacific, particularly in the Central Western Pacific, where convective activity may be more vigorous. Conversely, in weak MTI years, the monsoon trough's intensity diminishes, and its activity range contracts westward, predominantly shifting northwestward during years characterized by cold warm pool conditions, particularly affecting the South China Sea and the Philippine Islands. Furthermore, convective activity is notably more pronounced in the South China Sea and the West–Northwest Pacific region, indicating a weaker MTI and reduced monsoon trough activity, whereas during warm of warm pool years, the monsoon trough develops more eastward, reflecting a stronger monsoon trough.

The intensified cyclone activity within the monsoon trough can be explained by several interconnected factors. Key among these is the SST in the Western Pacific warm pool, which typically exceeds 28.5 °C beneath the monsoon trough, providing the necessary thermal foundation for cyclone development [4,11]. Moreover, the persistent presence of the monsoon trough over this warm pool during the summer months creates an environment characterized by favorable dynamic and thermal conditions, further promoting cyclone formation. First, the relative vorticity associated with atmospheric convergence and cyclonic characteristics within the monsoon trough is particularly strong at the 850 hPa level, supplying the rotational energy required for cyclone genesis. Second, the relatively low vertical wind shear within the monsoon trough enhances the stability of convective and circulation structures, facilitating the development and intensification of cyclones [11,21,32]. Additionally, the monsoon trough plays a critical role in supplying abundant water vapor, which is essential for the initiation and sustenance of cyclones. The continuous transport of warm, humid air toward the cyclone center supports robust convective activity. Simultaneously, the divergence at the 200 hPa level within the monsoon trough aids in the outward transport of heat and kinetic energy from the cyclone, supporting its ongoing development and intensification [28,29]. To obtain insights into the spatial and intensity variations in the Northwest Pacific Monsoon Trough, as well as the large-scale environmental factors influencing cyclone formation during the warm and cold periods of the Western Pacific warm pool, several atmospheric parameters must be considered (shown in Figure 19). The parameters considered included relative vorticity at 850 hPa, relative divergence at 200 hPa, vertical wind shear between 200 hPa and 850 hPa, and average relative humidity between 500 hPa and 700 hPa within the lower troposphere over the Northwest Pacific.

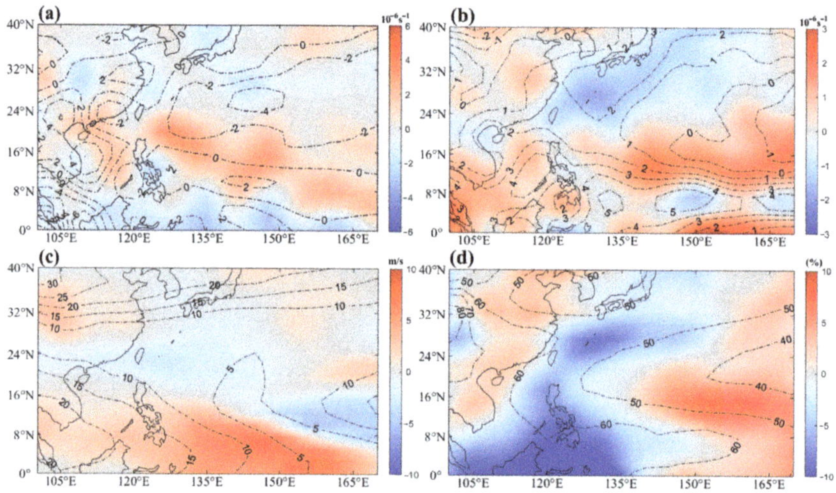

Figure 19. Abnormal distribution of environmental factors (colored) and climate mean state for warm pool heat status anomaly years from 1984 to 2022 ((**a**) is relative vorticity at 850 hPa, (**b**) is relative divergence at 250 hPa, (**c**) is wind shear between 850 hPa and 200 hPa, (**d**) is average relative humidity between 500 hPa and 700 hPa, with contour lines representing climate mean state).

Figure 19a illustrates the relative vorticity at 850 hPa during the anomalous years characterized by the warm pool's thermal conditions, in comparison to the climatological average. In the average state, positive relative vorticity exists in the central and eastern areas of the Northwest Pacific (8–16° N, 125–160° E). Positive anomalies also appear in this region and the eastern part of the South China Sea, signifying enhanced relative vorticity in colder years. It implies that the cyclonic relative vorticity at 850 hPa in the central to eastern regions of the Northwest Pacific is stronger in colder years than in warmer years, reflecting more intense airflow convergence in these areas under colder conditions. Figure 19b shows the relative divergence at 200 hPa during the abnormal years of the warm pool's thermal conditions relative to the climatological average. In the average state, positive divergence is observed in the central and eastern regions of the Northwest Pacific.

Additionally, positive divergence anomalies are present in the eastern part of the South China Sea and the eastern section of the Northwest Pacific (8–16° N, 125–170° E), indicating enhanced divergence in colder years. This indicates that the circulation divergence at 200 hPa over the South China Sea, as well as in the western and central-eastern regions of the Northwest Pacific, is enhanced during colder years, thereby promoting increased convective activity. The climatological mean of the vertical wind shear between 250 hPa and 850 hPa demonstrates minimal shear in the central and southeastern areas of the Northwest Pacific (Figure 19c). Negative anomalies were identified in the central and southeastern regions of the Western Pacific, suggesting a reduction in wind shear during colder years. This observation indicates that vertical wind shear in the central and southeastern areas of the Northwest Pacific is diminished in colder years. Furthermore, an analysis of the climatological mean of relative humidity between 500 hPa and 700 hPa is warranted to understand the implications of these findings (Figure 19d); there is sufficient moisture in the South China Sea and the southern part of the Western Pacific (5–12° N, 120–150° E). Negative anomalies are detected near the Philippine Islands and the surrounding sea areas, indicating reduced moisture content in colder years. On the contrary, positive anomalies are seen in the eastern part of the Western Pacific, suggesting adequate moisture availability in this region during colder years.

In summary, during the cold phase of the warm pool thermal conditions, anomalies in relative vorticity, positive relative divergence, and wind shear were predominantly observed over the eastern part of the South China Sea and the central to eastern regions of the Northwest Pacific. Additionally, a positive anomaly in relative humidity was primarily observed in the Northeastern Pacific Ocean. Most of these positive anomalies were concentrated in the maritime area bounded by 8–16° N and 125–160° E, corresponding to the active region of the monsoon trough during both warm and cold years. These findings indicate that convective activity within the operational range of the monsoon trough is notably vigorous, significantly enhancing the formation and intensification of cyclones. This phenomenon explains the increased frequency of cyclones occurring within the monsoon trough.

4. Conclusions

Through the analysis of the interannual variation characteristics of TC activity in the Northwest Pacific from 1984 to 2022, as well as the ACE and generation genesis, it was found that the frequency of TCs exhibits significant interannual variation characteristics, with a slight upward trend. The M-K test analysis revealed a mutation in cyclone frequency around 1996, and a mutation in ACE intensity occurred around 1998, with both showing a downward trend. By studying the thermal conditions of the warm pool and MTI, we found that the thermal conditions of the warm pool have been on the rise since 1996, with a significant increase after 2006, while MTI showed a downward trend after the mutation in 1996. Using wavelet analysis, we discovered that the frequency of TCs, ACE, warm pool SSTA, and MTI all exhibit similar low-scale periodic variations. Taking 2001–2005 as a dividing point, the significant periodicity of cyclone frequency in 2001 was 1–3 years, ACE had a significant periodicity of 1–6 years, warm pool SSTA had a periodicity of 1–3 years, and MTI had a periodicity of 1–4 years. After 2005, the significant periodicity of TCs frequency was 1–7 years, ACE had a significant periodicity of 1–4 years, the warm pool had a periodicity of 1–6 years, and MTI had a periodicity of 1–6 years. From the analysis of cyclone generation genesis over the 39 years, it was found that there are three main generation geneses, and the longitudinal position of the TC occurrence genesis has shifted westward.

Through our investigation into the correlation between the warm pool in the Western Pacific and the frequency of cyclone occurrences, we discovered that the origins of cyclones are frequently situated in regions of the ocean characterized by high sea surface temperatures. In years when the warm pool's thermal conditions are above or below average, we observed that during warmer (or cooler) years, the MTI may be weaker (or stronger). There exists a substantial negative correlation between the thermal conditions of the warm pool and MTI. In years characterized by elevated thermal conditions in the warm pool, the monsoon trough may extend as far as the South China Sea and the western region of the Northwestern Pacific at 130° E, while in cooler years, it can extend to the eastern part of the Northwestern Pacific at 160° E. This trend of monsoon trough activity is consistent with years of stronger or weaker MTI. In our study of the activity trends during years of stronger and weaker monsoon troughs, we found that the activity of the monsoon trough is generally accompanied by stronger convective activity, and the main generation areas for tropical cyclones are primarily located within the monsoon trough. In years characterized by a weaker monsoon trough, the primary regions for tropical cyclone formation are not limited to the monsoon trough itself; rather, they also more frequently arise within the northward airflow produced by the convergence of the northwesterly flow and the airflow originating from the Eastern Pacific.

Finally, we focused on analyzing the activity of environmental factors under different thermal conditions of the warm pool. In years when the warm pool is relatively cool, there are positive anomalies of relative vorticity and relative divergence over the eastern part of the South China Sea and the central and eastern regions of the Northwest Pacific. The positive anomaly of relative humidity is mainly concentrated over the eastern part of the Northwest Pacific. Most of these positive anomalies of environmental factors occur in the sea area between 8 and 16° N and 125 and 160° E, while wind shear shows a negative anomaly. The appearance of these anomalies indicates an enhancement of environmental factors in this region, which is more conducive to the development of convective activity. This area corresponds to the activity range of the monsoon trough during years when the warm pool is relatively cool. Within the monsoon trough's activity range, convective activity is particularly intense, and under strong convective conditions, the occurrence and development of cyclones are more favorable.

Author Contributions: Conceptualization, J.G. and J.S.; Methodology, X.H.; Validation, Y.F. and Y.C.; Data curation, Y.F.; Writing—original draft, S.W.; Writing—review & editing, J.G., X.H., J.S. and Y.C. All authors have read and agreed to the published version of the manuscript.

Funding: This research was funded by the Dalian Science and Technology Innovation Fund (2024JJ11PT007); the Dalian Science and Technology Program for Innovation Talents of Dalian (2022RJ06); the Science and Technology Program of Liaoning Province (2022JH2/101300222, 2022JH2/101300183); the Scientific Research Project of Education Department of Liaoning Province (LJ212410158039, LJ232410158056); and the Research Funds for Undergraduate Universities of Liaoning Province (2024JBPTZ001, 2024JBQNZ002).

Institutional Review Board Statement: Not applicable.

Informed Consent Statement: Not applicable.

Data Availability Statement: The data presented in this study are available upon request to the corresponding author.

Acknowledgments: We are thankful for the data support from the National Marine Scientific Data Center (Dalian), the National Science & Technology Infrastructure of China (http://odc.dlou.edu.cn/ (accessed on 24 March 2023)), and Oceanographic Data Center, Chinese Academy of Sciences (CODC, https://www.casodc.com/) for providing valuable data and information.

Conflicts of Interest: The authors declare no conflict of interest.

References

1. Briegel, L.M.; Frank, W.M. Large-Scale Influences on Tropical Cyclogenesis in the Western North Pacific. *Mon. Weather Rev.* **1997**, *125*, 1397–1413. [CrossRef]
2. Cheung, K.K.W. Large-Scale Environmental Parameters Associated with Tropical Cyclone Formations in the Western North Pacific. *J. Clim.* **2004**, *17*, 466–484. [CrossRef]
3. Harr, P.A.; Elsberry, R.L. Tropical Cyclone Track Characteristics as a Function of Large-Scale Circulation Anomalies. *Mon. Weather Rev.* **1991**, *119*, 1448–1468. [CrossRef]
4. Huang, W.; Dong, S. Long-Term and Inter-Annual Variations of Tropical Cyclones Affecting Taiwan Region. *Reg. Stud. Mar. Sci.* **2019**, *30*, 100721. [CrossRef]
5. Cho, Y.-M.; Lu, M.-M.; Sui, C.-H.; Solis, A.L.S.; Chen, M.-S. Decadal Changes of the Early Summer Asian Monsoon and the South China Sea Tropical Cyclones during 2001–2020. *Terr. Atmos. Ocean. Sci.* **2022**, *33*, 5. [CrossRef]
6. Raga, G.B.; Bracamontes-Ceballos, B.; Farfán, L.M.; Romero-Centeno, R. Landfalling Tropical Cyclones on the Pacific Coast of Mexico: 1850–2010. *Atmósfera* **2013**, *26*, 209–220. [CrossRef]
7. Hartmann, D.L.; Hendon, H.H.; Houze, R.A. Some Implications of the Mesoscale Circulations in Tropical Cloud Clusters for Large-Scale Dynamics and Climate. *J. Atmos. Sci.* **1984**, *41*, 113–121. [CrossRef]
8. Cornejo-Garrido, A.G.; Stone, P.H. On the Heat Balance of the Walker Circulation. *J. Atmos. Sci.* **1977**, *34*, 1155–1162. [CrossRef]
9. De Deckker, P. The Indo-Pacific Warm Pool: Critical to World Oceanography and World Climate. *Geosci. Lett.* **2016**, *3*, 20. [CrossRef]
10. Wang, C.; Enfield, D.B. The Tropical Western Hemisphere Warm Pool. *Geophys. Res. Lett.* **2001**, *28*, 1635–1638. [CrossRef]
11. Wang, C.; Lee, S.-K.; Enfield, D.B. Atlantic Warm Pool Acting as a Link between Atlantic Multidecadal Oscillation and Atlantic Tropical Cyclone Activity. *Geochem. Geophys. Geosyst.* **2008**, *9*, Q05V03. [CrossRef]
12. Emanuel, K. Increasing Destructiveness of Tropical Cyclones over the Past 30 Years. *Nature* **2005**, *436*, 686–688. [CrossRef]
13. Emanuel, K.; Sundararajan, R.; Williams, J. Hurricanes and Global Warming: Results from Downscaling IPCC AR4 Simulations. *Bull. Am. Meteorol. Soc.* **2008**, *89*, 347–368. [CrossRef]
14. Huang, R.-H.; Huangfu, J.-L.; Wu, L.; Feng, T.; Chen, G.-H. Research on the interannual and interdecadal variabilities of the monsoon trough and their impacts on tropical cyclone genesis over the western north pacific ocean. *J. Trop. Meteorol.* **2018**, *24*, 395–420. [CrossRef]
15. Chen, G.; Huang, R. Influence of Monsoon over the Warm Pool on Interannual Variation on Tropical Cyclone Activity over the Western North Pacific. *Adv. Atmos. Sci.* **2008**, *25*, 319–328. [CrossRef]
16. Wada, A.; Chan, J.C.L. Relationship between Typhoon Activity and Upper Ocean Heat Content. *Geophys. Res. Lett.* **2008**, *35*, L17603. [CrossRef]
17. Emanuel, K.A. The Dependence of Hurricane Intensity on Climate. *Nature* **1987**, *326*, 483–485. [CrossRef]
18. Zong, H.; Wu, L. Synoptic-Scale Influences on Tropical Cyclone Formation within the Western North Pacific Monsoon Trough. *Mon. Weather Rev.* **2015**, *143*, 3421–3433. [CrossRef]
19. Chen, G.-H.; Huang, R.-H. Role of Equatorial Wave Transitions in Tropical Cyclogenesis over the Western North Pacific. *Atmos. Ocean. Sci. Lett.* **2008**, *1*, 64–68. [CrossRef]
20. Miller, B.I. On the maximum intensity of hurricanes. *J. Atmos. Sci.* **1958**, *15*, 184–195. [CrossRef]
21. Li, C.Y.; Pan, J. The Interannual Variation of the South China Sea Summer Monsoon Trough and Its Impact. *Chin. J. Atmos. Sci.-Chin. Ed.* **2007**, *31*, 1049–1058. [CrossRef]
22. Gallina, G. Environmental Vertical Wind Shear and Tropical Cyclone Intensity Change Utilizing Enhanced Satellite Derived Wind Information. In Proceedings of the 25th Conference on Hurricanes and Tropical Meteorology, San Diego, CA, USA; 2002; pp. 172–173. Available online: https://ams.confex.com/ams/25HURR/webprogram/Paper35650.html (accessed on 1 March 2024).
23. Feng, T.; Chen, G.-H.; Huang, R.-H.; Shen, X.-Y. Large-Scale Circulation Patterns Favourable to Tropical Cyclogenesis over the Western North Pacific and Associated Barotropic Energy Conversions. *Int. J. Climatol.* **2014**, *34*, 216–227. [CrossRef]
24. Torrence, C.; Compo, G.P. A Practical Guide to Wavelet Analysis. *Bull. Am. Meteorol. Soc.* **1998**, *79*, 61–78. [CrossRef]
25. Polanco-Martínez, J.M.; Fernández-Macho, J.; Medina-Elizalde, M. Dynamic Wavelet Correlation Analysis for Multivariate Climate Time Series. *Sci. Rep.* **2020**, *10*, 21277. [CrossRef]
26. Ghafouri-Azar, M.; Lee, S.-I. Meteorological Influences on Reference Evapotranspiration in Different Geographical Regions. *Water* **2023**, *15*, 454. [CrossRef]
27. Gocic, M.; Trajkovic, S. Analysis of Trends in Reference Evapotranspiration Data in a Humid Climate. *Hydrol. Sci. J.* **2014**, *59*, 165–180. [CrossRef]

28. Zhang, X.; Wu, L.; Huangfu, J.; Fan, G.; Huang, R. Seasonal and Interannual Variability of the Western North Pacific Monsoon Trough and Its Relationship to Large-Scale Environmental Factors. *Clim. Environ. Res.* **2017**, *22*, 418–434. [CrossRef]
29. Fei, H.; Zhou, F.; Olson, P.J. Variations of the Atlantic and Pacific Blocking Anticyclones and Their Correlation in the Northern Hemisphere. *J. Ocean Univ. China* **2002**, *1*, 38–44. [CrossRef]
30. Bell, G.D.; Halpert, M.S.; Schnell, R.C.; Higgins, R.W.; Lawrimore, J.; Kousky, V.E.; Tinker, R.; Thiaw, W.; Chelliah, M.; Artusa, A. Climate Assessment for 1999. *Bull. Am. Meteorol. Soc.* **2000**, *81*, S1–S50. [CrossRef]
31. Chen, D.; Lian, T. Interaction of Western-Pacific Tropical Cyclones with El Niño Diversity. *Natl. Sci. Rev.* **2018**, *5*, 803–804. [CrossRef]
32. Fang, M.; Qu, L. Meridional Variation of the 1955–2003 Sea Level Anomalies in the Tropical Pacific Ocean Associated with El Niño Events. *J. Ocean Univ. China* **2007**, *6*, 332–338. [CrossRef]
33. Li, C.; Lu, R.; Chen, G. Promising Prediction of the Monsoon Trough and Its Implication for Tropical Cyclone Activity over the Western North Pacific. *Environ. Res. Lett.* **2017**, *12*, 074027. [CrossRef]
34. Huangfu, J.; Huang, R.; Chen, W. Relationship between the South China Sea Summer Monsoon Onset and Tropical Cyclone Genesis over the Western North Pacific. *Int. J. Climatol.* **2017**, *37*, 5206–5210. [CrossRef]

Journal of
Marine Science and Engineering

MDPI

Article

Tidal Current with Sediment Transport Analysis and Wind Turbine Foundation Pile Scour Trend Studies on the Central Bohai Sea

Xingtang Hu [1], Zhifeng Wang [2] and Xin Ma [1,*]

1 Physical Oceanography Laboratory, Ocean University of China, Qingdao 266100, China; 21220911091@stu.ouc.edu.cn
2 Department of Ocean Engineering, College of Engineering, Ocean University of China, Qingdao 266100, China; wzf1984@ouc.edu.cn
* Correspondence: maxin@ouc.edu.cn

Abstract: This study addresses the local scour problem of monopile foundations in the central Bohai Sea. This study integrates numerical simulations with experimental validation to conduct an in-depth analysis of the tidal current background, sediment transport, sediment sources, and scour characteristics over the past two decades. The analysis of ocean currents and sediment dynamics revealed that the monthly average tidal current speed in the majority of the study region is below 0.6 m·s^{-1}, with annual seabed erosion and accretion changes within 0.02 m, exhibiting minimal variation. The annual erosion and accretion changes in the wind farm areas are less than 0.01 m. The analysis of local scour around monopile foundations indicates that approximately 80% of the scour occurs during the initial phase. A comparative analysis of collar protection effectiveness indicates that the collar can effectively reduce scour depth by 50%, thereby demonstrating significant protective effects. However, the prevailing trend of scouring remains unaltered, indicating that collar protection has inherent limitations in regulating early-stage scouring. The findings of this study provide a theoretical basis for the design and protection of monopile foundations in the central Bohai Sea and offer a valuable reference point for the scour protection of wind turbine foundations in similar regions.

Keywords: current characteristics; sediment transport; local scour; collar protection

1. Introduction

The Bohai Sea is primarily constituted of three bays, which are as follows: Laizhou Bay, Bohai Bay, and Liaodong Bay. Furthermore, it includes the Bohai Mid-Basin and the Bohai Strait. The Bohai Sea is a semi-enclosed continental shelf sea with an average depth of 18 m. Moreover, 95% of the area is less than 30 m deep, with the maximum depth reaching approximately 70 m. The Bohai Mid-basin exhibits relatively abundant wind resources, with an average annual wind speed exceeding 6.0 m·s^{-1} [1]. The area is distinguished by a relatively flat topography, with water depths ranging from 15 to 30 m. Xu et al. determined that the majority of areas in the Bohai Sea susceptible to storm surges are situated in waters less than 10 m deep along the coastline [2]. Therefore, the Bohai Mid-basin, which is endowed with abundant wind resources, favorable water depths, and optimal topographical conditions, coupled with a relatively limited impact from storm surges, has emerged as an optimal locale for wind power development.

As offshore wind power continues to develop, research has demonstrated that monopile foundations are subjected to prolonged exposure to oceanic loads, including waves and currents. Such exposure results in the formation of a complex vortex flow field surrounding the foundations, comprising downward currents, horseshoe vortices, and wake vortices. Gazi et al. demonstrated that the formation of intense vortices around the piles can displace sediments around the foundations, leading to a gradual decline in bed elevation and the erosion of sediments [3]. This not only presents a potential risk to the stability of the wind turbine foundations, but also results in alterations to the topography and geomorphology of the original marine ecosystem. The issue of scouring around pile foundations is a highly complex one, warranting further investigation. The studies conducted by Breusers et al. and Ansari et al. have demonstrated that the depth of scouring is closely correlated with a number of factors, including water depth, sediment grain size, pile diameter, and Froude number [4,5]. Oliveto et al. indicated that when the water depth is substantial, surface vortices and horseshoe vortices do not interact with one another, and the local scour depth is primarily contingent on the pile diameter [6]. However, in situations where the water depth is relatively shallow, the influence of surface vortices is predominant, which in turn reduces the vortex-suction capability of horseshoe vortices, and consequently the local scour depth. Chiew et al. observed that when the ratio of pile diameter (D) to the median grain size (d50) is less than 50, the local scour is influenced by the sediment grain size [7]. Raudkivi et al. put forth the proposition that pile diameter represents a principal factor influencing scour depth, with scour depth increasing approximately linearly with pile diameter [8]. Jain observed that when the Froude number exceeds the critical Froude number, the scour depth initially declines and then increases with rising Froude number. The scour process surrounding pile foundations is influenced by the Froude number [9].

In examining the critical issues surrounding pile foundations, it is imperative to consider not only the immediate environmental impacts surrounding the piles themselves, but also the far-reaching effects of large-scale ocean currents and sediment on the scour process around pile groups at wind farms. The strength of ocean currents and tidal changes in the vicinity of the wind farm installation can have a significant impact on the extent of scour around pile foundations. Prior research has elucidated the tidal flow characteristics of the central Bohai Sea. Jiang et al. observed that, with the exception of the Bohai Strait, the ocean currents in the Bohai Sea are relatively weak, with the maximum surface velocity not exceeding 1.5 $m{\cdot}s^{-1}$ [10]. Moreover, the velocity of the flow in the central Bohai Sea is even lower. Bian et al. proposed that astronomical tides act as the primary driving force in the Bohai Sea, facilitating all mixing processes and the transportation of dissolved substances and suspended particles through residual currents [11]. In their study of sediment movement in the central Bohai Sea, Qiao et al. found that the distribution of suspended sediment concentration in the central Bohai Sea is not significantly correlated with the river discharge during the flood and dry seasons of the Yellow River, nor with the temperature and salinity distribution in winter and summer [12]. This is due to the fact that approximately 70–90% of the sediment discharged by the Yellow River is deposited in the vicinity of the river mouth, within a distance of 30 km from the coastline [13], while the remaining sediment is transported to the Yellow Sea via the Bohai Strait [14]. Liu et al. observed that the sediment from the Yellow River mouth predominantly migrates along two pathways: the western side of the shallow bank in the central Bohai Sea and along the Bohai Strait toward the Yellow Sea [15]. The winter season is distinguished by the prevalence of strong winds and waves, which contribute to the resuspension of sediment. Therefore, the winter season is the primary period for the transport of sediment from the Yellow River. Qiao et al. observed that the sedimentation rate in the central Bohai Sea and along the Shandong Peninsula is relatively low, with a range of 1.0 to 5.0 mm per year [16]. Notwithstanding the weak ocean currents in the central Bohai Sea, the sediment input from the Yellow River continues to exert a considerable influence on the stability of pile foundations. Therefore, a comprehensive investigation into the tidal flow characteristics, sediment transport, and deposition patterns in this region is essential for the long-term assessment of wind farm pile foundations.

The majority of hydrological data collected in the Bohai Sea is concentrated in the three bays and nearshore areas. However, there is a dearth of long-term observational data available for the central Bohai Sea, and the existing relevant research remains insufficient. In particular, the issue of scour around wind turbine pile foundations in the central Bohai Sea has not been adequately addressed or studied. This research employs numerical simulations of oceanographic environments and local scour analysis to examine the background of ocean currents, sediment movement characteristics, sediment sources, and the local scour and protection features of pile foundations in the central Bohai Sea. This study is structured into three principal sections. The initial two sections focus on the examination of the prevailing large-scale ocean current characteristics and the background of sediment movement. The third section addresses the local scour and protection features.

2. Ocean Current Characteristics of the Study Region

2.1. Governing Equations

The FVCOM was initially developed by Chen et al. [17], and subsequently underwent updates and optimizations at the University of Massachusetts and the Woods Hole Oceanographic Institution [18]. The software incorporates both non-hydrodynamic and hydrodynamic dynamics and utilizes unstructured grids, which are well suited for accurately representing complex coastlines, such as those with archipelagos [19]. The finite volume algorithm combines the flexibility of finite element geometric fitting with the computational efficiency of finite difference methods, rendering it an appropriate tool for investigating nearshore ocean processes, including tides, circulation, storm surges, waves, and sediment transport [20]. FVCOM can solve for ocean currents, temperature, salinity, and density based on the continuity, momentum, temperature, salinity, and density equations. The relevant equations are as follows:

(1) Continuity equation:

$$\frac{\partial \zeta}{\partial t}+\frac{\partial Du}{\partial x}+\frac{\partial Dv}{\partial y}+\frac{\partial w}{\partial \sigma}=0 \tag{1}$$

(2) Momentum equation:

$$\begin{aligned}&\frac{\partial uD}{\partial t}+\frac{\partial u^2D}{\partial x}+\frac{\partial uvD}{\partial y}+\frac{\partial u\omega}{\partial \sigma}-fvD=-gD\frac{\partial \zeta}{\partial x}-\frac{D}{\rho_0}\frac{\partial P_a}{\partial x}-\frac{1}{\rho_0}\left(\frac{\partial qD}{\partial x}+\frac{\partial qA_1}{\partial \sigma}\right)\\&-\frac{gD}{\rho_0}\left[\int_\sigma^0 D\left(\frac{\partial \rho}{\partial x}+\frac{\partial \rho}{\partial \sigma'}\frac{\partial \sigma'}{\partial x}\right)d\sigma'\right]+\frac{\partial}{\partial \sigma}\left(\frac{K_m}{D}\frac{\partial u}{\partial \sigma}\right)+DF_x\end{aligned} \tag{2}$$

$$\begin{aligned}&\frac{\partial vD}{\partial t}+\frac{\partial uvD}{\partial x}+\frac{\partial v^2D}{\partial y}+\frac{\partial v\omega}{\partial \sigma}+fuD=-gD\frac{\partial \zeta}{\partial y}-\frac{D}{\rho_0}\frac{\partial P_a}{\partial y}-\frac{1}{\rho_0}\left(\frac{\partial qD}{\partial y}+\frac{\partial qA_2}{\partial \sigma}\right)\\&-\frac{gD}{\rho_0}\left[\int_\sigma^0 D\left(\frac{\partial \rho}{\partial y}+\frac{\partial \rho}{\partial \sigma'}\frac{\partial \sigma'}{\partial y}\right)d\sigma'\right]+\frac{\partial}{\partial \sigma}\left(\frac{K_m}{D}\frac{\partial v}{\partial \sigma}\right)+DF_y\end{aligned} \tag{3}$$

(3) Temperature equation:

$$\frac{\partial TD}{\partial t}+\frac{\partial TuD}{\partial x}+\frac{\partial TvD}{\partial y}+\frac{\partial T\omega}{\partial \sigma}=\frac{1}{D}\frac{\partial}{\partial \sigma}\left(K_h\frac{\partial T}{\partial \sigma}\right)+DT_0+DF_T \tag{4}$$

(4) Salinity equation:

$$\frac{\partial SD}{\partial t} + \frac{\partial SuD}{\partial x} + \frac{\partial SvD}{\partial y} + \frac{\partial S\omega}{\partial \sigma} = \frac{1}{D}\frac{\partial}{\partial \sigma}\left(K_h \frac{\partial S}{\partial \sigma}\right) + DF_S \quad (5)$$

(5) Density equation:

$$\rho = \rho(T, S, p) \quad (6)$$

In the above equations, u, v, and ω represent the velocity components in the three-dimensional σ-coordinate system (x, y, σ). D is the total water depth, t is time, q is the non-hydrostatic pressure, g is the gravitational acceleration at the Earth's surface, p is the water pressure, P_a is the atmospheric pressure at the sea surface, F_x and F_y are the horizontal diffusion terms, K_m and K_h are the turbulent viscosity coefficient and the thermal turbulent diffusion coefficient in the vertical direction, respectively, T is the temperature, S is the salinity, T_0 is the initial seawater temperature, ρ_0 is the reference density, ρ is the seawater density, f is the Coriolis parameter, and F_T and F_S are the temperature and salinity diffusion terms in the three directions, respectively.

2.2. Model Setup and Validation

The study area for the model is situated between 37° and 40.0° N and 117.5° and 120° E, with the specific computational grid for the area illustrated in Figure 1. The locations of the measured tidal and tidal current data verification stations are illustrated in Figure 2a, while Figure 2b depicts the depth distribution and components of the Bohai Sea. The tidal verification at stations XD1 and XD2 was conducted over the course of 63 days, from 08:00 on 10 June 2023 to 08:00 on 11 July 2023. The tidal current verification at stations C1–C9 was conducted over the course of 48 h, from 16:00 on 12 April 2023 to 18:00 on 13 April 2023. The results of the tidal verification are presented in Figure 3, while the tidal current verification is illustrated in Figure 4. The model data demonstrate a high degree of correlation with the observed data.

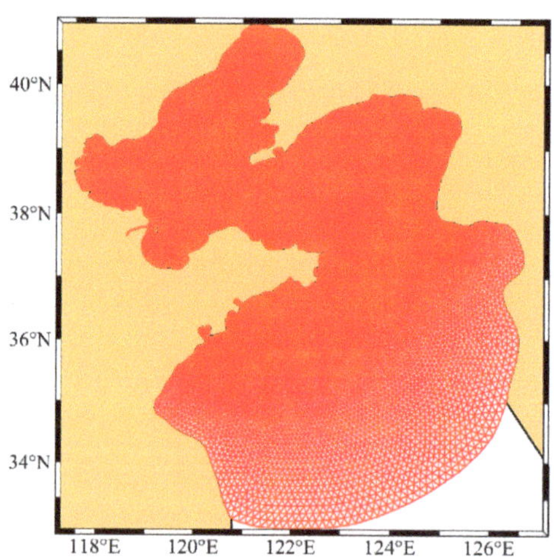

Figure 1. Schematic of the FVCOM model computational domain grid.

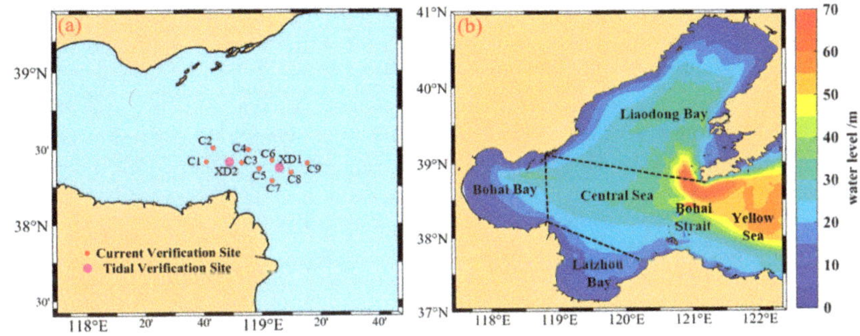

Figure 2. (a) Verification points of the study area; (b) Depth distribution of the study area.

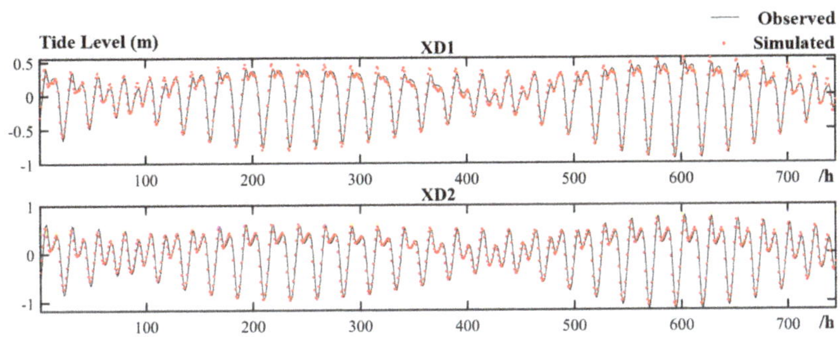

Figure 3. Tidal validation chart at XD1 and XD2 stations.

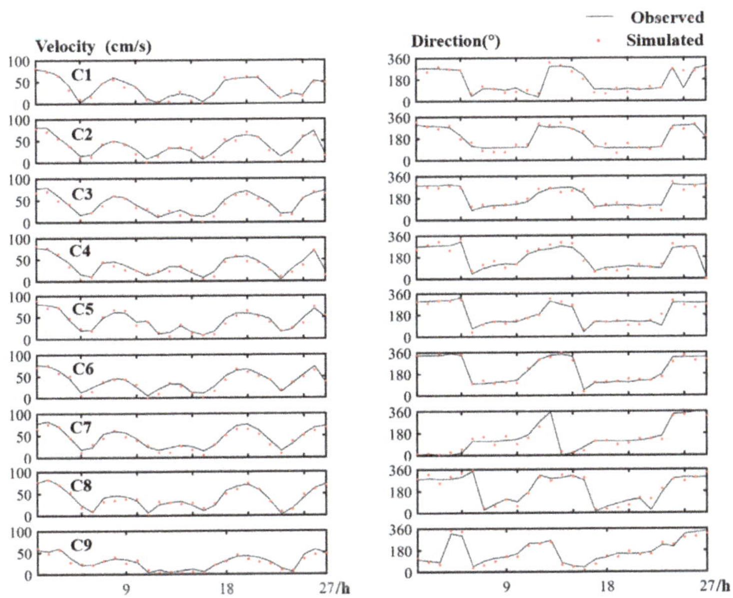

Figure 4. Velocity and direction validation at C1–C9 stations.

2.3. Current Field Characteristics of the Study Area

In order to analyze the flow field characteristics of the study area, to provide the tidal current background for the sediment erosion and deposition trends, and to offer data reference for the development and utilization of marine resources, as well as for the engineering design of both fixed and floating wind turbines in the area, this study processes and analyzes the simulated flow field data from 2004 to 2023.

Figures 5 and 6 illustrate the monthly averaged surface flood and ebb tidal current fields in the study area from 2004 to 2023. The locations of nine wind turbine foundation sites (S1–S9) are indicated on the maps. These are situated approximately 30 km north of Dongying Port, in a relatively flat area with a water depth of approximately 20 m, at a considerable distance from the area of sediment deposition in the vicinity of the Yellow River mouth. The analysis demonstrates that, with the exception of the regions proximate to the former Yellow River estuary and Dongying Port, where the monthly averaged flood tide current speed is typically between 0.7 m·s^{-1} and 1.0 m·s^{-1} due to the influence of geographical conditions, the monthly averaged flood and ebb tidal current speed in the majority of the study area is below 0.6 m·s^{-1}. The mean monthly surface flood and ebb tidal current speed at the wind turbine monopile foundation sites is approximately 0.4 m·s^{-1}. The monthly averaged flood tide current speed at the mouths of Liaodong Bay and Laizhou Bay is typically less than 0.3 m·s^{-1}, while the monthly averaged flood and ebb tidal current speed in the Bohai Bay mouth and in the vicinity of the Bohai Strait remains approximately 0.5 m·s^{-1}. The latitudinal differences are inferred to be primarily attributable to the wider mouths of Laizhou Bay and Liaodong Bay, where the direction of the bay mouth is almost perpendicular to the dominant flood and ebb tidal current. In contrast, the mouth of Bohai Bay is aligned with the direction of the dominant flood and ebb tidal current, resulting in a more concentrated hydrodynamic effect. In particular, in the western region of Bohai Bay, in the vicinity of the Bohai Strait, the streamlines converge, indicating an increase in current speed from east to west.

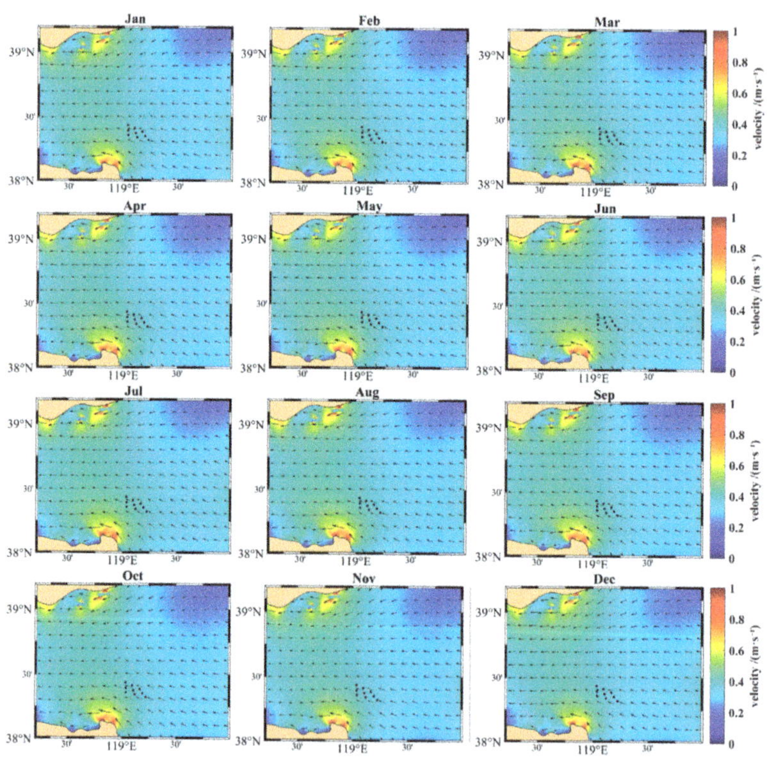

Figure 5. Monthly average surface flood tidal current field (2004–2023).

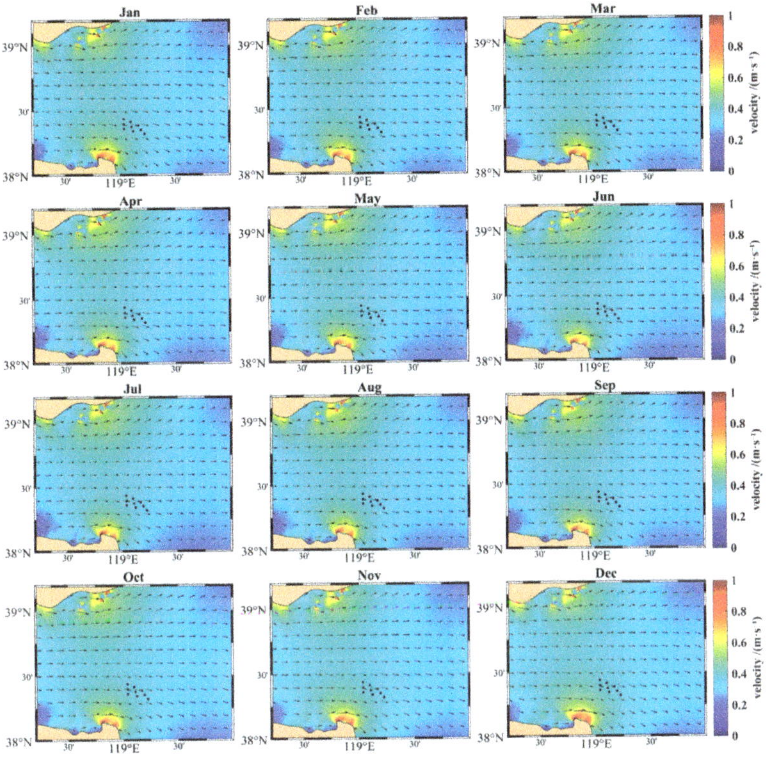

Figure 6. Monthly average surface ebb tidal current field (2004–2023).

The analysis demonstrates that during the flood tide, as the ocean current progresses towards the inner bays of the Bohai Three Bays, the flow velocity demonstrates a gradual increase due to the accumulation of water and localized changes in water depth. The tidal water advances rapidly inland, resulting in an increase in current speed. In contrast to the flood tide, the ebb tide is characterized by the reversal of the ocean current, with water flowing back from regions such as Bohai Bay, Laizhou Bay, and Liaodong Bay to the central Bohai Sea. This current eventually flows out through the Bohai Strait into the northern Yellow Sea. As a result of the reduction in water level and the absence of the accumulation effect observed during the flood tide, in addition to the considerable resistance exerted by the shoreline and the effect of water friction, the speed of the ebb tide is typically lower than that of the flood tide.

3. Sediment Transport in the Study Region

3.1. Sediment Transport Model

The FVCOM sediment module uses a concentration-based approach. The governing equation for suspended sediment transport is as follows:

$$\frac{\partial C_i}{\partial t} + \frac{\partial u C_i}{\partial x} + \frac{\partial v C_i}{\partial y} + \frac{\partial (w - w_i) C_i}{\partial \sigma} = \frac{\partial}{\partial x}\left(A_h \frac{\partial C_i}{\partial x}\right) + \frac{\partial}{\partial y}\left(A_h \frac{\partial C_i}{\partial y}\right) + \frac{\partial}{\partial z}\left(K_h \frac{\partial C_i}{\partial \sigma}\right) \quad (7)$$

In the above equation, the subscript i represents the i-th sediment group component, as specified by the user, C_i is the sediment concentration, w_i is the sediment settling velocity, and w represents the actual vertical velocity of sediment particles in the water. The user inputs parameters for each sediment group, including settling velocity, median particle diameter, density, and critical erosion stress. A_h represents the horizontal turbulent diffusion coefficient for sediment, and K_h is the vertical turbulent diffusion coefficient, which is calculated using the turbulence model.

At the surface, a no-flux boundary condition is applied for sediment concentration:

$$K_h \frac{\partial C_i}{\partial \sigma} = 0,\ \sigma = \zeta \tag{8}$$

At the bottom, the sediment flux is the difference between deposition and erosion:

$$K_h \frac{\partial C_i}{\partial \sigma} = E_i - D_i,\ \sigma = -H \tag{9}$$

The erosion rate E_i is calculated as follows:

$$E_i = \Delta t Q_i (1 - P_b) F_{bi} \left(\frac{\tau_b}{\tau_{ci}} - 1 \right) \tag{10}$$

In the above equation, H is the bottom depth, ζ is the height of the free surface, D_i represents the deposition rate, Q_i is the erosion flux, P_b is the bottom porosity, F_{bi} is the proportion of the sediment component, τ_b is the bottom shear stress, and τ_{ci} is the critical stress for sediment.

The settling of fine viscous sediment particles is represented by the settling probability P and settling flux WC as follows:

$$D = PWC \tag{11}$$

In the above equation, P is the settling probability, W is the settling velocity, and C is the sediment concentration.

The settling probability P is given by the formula proposed by Krone [21]:

$$P = 1 - \frac{\tau_b}{\tau_d}, \tau_b \leq \tau_d \tag{12}$$

$$P = 0, \tau_b > \tau_d \tag{13}$$

In the above equation, τ_d is the critical shear stress, and τ_b is the bottom shear stress.

In this model, the sediment transport resulting from the combined effect of waves and currents is calculated using the formula developed by Camenen and Larson [22]:

$$q_{bcw,i} = a_n \eta_i \rho_{s,i} \sqrt{(s_i - 1) g d_{50,i}^3} \sqrt{\theta_{cur}} \theta_{cw} \exp\left(-b_n \frac{\theta_{cr,i}}{\theta_{cw,max}} \right) \tag{14}$$

$$\tau_{cw} = \tau_c \left[1 + 1.2 \left(\frac{\tau_w}{\tau_c + \tau_w} \right)^{3.2} \right] \tag{15}$$

$$\tau_{cw,max} = \left[(\tau_{cw} + \tau_w \cos\varphi_{cw})^2 + (\tau_w \sin\varphi_{cw})^2 \right]^{0.5} \tag{16}$$

In this formula, a_n and b_n are empirical coefficients, with values of 0.03 and 3.0, respectively, η_i represents the concentration of the i-th sediment class, s_i denotes the specific gravity of the i-th sediment particles, and θ_{cur} is the Shields parameter under the influence of current alone; θ_{cw} and $\theta_{cw,max}$ represent the time-averaged and maximum Shields parameters under the combined effect of waves and currents, which can be calculated using Equations (15) and (16).

In the FVCOM model, due to sediment deposition and erosion, the thickness of each layer of the bed changes. At each time step, the active layer thickness z is first calculated.

The active layer thickness can be defined by the user or calculated according to the formula proposed by Harris and Wiberg [23]:

$$z = max[k_1(\tau_s - \overline{\tau_c})\rho_0, 0] + k_2 d_{50} \quad (17)$$

In this equation, τ_s is the bed shear stress, $\overline{\tau_c}$ is the average critical shear stress, and d_{50} is the median particle diameter, while k_1 = 0.007 and k_2 = 6 are empirical coefficients.

3.2. Suspended Sediment Concentration Validation

This study employs the FVCOM-SED sediment transport module, which, in conjunction with hydrodynamics, wave dynamics, and the coupling of sediment and topography, enables the comprehensive simulation of suspended sediment transport and bedload transport processes. The sediment model parameters are presented in Table 1.

Table 1. Sediment model parameter settings table.

Model Parameters	Model Settings			
Time Step (s)	1			
Suspended Load, Bedload Calculation	Both parameters are activated			
SD50 (mm)	0.005	0.010	0.015	0.025
Sediment Density (kg·m^{-3})	2650	2650	2650	2650
Sediment Settling Velocity (mm·s^{-1})	0.24	0.36	0.48	0.60
Erosion Rate (kg·m^{-2}·s^{-1})	2.0 × 10^{-4}	2.0 × 10^{-4}	2.0 × 10^{-4}	2.0 × 10^{-4}
Critical Shear Stress for Scour (N·m^{-2})	0.16	0.25	0.30	0.40
Critical Shear Stress for Deposition (N·m^{-2})	0.15	0.20	0.25	0.35
Bed Layer Thickness (m)	1.00	1.00	1.00	1.00
Porosity	0.50	0.50	0.50	0.50

In order to ascertain the veracity of the sediment model, Figure 7 illustrates the validation of the suspended sediment concentration (SSC, $SSC = \sum_i C_i$) at stations C1 to C9. A comparison between the model data and the measured data indicates a high degree of correlation.

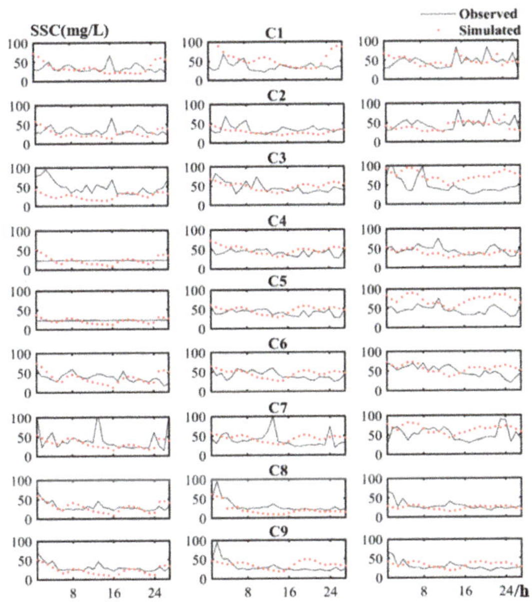

Figure 7. Validation of SSC (surface, middle, bottom layers).

3.3. Suspended Sediment Distribution

The sediments in the Bohai Sea are primarily derived from riverine inputs, with approximately 75% originating from terrestrial sources and the remaining 25% resulting from atmospheric deposition, coastal erosion, and resuspension within the marine environment [16]. The Bohai Sea is fed by over 40 rivers, with a total annual water volume of 68.5×10^9 m^3 and a suspended sediment load of 1.1×10^9 t. The Yellow River is the largest contributor, with an annual flow of 4.2×10^{10} m^3 and an annual sediment load of 1.0×10^9 t, accounting for more than 50% of the total freshwater inflow and over 90% of the total sediment load. It is noteworthy that there are significant seasonal variations in the sediment and water discharge from the Yellow River. In addition, the Luan River, Liao River, and Hai River also contribute a significant quantity of sediment to the Bohai Sea [1]. This section presents an overview of the sources of sediment in the study area. The initial source is the Yellow River estuary, where the Yellow River has transported considerable quantities of sediment into the sea over time, resulting in the continuous deposition and formation of the expansive Yellow River Delta. The second source comprises several rivers along the coastline in the vicinity of Tangshan, including the Luan River and Su River, which contribute a considerable amount of sediment. Upon entering the sea, the water flow from these rivers experiences a notable reduction in velocity, which results in the gradual deposition of sediment and other materials on the seabed. This study demonstrates that when ocean currents flow into the northern Tangshan offshore island group in Bohai Bay and the southern Yellow River Delta area, the flow velocity increases. The two regions have developed into discrete areas of elevated suspended sediment concentration as a result of the prolonged deposition of considerable quantities of sediment transported by the Yellow River, the Luan River, and the Su River.

Figure 8 illustrates the distribution of SSC during the flood and ebb tide periods for the surface, middle, and bottom layers at the wind turbine foundation sites in the study area in 2023. In consideration of these observations, it can be postulated that the suspended sediment concentration in the wind farm area is predominantly influenced by the elevated suspended sediment concentration region of the Yellow River Delta.

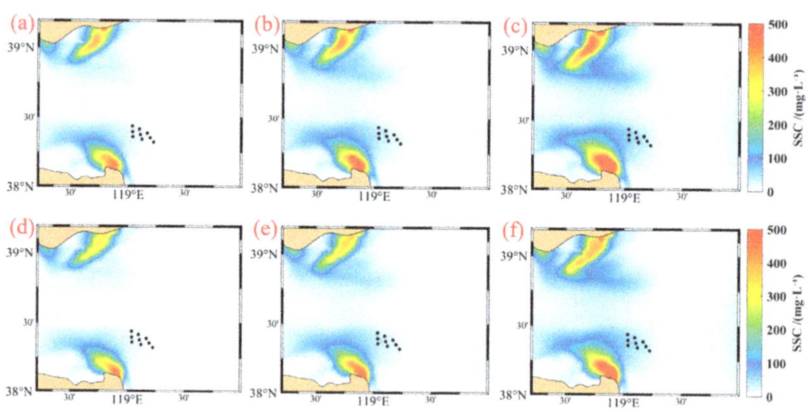

Figure 8. Annual average SSC distribution in the 2023 simulation. ((**a–c**): flood tide, (**d–f**): ebb tide, surface, middle, bottom layers).

3.4. Suspended Sediment Transport

The annual sediment discharge in the study area is substantial, and the sediment conditions are complex. The interaction of coastal, tidal, and ocean currents gives rise to the transportation, deposition, and redistribution of sediment on the seabed, which has a significant impact on the research and subsequent maintenance of monopile foundations [16]. In accordance with the preceding analysis, the study area is subject to higher ocean current velocities during the summer months. Figure 9 illustrates the fluctuations in SSC throughout the flood and ebb tide phases in August 2023.

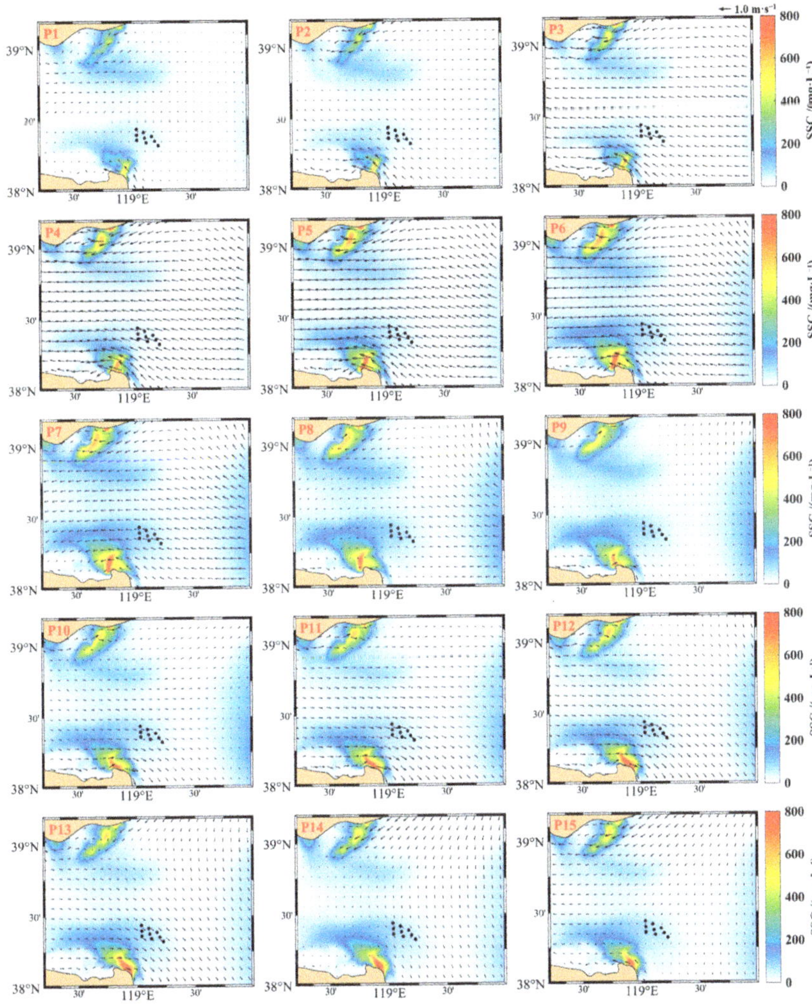

Figure 9. SSC sequence during the spring tide in August 2023 simulation.

As illustrated in the diagram, during the periods of flood and ebb tides, the ocean current in the vicinity of the shorelines at Dongying Port and Yellow River Port converges, with flow speeds exceeding 1.0 $m \cdot s^{-1}$. This results in local scouring and a notable increase in the SSC in the northern Dongying Port region. The direction of the flood and ebb tides in the eastern Dongying Port area is from south-west to west-north-west. During the period of flood tide, the center of the SSC is situated on the north-western side of the northern Dongying Port area, with concentrations reaching up to 800 mg/L, which then spread in a north-westerly direction to reach approximately 400 mg/L. During the period of ebb tide, the center of the SSC shifts in a southeasterly direction in accordance with the tidal flow, and the concentration typically decreases to below 400 mg/L. In comparison to the flood and ebb tidal currents directions, the SSC at the wind farm sites is more susceptible to fluctuations in flow speed. This is likely attributable to the fact that the primary source of suspended sediment is the movement of sediment caused by flow velocity, rather than direct input from the high-concentration SSC centers.

3.5. Seabed Stability Analysis

Figure 10 illustrates the measured annual erosion and sedimentation intensity in the vicinity of the wind farm from 1991 to 2022. Figure 11 depicts the simulated annual erosion and sedimentation intensity for the study area. The annual erosion and sedimentation variation map indicates that, with the exception of the northern offshore area of Dongying Port, where the annual erosion depth reaches 0.2 m, and the northeastern side of Dongying Port, where the annual sedimentation depth reaches 0.3 m, the majority of the study region experiences seabed annual erosion and sedimentation changes within 0.02 m. This corroborates the hypothesis that the annual erosion and sedimentation variation at the wind farm sites is negligible, within 0.01 m. During the ebb and flood tides, the ocean current along the shorelines of Dongying Port and Yellow River Port converges, with flow speeds exceeding 1.0 m$\cdot$s^{-1}, resulting in intense scouring and a notable increase in the suspended sediment concentration in the northern Dongying Port area.

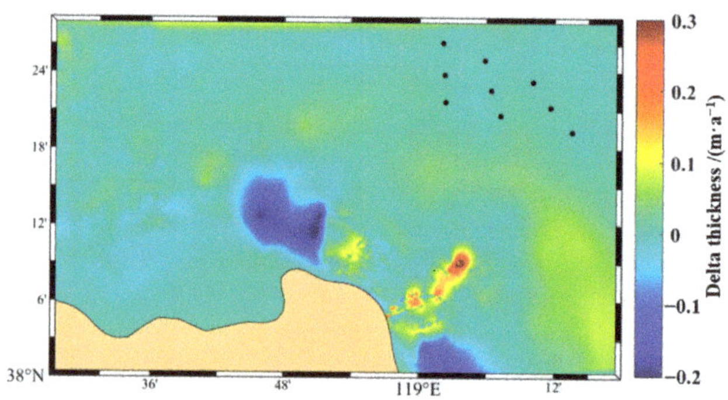

Figure 10. Measured annual erosion and deposition changes.

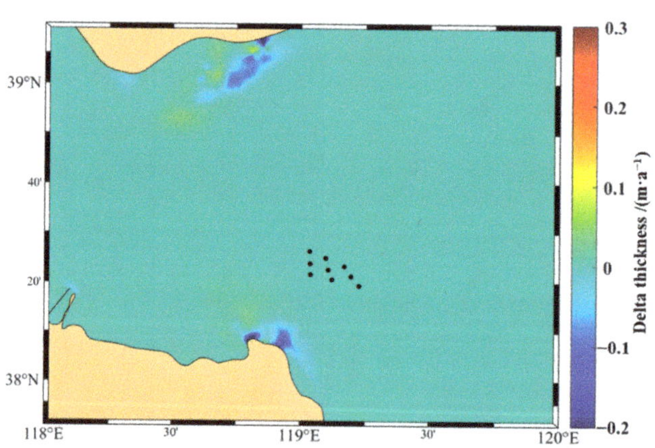

Figure 11. Simulated annual erosion and deposition changes.

Table 2 presents the results of the measured and simulated annual scour and deposition changes at the wind turbine foundation sites, designated S1 to S9.

Table 2. Annual erosion and deposition changes (m).

Site	Measured	Simulated
S1	0.0016	0.0027
S2	0.0021	0.0027
S3	0.0054	0.0540
S4	0.0035	0.0035
S5	0.0024	0.0032
S6	0.0078	0.0063
S7	0.0085	0.0103
S8	0.0088	0.0087
S9	0.0022	0.0048

4. Local Scour Around Wind Turbine Foundation Piles

4.1. Governing Equations

FLOW-3D (Version 11.2) is three-dimensional fluid dynamics simulation software based on the finite volume method. It provides flexibility and practicality for porous media flow and a simple setup for complex structures [24]. It has powerful multi-scale, multi-physics coupling capabilities and accurately and efficiently solves coupled problems [25].

The governing equations include the continuity equation and the RANS equation. The continuity equation is expressed as follows:

$$\frac{\partial \rho}{\partial t}+\frac{\partial(\rho u)}{\partial x}+\frac{\partial(\rho v)}{\partial y}+\frac{\partial(\rho w)}{\partial z}=0 \tag{18}$$

The RANS equation is expressed as follows:

$$\frac{\partial(\rho u)}{\partial t}+div(\rho u\overline{u})=div(\mu \mathrm{grad}(u))-\frac{\partial p}{\partial x}+F_u \tag{19}$$

$$\frac{\partial(\rho v)}{\partial t}+div(\rho v\overline{u})=div(\mu \mathrm{grad}(v))-\frac{\partial p}{\partial y}+F_v \tag{20}$$

$$\frac{\partial(\rho w)}{\partial t}+div(\rho w\overline{u})=div(\mu \mathrm{grad}(w))-\frac{\partial p}{\partial z}+F_w \tag{21}$$

$$div(\overline{u})=\frac{\partial u}{\partial x}+\frac{\partial v}{\partial y}+\frac{\partial w}{\partial z} \tag{22}$$

$$grad()=\frac{\partial()}{\partial x}+\frac{\partial()}{\partial y}+\frac{\partial()}{\partial z} \tag{23}$$

In the equation, ρ is the fluid density, t is time, and u, v, and w are the velocity components in the x, y, and z directions, respectively. p is pressure, F_u, F_v, and F_w are the components of the external forces in the x, y, and z directions, $\overline{u}$ is the fluid velocity vector, and μ is the dynamic viscosity coefficient.

The FLOW-3D methodology for determining sediment entrainment involves the evaluation of critical shear stress. The Shields parameter is calculated using the Soulsby–Whitehouse formula, which is expressed as follows:

$$\theta_{cr}=\frac{0.3}{1+1.2d_*}+0.055[1-\exp(-0.02d_*)] \tag{24}$$

$$d_* = d_{50}\left[\frac{\rho(\rho_s-\rho)\|g\|}{\mu^2}\right]^{1/3} \tag{25}$$

In light of the impact of the actual bed slope on sediment entrainment, the following modification is made to Equation (24):

$$\theta'_{cr} = \theta_{cr} \frac{\left(\tan^2\varphi\cos^2\beta - \sin^2\psi\sin^2\beta\right)^{0.5} + \cos\psi \cdot \sin\beta}{\tan\varphi} \tag{26}$$

In the equation, d_* is the dimensionless particle diameter parameter; d_{50} is the median particle diameter of the sediment; ρ is the fluid density; ρ_s is the sediment density; μ is the dynamic viscosity coefficient; g is the acceleration due to gravity; φ is the underwater repose angle of the sediment, typically assumed to be 32°; β is the bed slope; and ψ is the angle between the flow direction and the upslope direction of the bed.

When the local Shields parameter θ_i exceeds the critical Shields parameter θ'_{cr}, sediment particles will be entrained. The equation for calculating the local Shields parameter θ_i is expressed as follows:

$$\theta_i = \frac{\tau}{\parallel g \parallel d_i(\rho_i - \rho)} \tag{27}$$

In the equation, τ is the bed shear stress; d_i is the particle diameter of the i-th sediment class; and ρ_i is the density of the i-th sediment particle.

4.2. Turbulence Model

The presence of pile foundations has been observed to increase turbulence around the piles, generating turbulence that is characterized by high-frequency oscillations and vortices of varying energy levels. In their analysis of horseshoe vortex turbulence, Dey et al. observed that as the scour hole develops, the core area of vertical velocity increases, accompanied by a rise in both the core turbulence intensity of the horseshoe vortex and the Reynolds stress [26]. Turbulence gives rise to localized high shear forces at the bed, which drive sediment motion and increase the suspended load of sediment. Zhang et al. demonstrated that the RNG k-ε turbulence model is highly accurate and can significantly reduce computation time [27]. Omara et al. asserted that the accurate calculation of near-wall shear stress is essential for predicting sediment scour [28]. They demonstrated the effectiveness of the RNG k-ε model in estimating maximum scour depth. Jalal et al. compared the RNG k-ε model with experimental data, finding that the error rate for the maximum scour hole depth was 10% [29]. Nazari et al. employed the RNG k-ε model to solve the turbulence field, noting that large eddy simulation might provide superior results [30], but at the cost of longer computation times [31]. Nielsen et al. employed the RNG k-ε vortex model in FLOW-3D and observed that the simulated fluid velocity distribution, shear stress distribution, and scour protection layer settlement were in good agreement [24]. The RNG model demonstrates efficacy in scour simulations, particularly in scenarios where turbulence is generated by fluid flow through control structures.

The RNG k-ε model was derived by Yakhot et al. through the application of the Gauss statistical method to the unsteady Navier–Stokes equations in the equilibrium state, resulting in the formulation of new turbulence equations [32,33]. The model takes into account the non-uniformity and anisotropy of the flow in terms of turbulence kinetic energy and its dissipation rate, thereby enhancing its adaptability to different flow conditions.

The equations for turbulence kinetic energy and its dissipation rate are as follows:

$$\frac{\partial k_T}{\partial t} + \frac{1}{V_F} F_{x,y,z} = P_T + G_T + DIff_{k_T} - \varepsilon_T \tag{28}$$

$$\frac{\partial k_T}{\partial t} + \frac{1}{V_F} F_{x,y,z} = \frac{1.42 \cdot \varepsilon_T}{k_T}(P_T + 0.2 \cdot G_T) + DIff_\varepsilon - 1.68\frac{\varepsilon_T^2}{k_T} \tag{29}$$

$$F_{x,y,z} = uA_x\frac{\partial k_T}{\partial x} + vA_y\frac{\partial k_T}{\partial y} + wA_z\frac{\partial k_T}{\partial z} \tag{30}$$

$$k_T = \frac{1}{2}\left(\overline{u}'^2 + \overline{v}'^2 + \overline{w}'^2\right) \tag{31}$$

$$P_T = \left(\frac{\mu}{\rho V_F}\right)\begin{cases} 2A_x\left(\frac{\partial u}{\partial y}\right)^2 + 2A_y\left(\frac{\partial v}{\partial y} + \xi\frac{u}{x}\right)^2 + 2A_z\left(\frac{\partial w}{\partial z}\right)^2 \\ +\left(\frac{\partial v}{\partial x} + \frac{\partial u}{\partial y} - \xi\frac{v}{x}\right)\left[A_z\frac{\partial v}{\partial x} + A_y\left(\frac{\partial u}{\partial y} - \xi\frac{v}{x}\right)\right] \\ +\left(\frac{\partial u}{\partial z} + \frac{\partial w}{\partial x}\right)\left(A_z\frac{\partial u}{\partial z} + A_x\frac{\partial w}{\partial x}\right) \\ +\left(\frac{\partial v}{\partial z} + R\frac{\partial w}{\partial y}\right)\left(A_z\frac{\partial v}{\partial z} + A_y\frac{\partial w}{\partial y}\right) \end{cases} \tag{32}$$

$$G_T = -CRHO\left(\frac{\mu}{\rho^3}\right)\left(\frac{\partial\rho}{\partial x}\frac{\partial p}{\partial x} + R^2\frac{\partial\rho}{\partial y}\frac{\partial p}{\partial y} + \frac{\partial\rho}{\partial z}\frac{\partial p}{\partial z}\right) \tag{33}$$

$$\begin{aligned} DIff_{k_T} &= \frac{1}{V_F}\left\{\frac{\partial}{\partial x}\left(\nu_k A_x\frac{\partial k_T}{\partial x}\right) + R\frac{\partial}{\partial y}\left(\nu_k A_y R\frac{\partial k_T}{\partial y}\right) + \frac{\partial}{\partial z}\left(\nu_k A_z\frac{\partial k_T}{\partial z}\right) + \xi\frac{\nu_k A_x k_T}{x}\right\} \\ DIff_\varepsilon &= \frac{1}{V_F}\left\{\frac{\partial}{\partial x}\left(\nu_\varepsilon A_x\frac{\partial \varepsilon_T}{\partial x}\right) + R\frac{\partial}{\partial y}\left(\nu_\varepsilon A_y R\frac{\partial \varepsilon_T}{\partial y}\right) + \frac{\partial}{\partial z}\left(\nu_\varepsilon A_z\frac{\partial \varepsilon_T}{\partial z}\right) + \xi\frac{\nu_\varepsilon A_x \varepsilon_T}{x}\right\} \end{aligned} \tag{34}$$

$$\nu_k = \rho C_\mu\frac{k_T^2}{\varepsilon_T} \tag{35}$$

$$\varepsilon_T = 0.085^{3/4}\frac{k_T^{3/2}}{TLEN} \tag{36}$$

In the equation, k_T is the turbulent kinetic energy, V_F is the fluid volume fraction, A_x, A_y, and A_z are the components of the fluid fraction in the x, y, and x directions, respectively, u, v, and w are the components of the velocity vector in the x, y, and x directions, respectively, P_T is the turbulent kinetic energy production term, ξ is the default value in the Cartesian coordinate system, which is 0, G_T is the buoyancy production term, $DIff_{k_T}$ is the turbulence diffusion term for k_T, $DIff_\varepsilon$ is the turbulence dissipation diffusion term, ν_ε is the turbulence viscosity coefficient, C_μ is the empirical constant, taken as 0.0845, ε_T is the turbulence kinetic energy dissipation term, and $TLEN$ is the turbulence length scale.

4.3. Physical Model Validation

This study employs the RANS equations, the Reynolds–Gaussian k-ε turbulence model, and the VOF algorithm to solve for three-dimensional incompressible viscous fluid flow.

Flow model validation is conducted based on the physical experiments of Roulund et al. on flow around vertical cylindrical pile foundations [34]. The findings of experimental studies conducted by Chiew et al. indicate that the impact of side walls can be disregarded when the width of the computational domain exceeds 8D [7]. The computational domain is defined with dimensions of 20.0 m by 10.0 m by 0.6 m, and the mesh size is set to 0.025 m. The water depth is 54.0 cm, the pile diameter is 10.0 cm, and the average flow velocity is 32.6 cm. Figure 12a illustrates the horizontal flow velocity around the pile, while Figure 12b depicts the vertical flow velocity around the pile, both of which demonstrate satisfactory validation results.

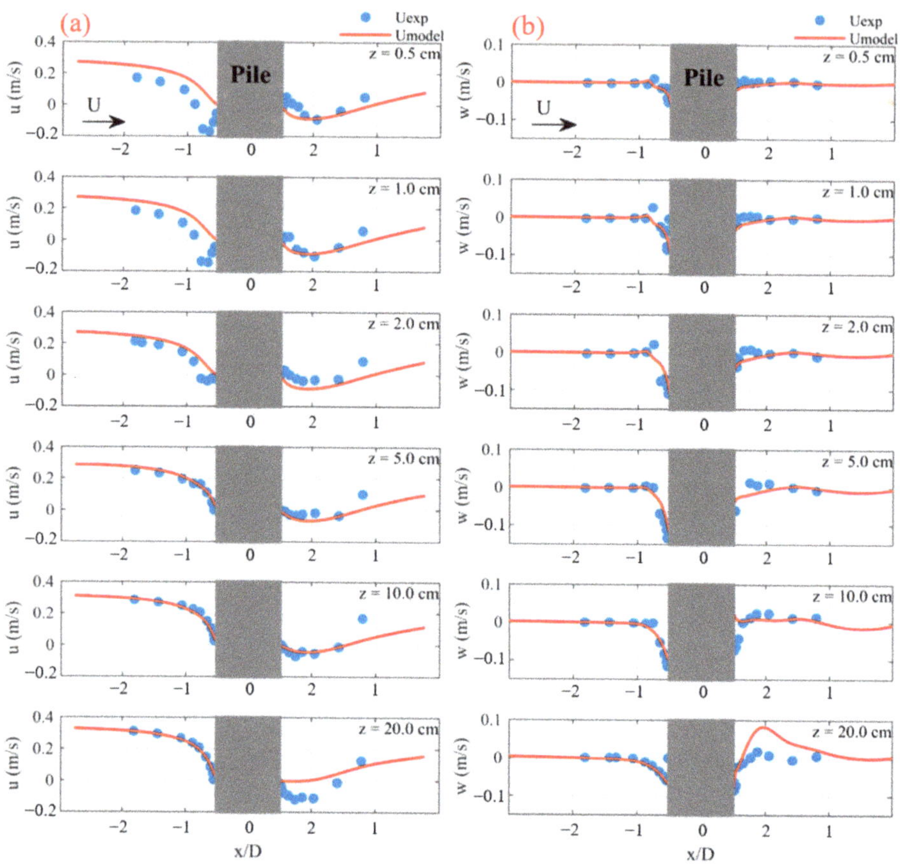

Figure 12. Velocity validation around the pile—(**a**) Comparison of horizontal velocity in the X direction; (**b**) Comparison of vertical velocity in the Z direction.

The physical experiment for validating local scour is conducted in a computational domain with grid dimensions of 8.0 m × 2.0 m × 0.5 m, with a minimum grid size of 0.5 cm. The water depth is 40.0 cm, the pile diameter is 10.0 cm, the average flow velocity is 28.6 cm·s^{-1}, the critical Shields number is 0.05, and the median grain size is 0.29 mm. The local scour depth verification in Figure 13 shows excellent agreement between the model and experimental data, indicating high precision and reliability.

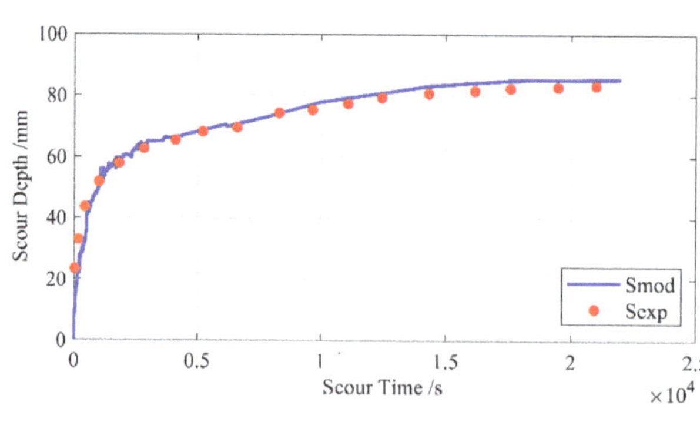

Figure 13. Cylindrical local scour model validation.

4.4. Wind Turbine Pile Foundation Model Parameter Setup

In this study, the RANS equations, RNG k-ε turbulence model, and VOF algorithm are utilized to solve for three-dimensional incompressible viscous flow. The wind turbine foundation has a pile diameter of 9.8 m, and the computational domain extends 150.0 m in width, where the effects of boundary walls are negligible. To ensure a stable flow velocity profile with a logarithmic distribution in front of the pile, the distance from the pile center to the velocity inlet is 600.0 m, and the distance from the pile center to the lateral boundaries is 140.1 m. The water depth at the wind farm site is 23.02 m, and the computational domain dimensions are 800 m × 150 m × 25 m. The inlet boundary on the left is set as a wave boundary condition, while the right side is designated as an outflow boundary. The lateral and bottom boundaries are wall boundary conditions, and the top boundary is a symmetry condition. The model boundary conditions are depicted in Figure 14, and the local scour model parameters are listed in Table 3.

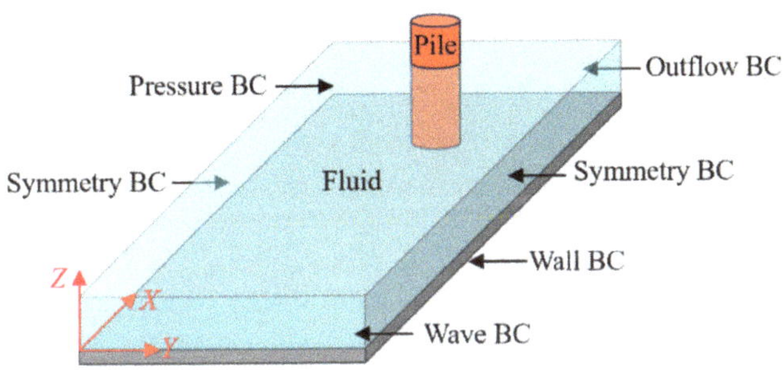

Figure 14. Schematic of model boundary conditions.

Table 3. Local scour model condition parameters.

Condition	Parameter
Water Depth (m)	23.02
Pile Diameter (m)	9.8
Average Flow Velocity (m·s^{-1})	1.13
Wave Period (s)	7.71
Wave Height (m)	4.28
Critical Shields Number	0.04
Bed Roughness Height (m)	0.0011
Median Particle Diameter d_{50} (mm)	0.0115

4.5. Monopile Surrounding Flow Field Characteristics

The vortex system in the vicinity of the wind turbine pile foundation represents the fundamental mechanism of local scour [4]. The application of LDV and PIV has enabled the in-depth and comprehensive study of the turbulent flow field around the pile foundation, with the identification of four specific types of turbulent flow: downward flow, horseshoe vortices, wake vortices, and contracted streamlines [35–37]. Figure 15 illustrates the local flow fields in the XY bottom layer, XZ, and YZ cross-sections of the pile foundation, elucidating the distribution of downward flow, horseshoe vortices, wake vortices, and contracted streamlines around the pile.

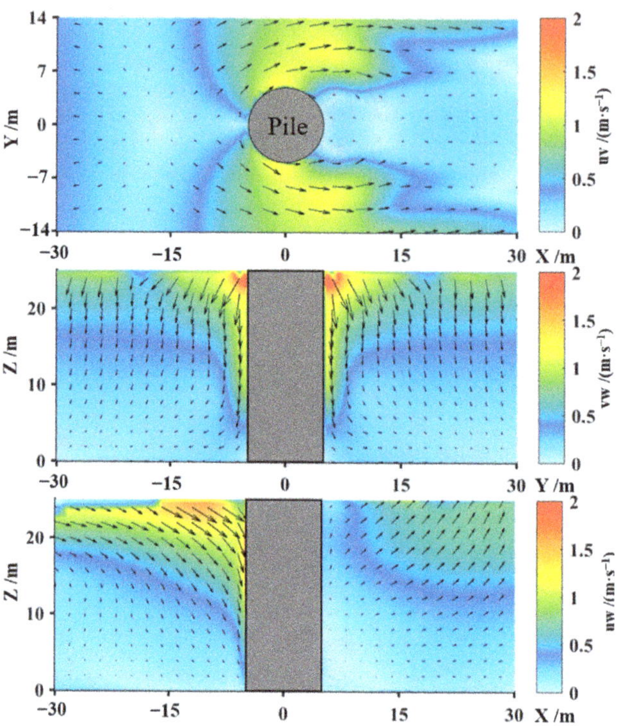

Figure 15. Monopile surrounding flow field in the XY, XZ, and YZ sections.

4.6. Local Scour and Protection Process Characteristics

The critical Shields parameter θ_{cr} is an important physical quantity related to sediment movement and flow velocity. Soulsby et al. proposed a method for calculating the critical Shields parameter θ_{cr}, and the formula is as follows [38]:

$$D_* = d_{50}\left[\left(\frac{\rho_s}{\rho} - 1\right)g/v^2\right]^{1/3} \tag{37}$$

$$\theta_{cr} = \frac{0.30}{1+1.2D_*} + 0.055[1 - \exp(-0.020D_*)] \tag{38}$$

$$\theta = \frac{\rho C_D V^2}{\rho g\left(\frac{\rho_s}{\rho} - 1\right)d_{50}} \tag{39}$$

$$C_D = \left\{\kappa/[\text{In}(z_{os}/h) + 1]\right\}^2 \tag{40}$$

In the equation, d_{50} is the median particle diameter of the sediment, ρ_s is the particle density of the sediment, ρ is the water density, v is the kinematic viscosity, and κ is the Karman constant, taken as 0.4. $z_{os} = d_{50}/12$. By substituting the parameters into the formula above, the critical Shields parameter $\theta_{cr} < \theta$ is obtained, indicating live-bed scour. The equilibrium scour depth is reached when the amount of sediment removed from the scour hole equals the amount entering the hole. This process occurs more quickly than the equilibrium under clear-water scour conditions [39].

In consideration of scour protection measures, the collar protection is selected in this section due to its extensive application, straightforward structure, and minimal maintenance costs. The collar has been demonstrated to significantly reduce the scour effect by weakening the horseshoe vortex and the wake vortices around the foundation, thereby reducing the scour depth around the foundation. Bazilevs et al. investigated the influence of collar protection on the evolution of scour times [40]. The study demonstrated that, in addition to reducing the maximum scour depth, the collar also delays the lateral development of the scour hole. Zhu et al. investigated the equilibrium scour conditions with collar protection and determined that the efficacy of the protection is primarily contingent upon the installation height and outer diameter of the collar [41]. She et al. conducted experimental research which demonstrated that collars with a diameter exceeding three times that of the pile were ineffective in preventing sediment scour [42]. The optimal protection effect was observed when the collar was positioned slightly below the seabed surface. When the outer diameter of the collar was 2.3 times the diameter of the foundation, the protection effect reached its optimal value. This ratio is employed in this study.

Figure 16 depicts the variation curve of the maximum scour depth in the vicinity of the pile foundation. The scour depth on the upstream side undergoes a two-stage change over time, comprising a rapid scouring phase and a subsequent stabilization phase. The aforementioned process is influenced by a number of factors, including the dimensions of the pile foundation, the distribution of sediment particles, and the velocity of the flow. The figure illustrates that following approximately 1000 s, the scour depth reaches equilibrium, with approximately 80% of the scour occurring during the initial 500 s. In the absence of collar protection, the maximum scour depth for the wind turbine pile foundation is 4.23 m. In contrast, with collar protection, the maximum scour depth is reduced to 2.12 m, indicating that the collar protection effectively reduces the original scour depth by 50%. The results clearly demonstrate the significant protective effect of the collar, indicating its effectiveness in reducing scour. Nevertheless, the overall tendency of scouring remains unchanged, indicating that the collar's influence on the initial stages of scouring still has some limitations.

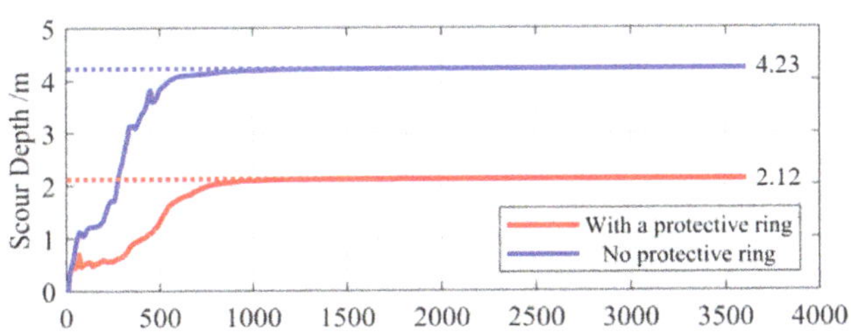

Figure 16. Scour depth process curve for the monopile foundation.

Figure 17a,b demonstrate the local scour conditions at the equilibrium scour stage with and without collar protection. In this study, the scour condition is characterized as a dynamic bed scour, whereby the sediment carried by the flow does not accumulate behind the pile foundation, but rather is swept away with the flow. The greatest depth of scouring is observed on both sides of the pile foundation, particularly in the vicinity of the upstream side. This phenomenon can be attributed to the formation of a dominant horseshoe vortex, which is a consequence of the strong flow and results in an increased scour rate. In the wake zone of the pile foundation, the formation of separated vortices may result in the emergence of smaller vortices. It is hypothesized that the movement of these small vortices may be a contributing factor to the expansion of the scour region. The application of collar protection results in a notable shift in the position of the maximum scour depth, with the outer diameter of the collar becoming the new focal point. This shift demonstrates that the collar not only reduces the scour depth, but also effectively guides the flow, thereby altering the distribution of the scour area and optimizing the protection effect of the pile foundation. To provide a more intuitive representation of the morphology and location of the local scour hole around the pile foundation, Figure 18 presents a three-dimensional illustration of the local scour around the pile foundation with and without collar protection at the equilibrium stage.

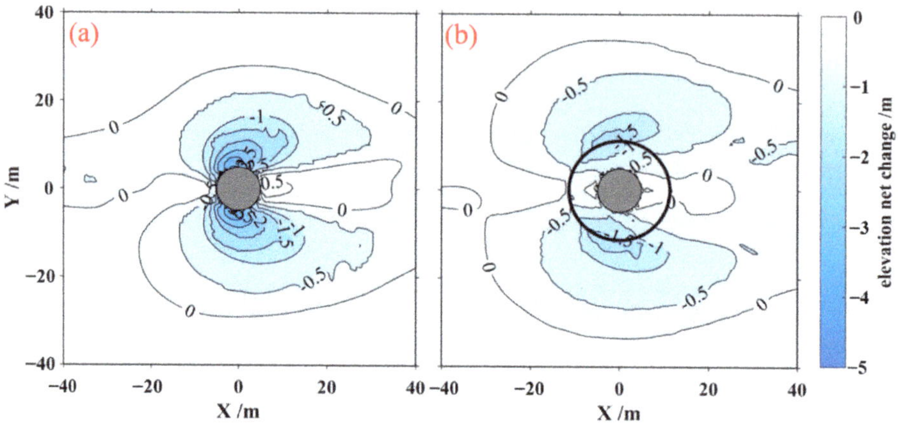

Figure 17. Local scour around the monopile foundation at equilibrium time: (**a**) without collar protection; (**b**) with collar protection.

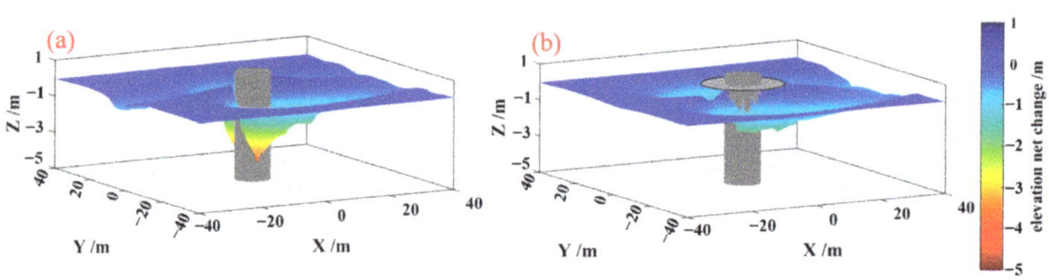

Figure 18. Three-dimensional local scour around the monopile foundation at equilibrium time: (**a**) without collar protection; (**b**) with collar protection.

5. Conclusions

This study employs a multidisciplinary approach, integrating numerical simulations and an analysis of ocean currents, sediment transport, and monopile scour. This comprehensive and in-depth analysis provides insights into the current and sediment background in the study area, as well as the scour protection characteristics of monopile foundations. The specific results of the analysis are as follows:

The analysis of tidal currents reveals that, with the exception of the areas in the vicinity of the Old Yellow River mouth and Dongying Port, where the tidal current speed ranges from 0.7 to 1.0 m·s^{-1} due to geographic conditions, the monthly average tidal current speed in the majority of the study region is below 0.6 m·s^{-1}. The monthly average surface tidal current speed at the location of the wind farm pile foundation is approximately 0.4 m·s^{-1}. The flood tide direction is around 300°, and the ebb tide direction is around 110°.

From the perspective of sediment transport, the Yellow River is responsible for the substantial delivery of sediment into the ocean through its mouth, where it continues to deposit and accumulate. The sediment is notably loose in consistency. This demonstrates that the suspended sediment concentration at the wind farm site is predominantly influenced by the elevated suspended sediment concentration regions in the vicinity of the Yellow River mouth. The concentration of suspended sediment is typically below 100 mg/L. In the majority of locations within the study region, the annual seabed erosion and deposition changes are within 0.02 m, with minimal annual erosion and deposition variations. The wind farm site exhibits an annual erosion and deposition change of less than 0.01 m.

The analysis of local scour around the pile foundation indicates that approximately 80% of the scour occurs in the initial stage. In the absence of collar protection, the maximum scour depth of the wind turbine pile foundation is observed to reach 4.23 m. However, the implementation of collar protection has been observed to result in a reduction in the maximum scour depth to 2.12 m, effectively decreasing the scour depth by 50%. This evidence illustrates the substantial protective impact of the collar in mitigating scour. Nevertheless, the prevailing trend of scouring remains consistent, suggesting that the collar still exhibits certain constraints in regulating the early-stage scour process. The application of collar protection results in a notable shift in the location of the maximum scour depth, with the outer diameter of the collar becoming the primary area of concern. This alteration demonstrates that the collar not only diminishes the depth of scour but also effectively channels the flow of scour water, modifying the distribution area of the scour and thereby optimizing the protection of the pile foundation. Furthermore, it is possible that the collar may result in an expansion of the scour area, potentially due to vortex shedding in the wake zone of the pile foundation, where larger vortices break into smaller ones. It is possible that these larger vortices may be a contributing factor to the observed expansion of the scour area.

Author Contributions: Conceptualization, X.H. and Z.W.; methodology, X.H. and Z.W.; software, X.H.; validation, X.H.; formal analysis, X.H., Z.W. and X.M.; investigation, X.H.; resources, Z.W. and X.M.; data curation, X.H.; writing—original draft preparation, X.H.; writing—review and editing, Z.W. and X.M.; visualization, X.H.; supervision, Z.W.; project administration, Z.W. and X.M.; funding acquisition, Z.W. and X.M. All authors have read and agreed to the published version of the manuscript.

Funding: This study was supported financially by the National Key Research and Development Program of China (2022YFC3104205) and the National Natural Science Foundation of China (42377457).

Institutional Review Board Statement: Not applicable.

Informed Consent Statement: Informed consent was obtained from all subjects involved in the study.

Data Availability Statement: Due to privacy, the data supporting the results of this study are not publicly available. However, they can be accessed upon reasonable request from the corresponding author.

Acknowledgments: We would like to thank Marine Big Data Center of Institute for Advanced Ocean Study of Ocean University of China for providing data conversion and storage, and the professional and technical services provided by Yujie Dong.

Conflicts of Interest: The authors declare no conflict of interest.

References

1. Song, J.; Duan, L. The Bohai Sea. In *World Seas: An Environmental Evaluation*; Elsevier: Amsterdam, The Netherlands, 2019; pp. 377–394.
2. Xu, Y.; Gao, H.; Wei, X.; Zhu, J. The Effects of Reclamation Activity and Yellow River Runoff on Coastline and Area of the Laizhou Bay, China. *J. Ocean Univ. China* **2021**, *20*, 729–739. [CrossRef]

3. Gazi, A.H.; Afzal, M.S. A review on hydrodynamics of horseshoe vortex at a vertical cylinder mounted on a flat bed and its implication to scour at a cylinder. *Acta Geophys.* **2020**, *68*, 861–875. [CrossRef]
4. Breusers, H.N.C.; Nicollet, G.; Shen, H.W. Local Scour Around Cylindrical Piers. *J. Hydraul. Res.* **1977**, *15*, 211–252. [CrossRef]
5. Ansari, S.A.; Kothyari, U.C.; Ranga Raju, K.G. Influence of cohesion on scour around bridge piers. *J. Hydraul. Res.* **2002**, *40*, 717–729. [CrossRef]
6. Oliveto, G.; Hager, W.H. Further Results to Time-Dependent Local Scour at Bridge Elements. *J. Hydraul. Eng.* **2005**, *131*, 97–105. [CrossRef]
7. Chiew, Y.M.; Melville, B.W. Local scour around bridge piers. *J. Hydraul. Res.* **1987**, *25*, 15–26. [CrossRef]
8. Raudkivi, A.J.; Ettema, R. Clear-Water Scour at Cylindrical Piers. *J. Hydraul. Eng.* **1983**, *109*, 338–350. [CrossRef]
9. Jain, S.C.; Fischer, E.E. *Scour Around Circular Bridge Piers at High Froude Numbers*; US Department of. Transportation, Federal Highway Administration: Washington, DC, USA, 1979.
10. Jiang, W.; Pohlmann, T.; Sündermann, J.; Feng, S. A modelling study of SPM transport in the Bohai Sea. *J. Mar. Syst.* **2000**, *24*, 175–200. [CrossRef]
11. Bian, C.; Jiang, W.; Pohlmann, T.; Sündermann, J. Hydrography-Physical Description of the Bohai Sea. *J. Coast. Res.* **2016**, *74*, 1–12. [CrossRef]
12. Qiao, L.; Zhong, Y.; Wang, N.; Zhao, K.; Huang, L.; Wang, Z. Seasonal transportation and deposition of the suspended sediments in the Bohai Sea and Yellow Sea and the related mechanisms. *Ocean Dyn.* **2016**, *66*, 751–766. [CrossRef]
13. Saito, Y.; Yang, Z.; Hori, K. The Huanghe (Yellow River) and Changjiang (Yangtze River) deltas: A review on their characteristics, evolution and sediment discharge during the Holocene. *Geomorphology* **2001**, *41*, 219–231. [CrossRef]
14. Zhou, Z.; Bian, C.; Wang, C.; Jiang, W.; Bi, R. Quantitative Assessment on Multiple Timescale Features and Dynamics of Sea Surface Suspended Sediment Concentration Using Remote Sensing Data. *J. Geophys. Res. Ocean.* **2017**, *122*, 8739–8752. [CrossRef]
15. Liu, X.; Qiao, L.; Zhong, Y.; Wan, X.; Xue, W.; Liu, P. Pathways of suspended sediments transported from the Yellow River mouth to the Bohai Sea and Yellow Sea. *Estuar. Coast. Shelf Sci.* **2020**, *236*, 106639. [CrossRef]
16. Qiao, S.; Shi, X.; Wang, G.; Zhou, L.; Hu, B.; Hu, L.; Liu, S. Sediment accumulation and budget in the Bohai Sea, Yellow Sea and East China Sea. *Mar. Geol.* **2017**, *390*, 270–281. [CrossRef]
17. Chen, C.; Liu, H.; Beardsley, R.C. An Unstructured Grid, Finite-Volume, Three-Dimensional, Primitive Equations Ocean Model: Application to Coastal Ocean and Estuaries. *J. Atmos. Ocean. Technol.* **2003**, *20*, 159–186. [CrossRef]
18. Chen, C.; Cowles, G. An Unstructured-Grid Finite-Volume Coastal Ocean Model (FVCOM) System. *Oceanography* **2006**, *19*, 78–89. [CrossRef]
19. Jin, Z.; Zuo, C.; Wang, Z. Impact of Phase III Project of Maji Mountain Port on sediment siltation in adjacent sea area. *Acta Oceanol. Sin.* **2017**, *36*, 111–118. [CrossRef]
20. Wu, L.; Chen, C.; Guo, P.; Shi, M.; Qi, J.; Ge, J. A FVCOM-based unstructured grid wave, current, sediment transport model, I. Model description and validation. *J. Ocean Univ. China* **2011**, *10*, 1–8. [CrossRef]
21. Krone, R.B. *Flume Studies of the Transport of Sediment in Estuarial Shoaling Processes: Final Report*; Hydraulic Engineering Laboratory and Sanitary Engineering Research Laboratory, University of California: Berkeley, CA, USA, 1962.
22. Camenen, B.; Larson, M. A general formula for non-cohesive bed load sediment transport. *Estuar. Coast. Shelf Sci.* **2005**, *63*, 249–260.
23. Harris, C.K.; Wiberg, P.L. A two-dimensional, time-dependent model of suspended sediment transport and bed reworking for continental shelves. *Comput. Geosci.* **2001**, *27*, 675–690. [CrossRef]
24. Nielsen, A.W.; Liu, X.; Sumer, B.M.; Fredsøe, J. Flow and bed shear stresses in scour protections around a pile in a current. *Coast. Eng.* **2013**, *72*, 20–38. [CrossRef]
25. Abdelaziz, S.; Bui, M.D.; Rutschmann, P. Numerical simulation of scour development due to submerged horizontal jet. In *River Flow: Process of the International Conference on Fluvial Hydraulics*; CRC Press: Boca Raton, FL, USA, 2010.
26. Dey, S.; Raikar, R.V. Characteristics of Horseshoe Vortex in Developing Scour Holes at Piers. *J. Hydraul. Eng.* **2007**, *133*, 399–413. [CrossRef]
27. Zhang, Q.; Zhou, X.-L.; Wang, J.-H. Numerical investigation of local scour around three adjacent piles with different arrangements under current. *Ocean Eng.* **2017**, *142*, 625–638. [CrossRef]
28. Omara, H.; Tawfik, A. Numerical study of local scour around bridge piers. *IOP Conf. Ser. Earth Environ. Sci.* **2018**, *151*, 012013. [CrossRef]
29. Jalal, H.K.; Hassan, W.H. Three-dimensional numerical simulation of local scour around circular bridge pier using Flow-3D software. *IOP Conf. Ser. Mater. Sci. Eng.* **1979**, *745*, 012150. [CrossRef]
30. Nazari-Sharabian, M.; Nazari-Sharabian, A.; Karakouzian, M.; Karami, M. Sacrificial Piles as Scour Countermeasures in River Bridges a Numerical Study using FLOW-3D. *Civ. Eng. J.* **2020**, *6*, 1091–1103. [CrossRef]
31. Aminoroayaie Yamini, O.; Mousavi, S.H.; Kavianpour, M.R.; Movahedi, A. Numerical modeling of sediment scouring phenomenon around the offshore wind turbine pile in marine environment. *Environ. Earth Sci.* **2018**, *77*, 776. [CrossRef]

32. Yakhot, V.; Orszag, S.A. Renormalization group analysis of turbulence. *I. Basic Theory. J. Sci. Comput.* **1986**, *1*, 3–51. [CrossRef] [PubMed]
33. Yakhot, V.; Smith, L.M. The Renormalization Group, the e-Expansion and Derivation of Turbulence Models. *J. Sci. Comput.* **1992**, *7*, 35–61. [CrossRef]
34. Roulund, A.; Sumer, B.M.; Fredsøe, J.; Michelsen, J. Numerical and experimental investigation of flow and scour around a circular pile. *J. Fluid Mech.* **2005**, *534*, 351–401. [CrossRef]
35. Guan, D.; Chiew, Y.-M.; Wei, M.; Hsieh, S.-C. Characterization of horseshoe vortex in a developing scour hole at a cylindrical bridge pier. *Int. J. Sediment Res.* **2019**, *34*, 118–124. [CrossRef]
36. Jia, Y.; Altinakar, M.; Guney, M.S. Three-dimensional numerical simulations of local scouring around bridge piers. *J. Hydraul. Res.* **2018**, *56*, 351–366. [CrossRef]
37. OuYang, H.; Dai, G.; Gao, L.; Zhu, W.; Du, S.; Gong, W. Local scour characteristics of monopile foundation and scour protection of cement-improved soil in marine environment—Laboratory and site investigation. *Ocean Eng.* **2022**, *255*, 111443. [CrossRef]
38. Soulsby, R. Dynamics of marine sands: A manual for practical applications. *Oceanogr. Lit. Rev.* **2019**, *9*, 947.
39. Akhlaghi, E.; Babarsad, M.S.; Derikvand, E.; Abedini, M. Assessment the Effects of Different Parameters to Rate Scour around Single Piers and Pile Groups: A Review. *Arch. Comput. Methods Eng.* **2010**, *27*, 183–197. [CrossRef]
40. Bazilevs, Y.; Takizawa, K.; Tezduyar, T.E. Challenges and Directions in Computational Fluid–Structure Interaction. *Math. Models Methods Appl. Sci.* **2013**, *23*, 215–221. [CrossRef]
41. Zhu, H.P.; Zhou, Z.Y.; Yang, R.Y.; Yu, A.B. Discrete particle simulation of particulate systems: Theoretical developments. *Chem. Eng. Sci.* **2007**, *62*, 3378–3396. [CrossRef]
42. She, K.; Created, C.A.; Easson, W.J. Experimental study of three-dimensional breaking wave kinematics. *Appl. Ocean Res.* **1997**, *19*, 329–343. [CrossRef]

Journal of
Marine Science and Engineering

MDPI

Article

Response of Subsurface Chlorophyll Maximum Depth to Evolution of Mesoscale Eddies in Kuroshio–Oyashio Confluence Region

Ziwei Chuang, Chunling Zhang *, Jiahui Fan and Huangxin Yang

College of Oceanography and Ecological Science, Shanghai Ocean University, Shanghai 201306, China; czw15161290639@163.com (Z.C.); m230200646@st.shou.edu.cn (J.F.); m240300832@st.shou.edu.cn (H.Y.)
* Correspondence: clzhang@shou.edu.cn; Tel.: +86-21-6190-8337

Abstract: The subsurface chlorophyll maximum depth (SCMD) is an indicator of the spatial activity of marine organisms and changes in the ecological environment. Ubiquitous mesoscale eddies are among the important factors regulating the Kuroshio–Oyashio confluence region. In this study, we use satellite altimeter observations and high-resolution reanalysis data to explore seasonal variations in the SCMD and its responses to different types of eddies based on methods of composite averaging and normalization. The results show that variations in the SCMD induced by the evolution of the eddies were prominent in the summer and autumn. The monopoles of the SCMD exhibited internally shallow and externally deep features in the cyclonic eddies (CEs), while the contrary trend was observed in the anticyclonic eddies (ACEs). The SCMD was positively correlated with the intensity of the eddies and sea surface temperature, and was negatively correlated with the depth of the mixed layer. These correlations were more pronounced in the CEs (summer) and ACEs (autumn). Both the CEs and ACEs prompted the westward transport of chlorophyll-a (Chl-A), where ACEs transported it over a longer distance than the CEs.

Keywords: eddies; subsurface chlorophyll maximum depth; evolution; Kuroshio–Oyashio confluence region

1. Introduction

The Kuroshio–Oyashio confluence region features complex changes in the dynamic environment of the ocean and strong sea–air interactions [1,2]. The intense mixing resulting from the convergence of cold and warm currents is conducive to marine life, because of which this region is a key fishing ground for cephalopods [3]. The subsurface chlorophyll maximum depth (SCMD) is an important indicator for assessing the primary productivity of the ocean, as well as the spatial activity of marine organisms and changes in their ecological environment. The SCMD of seawater is primarily influenced by dynamic processes, in addition to the natural conditions of photosynthesis. Mesoscale eddies are a common form of movement of seawater in the Kuroshio–Oyashio confluence region, and influence the position of the thermocline, supply of nutrients, and distribution of algae to the SCMD through their dynamic effects [4]. Their internal vertical circulation regulates the vertical structure of the local environment through mixing and pumping [5,6]. Cyclonic eddies (CEs) mostly originate from the Oyashio meanders and carry productive subarctic water, while anticyclonic eddies (ACEs) originate from the Kuroshio meanders and carry oligotrophic subtropical water [7]. ACEs are more frequent than CEs, and have a longer lifetime [8]. The large amount of kinetic energy carried by the nonlinear movement of the eddies promotes the horizontal transport and redistribution of marine materials and resources [9,10]. A comprehensive analysis of the response-based relationship between the SCMD and the spatiotemporal evolution of these eddies can provide a firm scientific basis for understanding the dynamic mechanisms of marine ecosystems and changes in fishing grounds.

The concentration of chlorophyll-a (Chl-A) in seawater reflects the primary productivity of marine areas. A key focus in the relevant research is on its interactions with dynamic processes in the ocean. Recent studies have revealed that about half of all Chl-A is trapped by eddies in the global ocean [11]. The rotational characteristics and advection of eddies play an essential role in stirring the horizontal distribution of Chl-A [12,13]. Mesoscale eddies influence the distribution of Chl-A through vertical pumping, Ekman pumping, eddy advection, and submesoscale pumping [5]. The upwelling in CEs transports nutrient-rich water to the euphotic zone to enhance Chl-A photosynthesis and promote the growth of marine phytoplankton. This can increase primary productivity by 10% to 50%. By contrast, the downwelling in ACEs moves surface water away from the euphotic zone, resulting in reduced productivity [14–17]. Mesoscale eddies generally influence changes in Chl-A through multiple mechanisms [6,13,18–20].

Chl-A exhibits a vertical distribution, and its maximum values typically occur in the subsurface layer [21–23]. The mechanisms sustaining the subsurface chlorophyll maximum (SCM) include the local rate of maximum growth of phytoplankton, their adaptation to light, and swimming behavior [24]. The concentration of Chl-A within the SCMD accounts for 30–70% of the primary productivity of the entire water column. As a major dynamic factor influencing it, mesoscale eddies have complex and significant effects on the SCMD. CEs tend to reduce the SCMD, which results in an increased vertical gradient of Chl-A, while ACEs have the opposite effect on it [25–27]. Most relevant research has relied on observations from ships or BGC–Argo to investigate variations in the SCMD within individual eddies because only a limited number of observational profiles of Chl-A are available. Research has shown that the average SCMD increases by 15 m in CEs and decreases by 20 m in ACEs [28]. Moreover, the maximum Chl-A concentration in CEs is 62% higher than the background level, while it constitutes only 54% of the background in ACEs [25,28]. Some studies have used composite analysis to demonstrate that CEs can enhance subsurface chlorophyll concentrations and the biomass of deep phytoplankton by providing optimal conditions of light and nutrients at a global scale [29].

Physical–biogeochemical models and data for objective reanalysis have proven to be valuable tools in recent years for analyzing the response of subsurface Chl-A to different types of eddies [30]. The composition of communities of phytoplankton varies significantly with different types of eddies, and in different stages of eddy movement [4]. Therefore, further investigation is needed to examine the vertical structure of Chl-A and the response of the SCMD to the evolution of eddies. In this study, we use high-resolution reanalysis data on Chl-A to discuss the spatiotemporal variations in the SCMD under the evolution of different kinds of eddies in the Kuroshio–Oyashio confluence region. The remainder of this paper is structured as follows: we detail the data and methods used in this research in Section 2. Section 3 describes seasonal variations in Chl-A and the SCMD in different stages of the evolution of eddies, and the response relationship between the latter and the SCMD. The characteristics of the maximum Chl-A induced by the evolution of the eddies are discussed in Section 4, while the conclusions of this study are summarized in Section 5.

2. Data and Methods

2.1. Data

We used data on five clusters of materials in this study: sea surface currents, subsurface Chl-A content, sea level anomalies (SLAs), sea surface temperature (SST), and potential temperature. All datasets were obtained from the Copernicus Marine Service (http://marine.copernicus.eu/ (accessed on 3 November 2023)). Data on the daily sea surface current were obtained at a spatial resolution of 0.25° × 0.25° by inverting satellite altimeter observations, and were used to detect mesoscale eddies. The daily reanalysis products of the subsurface Chl-A relied on a physical–biogeochemical model based on the Nucleus for European Modeling of the Ocean (NEMO) version 3.6, and the biogeochemical Pelagic Interaction Scheme for Carbon and Ecosystem Studies (PISCES) model (https://data.marine.copernicus.eu/product/GLOBAL_MULTIYEAR_BGC_001_029/services (accessed on 16 November 2023)). Previous studies have compared observational data with the chlorophyll profiles generated by this model, and their results have shown that it can accurately reflect the seasonal characteristics of chlorophyll in the ocean [31]. It has also been used to examine the relationship between eddies and primary productivity [32]. This product has been validated against climatic data and subjected to root mean-squared error testing [33]. It has a spatial resolution of 0.25° × 0.25°, and spans over a vertical range of 0.5 to 5902 m, divided into 75 layers with unequal intervals.

Data on the SLA and SST were obtained from the same source as the data on the currents. The daily potential temperature used to compute the mixing layer depth (MLD) was sourced from the reanalysis products provided by the dynamic model of the ocean (https://data.marine.copernicus.eu/product/GLOBAL_MULTIYEAR_PHY_001_030/services (accessed on 7 January 2024)) at a spatial resolution of 0.083° × 0.083°. The MLD was defined as the depth of the ocean at which the density increased in comparison with that at a depth of 10 m, and corresponded to a decline of 0.2 °C in the temperature under standard atmospheric pressure [4,34]. We used data from the Kuroshio–Oyashio confluence region (145° E–170° E, 35° N–45° N) spanning from 1 January 2009 to 31 December 2022. Figure 1a illustrates the bathymetric topography and background system of currents, while Figure 1b shows the spatial distribution of sea surface currents corresponding to surface Chl-A content on 3 June 2020.

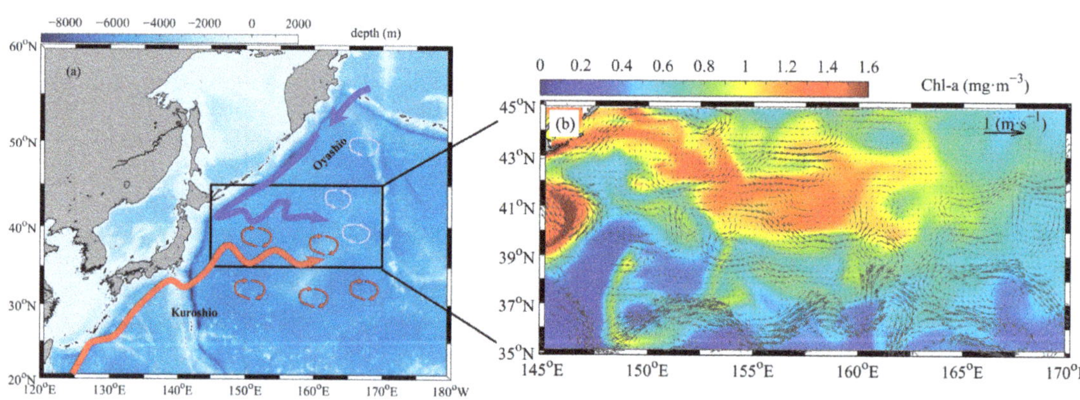

Figure 1. Research area (**a**) and sea surface current corresponding to Chl-A on 3 June 2020 (**b**). The black frame in panel (**a**) denotes the research area. Blue and red arrows and circles represent CEs and ACEs, respectively.

The BGC–Argo-based observational profiles of Chl-A were derived from the China Argo Real-time Data Center (https://www.argo.org.cn/ (accessed on 10 December 2024)). A total of 2609 profiles were collected for the study area, and were subjected to data processing and quality control at the China Argo Real-time Data Center [35]. Data on the location and time of each eddy were used to match the vertical profiles obtained from Argo floats within the eddies. Data from the Argo floats were normalized into a standard vertical profile in different seasons, and this provided the characteristics of the distribution of Chl-A within the eddies in the Kuroshio–Oyashio confluence region.

2.2. Methods

2.2.1. Detection of Mesoscale Eddies

We used the flow vector method, proposed by Nencioli et al. [36], to detect and identify mesoscale eddies. This technique is based on the geometric characteristics of the velocity field of the eddy. The eddy centers are determined by the following four constraints: (1) The velocity component v along the east–west (EW) section should have opposite numerical signs on either side of the eddy center, and its magnitude should increase away from the center. (2) The velocity u component along the north–south (NS) section should have opposite numerical signs on either side of the eddy center, its magnitude should increase away from the center, and the sense of rotation of u needs to be identical to that of v. (3) The magnitude of velocity has a local minimum at the eddy center. (4) The directions of the velocity vectors around the eddy center must change with a constant sense of rotation, while the directions of two adjacent velocity vectors must lie within the same quadrant or in adjacent quadrants.

The boundaries of the eddy were defined as the outermost closed streamline surrounding its center, and its radius was assumed to be the average distance between each edge point and the eddy center. Mesoscale eddies were detected every day. We recorded the number of eddies, their types, the position of the eddy center, its mean radius, and the position of its boundary. We used the method developed by Liu et al. [37] to calculate the curves of average variations in the radius of the eddy, eddy kinetic energy (EKE), vorticity, and rate of deformation in the research area during the normalized lifetime of the eddy. The lifetime of each eddy was normalized and systematically divided into three stages of evolution: a developmental stage (the first 30% of its lifetime), a stable stage (30–70% of its lifetime), and a decay stage (the final 30% of its lifetime) [37,38].

We used the radius of the eddies, the division of their evolutionary stages, and their horizontal motion to focus only on eddies with a lifetime exceeding 14 days. The physical characteristics of the tracked eddies were analyzed, and their corresponding amplitude (*APD*) was defined as the difference between the mean SLA along the edge of the eddy (h_0) and the eddy center (h_c):

$$APD = |\ h_0 - h_c\ | \tag{1}$$

2.2.2. Composite Analysis of Chl-A and SCMD

To eliminate differences in the Chl-A concentration across eddies of different types and spatiotemporal scales, we normalized and averaged the Chl-A content within the spatial scale of the eddies [13,39]. We first normalized their spatial scale. The Chl-A concentration of mesoscale eddies was first normalized based on their mean radius (R). To identify each eddy, we extracted a circular area from its center that was three times its radius (3R), and projected it onto a uniform grid with a spatial scale ranging from −3 to 3 in both the meridional and zonal directions (the region within 3R also facilitated a comparison highlighting the differences between the center of the eddy and its outer edge). Following this, we composited snapshots of the Chl-A concentrations of all the normalized eddies to obtain their mean spatial distribution. The temporal scale of the eddies was normalized based on their lifetime. The relative lifetime (T) on each day (D) during the evolution of each eddy was determined by Formula (2). Finally, the Chl-A concentration of each layer was superimposed and averaged, according to different stages of its normalized lifetime, over the 3R spatial scale. We used the surface radius to match the data obtained from deep below the ocean surface. The composite of the SCMD was obtained by using a similar method:

$$T = (D - D_{min})/(D_{max} - D_{min}) \tag{2}$$

where T is the normalized lifetime of the eddy, D is the ordinal of its lifetime, and D_{min} and D_{max} represent the times of generation and extinction of the eddy, respectively.

The Chl-A concentration in the ocean typically exhibits significant seasonal variations, because of which calculating anomalies in the Chl-A concentration can reveal the influence of the eddies on it [30]. We used the average Chl-A concentration inside a rectangular region within 5R as the background concentration, and then calculated the difference at this depth on each day. This yielded in the difference between the internal Chl-A concentration of each eddy and the background. We also conducted a composite analysis of all eddies according to their evolutionary stages, with the aim of revealing the relative changes in their evolution.

2.2.3. Monopole of SCMD

The primary dynamic mechanisms that influence the distribution of Chl-A in eddies include their stirring, trapping, pumping, and eddy–Ekman pumping [5,13]. All these mechanisms are expected to generate monopole-like anomalies in Chl-A, with the exception of eddy stirring [25]. The anomalies in Chl-A are predominantly positive in CEs and negative in ACEs [13]. In case of the stirring of the eddy, its interior Chl-A generally exhibits an asymmetric structure. Therefore, the structure of Chl-A trapped by eddies can be viewed as the superposition of a monopole and a dipole [5,19,25]. Extraction of monopole eddy structure is equivalent to performing a radial average based on the distance from the eddy center, with the aim of highlighting the differences in SCMD between the eddy core and its periphery. We first applied polar coordinate transformation to obtain the relative distance between the SCMD and the eddy center, and then decomposed the SCMD into a monopole corresponding to different types and stages of evolution of the eddy:

$$\mathrm{SCMD}_{(r,\theta)} = \overline{\mathrm{SCMD}_{(r)}} \tag{3}$$

where r is the relative distance between the eddy center and the SCMD within the normalized composite region, and θ is the angle of rotation in the polar coordinate system. The overbar denotes the radial average, and we obtained the composite average of each part of all eddies.

2.2.4. Transport Distance of SCM

We used the calculated physical parameters of the eddies to divide the evolution of each into three stages. The position of the eddy center on the first day was used as the starting point of each stage, while the coordinates of the eddy center on the last day were used as the endpoint. The relative azimuth and transport distance were then calculated. The SCM trapped by each eddy was assumed to be the average value within the average radius of the given stage. To comprehensively assess the distribution and movement of the mesoscale eddies, we divided the directions of their movement into 16 azimuths, with each direction separated by 20 km. We then used these azimuths to divide the study area into several grids. Finally, we calculated the average SCM of all eddies within each grid.

3. Results

3.1. Properties of Chl-A Induced by Eddy Evolution

Owing to the dynamic differences in eddy evolution, the properties of Chl-A induced by different types of eddies in varying stages of evolution varied significantly. We identified a total of 725 CEs and 1115 ACEs with lifetimes exceeding 14 days. The CEs were most common in the summer, while the ACEs tended to form in the spring and summer (Figure 2). A majority of the CEs had a lifetime of two to four weeks, while the ACEs generally had a longer lifetime, with 510 of them lasting over four weeks. The seasonal *APD* was the strongest in the summer, on average.

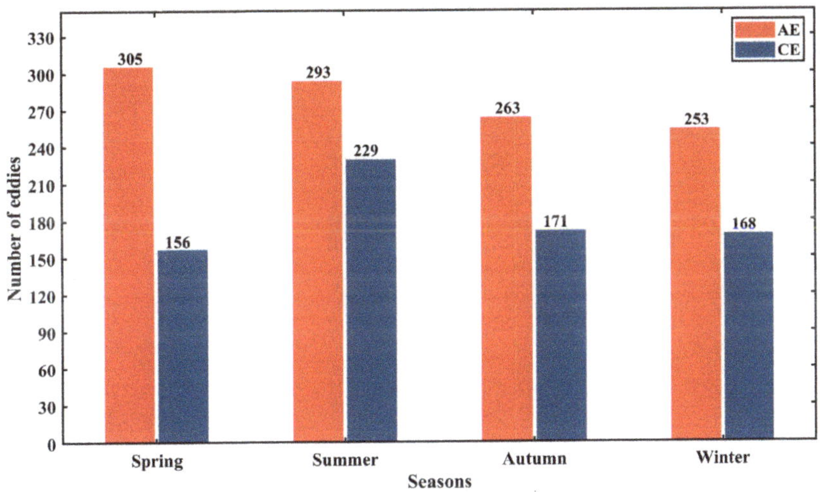

Figure 2. The number of eddies generated in different seasons.

The anomalies in Chl-A were obtained by subtracting the spatial average within the region of the eddy (Figure 3). Both the CEs and ACEs exhibited significant changes in Chl-A at depths of 0–40 m, with anomalies ranging from −0.03 to 0.03 mg/m^3. The anomalies in Chl-A exhibited a horizontal distribution that was characterized by "higher in the north and lower in the south." This was consistent with the results of comparison of the BGC–Argo profile reported by Chen et al. [40]. By contrast, the anomalies in Chl-A below 40 m were higher in the south and lower in the north, and had a more uniform distribution (−0.015–0.015 mg/m^3). The increase in Chl-A within 1R reached its peak around 60 m. During the evolution of the eddies, the spatial differences in Chl-A concentration in the stable stage were more pronounced than those in the developmental and decay stages at depths shallower than 100 m.

High values of positive anomalies in Chl-A (higher than 0.01 mg/m^3 on the surface, 0.015 mg/m^3 at a depth of 20 m, and higher than 0.01 mg/m^3 at a depth of 40 m) gradually expanded from the edges of the eddy to its center at a depth shallower than 40 m from the developmental to the stable stage of evolution of CEs (Figure 3(a1–a3)). This tendency weakened in the stage of decay of CEs. The vertical distribution of positive anomalies in Chl-A expanded in a roughly counterclockwise manner in each stage of evolution of the CEs, which reflected their nonlinear rotational characteristic. The high values of positive anomalies in Chl-A at 0–40 m (higher than 0.015 mg/m^3 on the surface, 0.015 mg/m^3 at a depth of 20 m, and higher than 0.015 mg/m^3 at 40 m) gradually expanded from the center of the eddy to its northwest edge for the ACEs (Figure 3(b1–b3)). This expansion spanned from the edge of 1R to 2R from the surface to a depth of 20 m. The anomalies in Chl-A over

3R at 60 m exhibited a more pronounced positive increase compared with those in the CEs. Moreover, the range of anomalies in Chl-A that were higher than 0.005 mg/m^3 continually decreased during the evolution of the ACEs at 80–100 m, and were gradually concentrated between 2R and 3R in the southern part of the eddy.

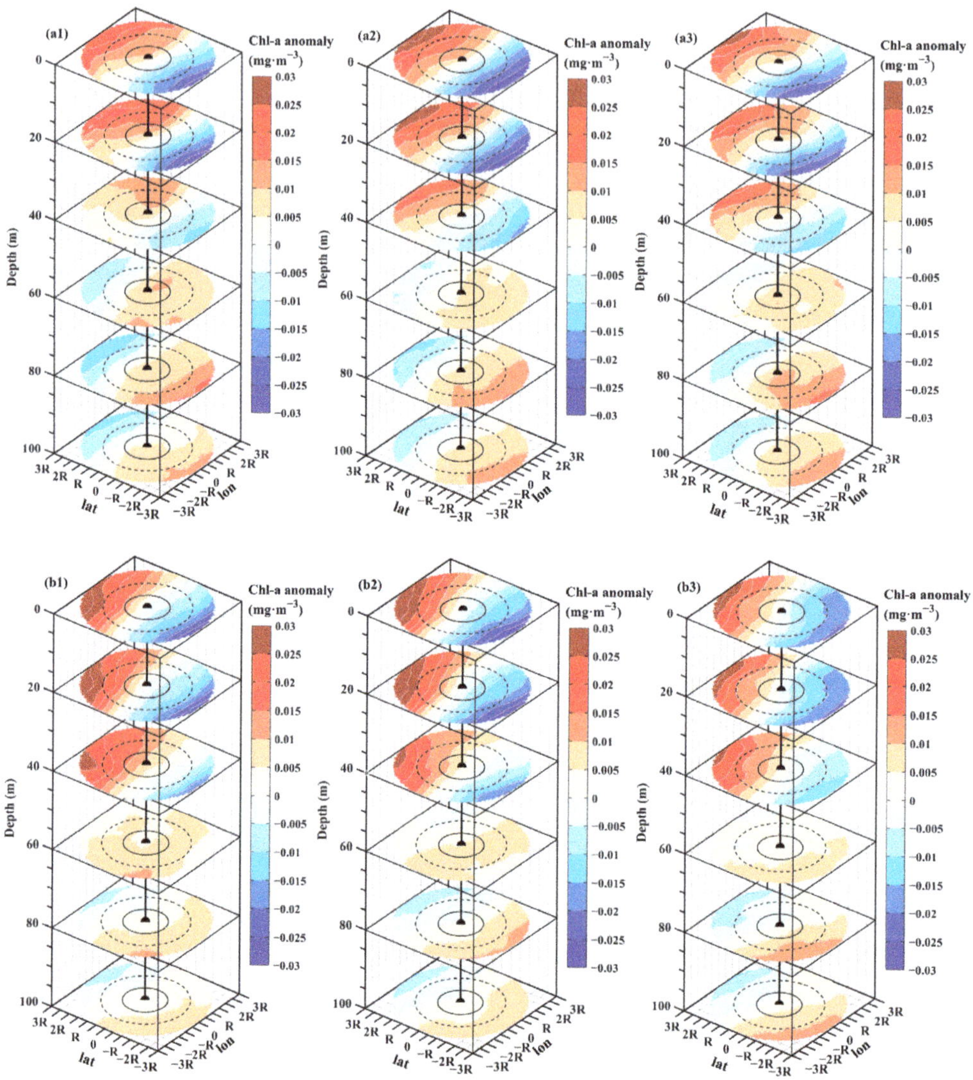

Figure 3. Anomalies in Chl-A over an area three times the radius of the eddy during different stages of its evolution. (**a1**–**a3**) represent anomalies in Chl-A corresponding to the stages of development, stability, and decay of CEs, while (**b1**–**b3**) represent the corresponding anomalies for ACEs. The solid black line represents the mean radius of the eddy, and the dashed black line represents twice its radius. The eastward and northward directions were set as the positive axes of the longitude and latitude, respectively. The anomalies in Chl-A concentration were obtained by subtracting the spatial average value within the region of the eddy, to obtain the composite average of each part for all eddies.

Figure 4 shows the structure of the composite section of Chl-A along the meridian in response to seasonal variations. The vertical distributions of Chl-A in CEs and ACEs were similar to each other: Its concentration was the highest in spring, followed by winter and summer, with the lowest concentrations occurring in autumn. The highest Chl-A concentration in spring was >0.85 mg/m^3, with a uniform distribution at a depth of 0–20 m within the eddy owing to surface mixing (Figure 4(a1,b1)), and gradually decreasing with the depth. The vertical distribution of Chl-A in winter was similar to that in spring, except that the mixing of the upper layer deepened to 60 m. High values of Chl-A were identified at depths of 30–70 m in the summer and autumn.

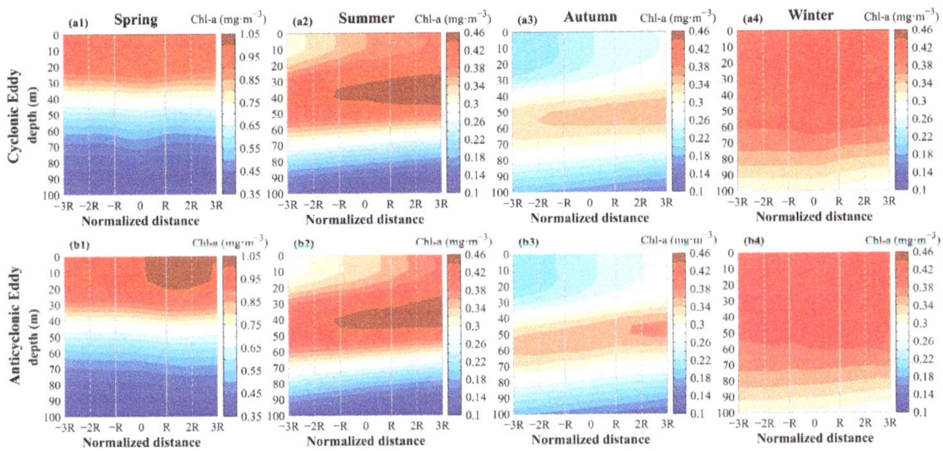

Figure 4. The south–north sections of Chl-A concentration along the meridians of centers of the CEs (**a1**–**a4**) and ACEs (**b1**–**b4**) in different seasons. The positive radius is directed northward and the negative radius southward. The solid and dashed lines represent the mean radius and twice the mean radius, respectively.

High values of Chl-A concentration were distributed in the north of the CEs within 3R in winter (Figure 4(a4)). Similarly, higher Chl-A concentrations were identified in the north of the center of the ACEs within 2R–3R (Figure 4(b4)). Curves of the contour lines of Chl-A moved downward to 40 m at the center of the CEs, reflecting the increase in higher values of Chl-A deeper below the surface (Figure 4(a1,a4)). By contrast, the contour lines of Chl-A at the center of the ACEs exhibited a slight increase, with low values of Chl-A increasing closer to the surface (Figure 4(b1,b4)). Both the CEs and ACEs yielded the maximum Chl-A concentration (>0.38 mg/m^3) in the subsurface layer (30–50 m) in the summer (Figure 4(a2,b2)). Higher Chl-A values were concentrated in the north of the eddies. The SCM (>0.3 mg/m^3) was approximately in the range of depths of 50–70 m in autumn (Figure 4(a3,b3). The vertical change in the Chl-A concentration was thus more pronounced than that in the summer.

3.2. Properties of SCMD Induced by Eddy Evolution

Figure 4 shows that the SCMD within the CEs and ACEs exhibited significant seasonal variations, while Figure 5 shows the spatiotemporal features of the composite SCMD within an area three times the radius of the eddy. The SCMD was 40–50 m in the summer, which is shallower than that in autumn (48–60 m), and was shallower in the spring and winter by 10 m. Its spatial distribution within the eddies typically followed a pattern of deeper in the south and shallower in the north across the four seasons. The maximum depths were concentrated in the south of the eddies, between 2R and 3R, and were particularly pronounced during summer and autumn. By contrast, the range of the SCMD in the ACEs was greater than that in the CEs. The mean SCMD of ACEs shows no significant difference compared to that of CEs in both summer (Figure 5(a2,b2)) and autumn (Figure 5(a3,b3)). The SCMD within CEs was spatially distributed in the upper layers of seawater in spring and winter (<6 m).

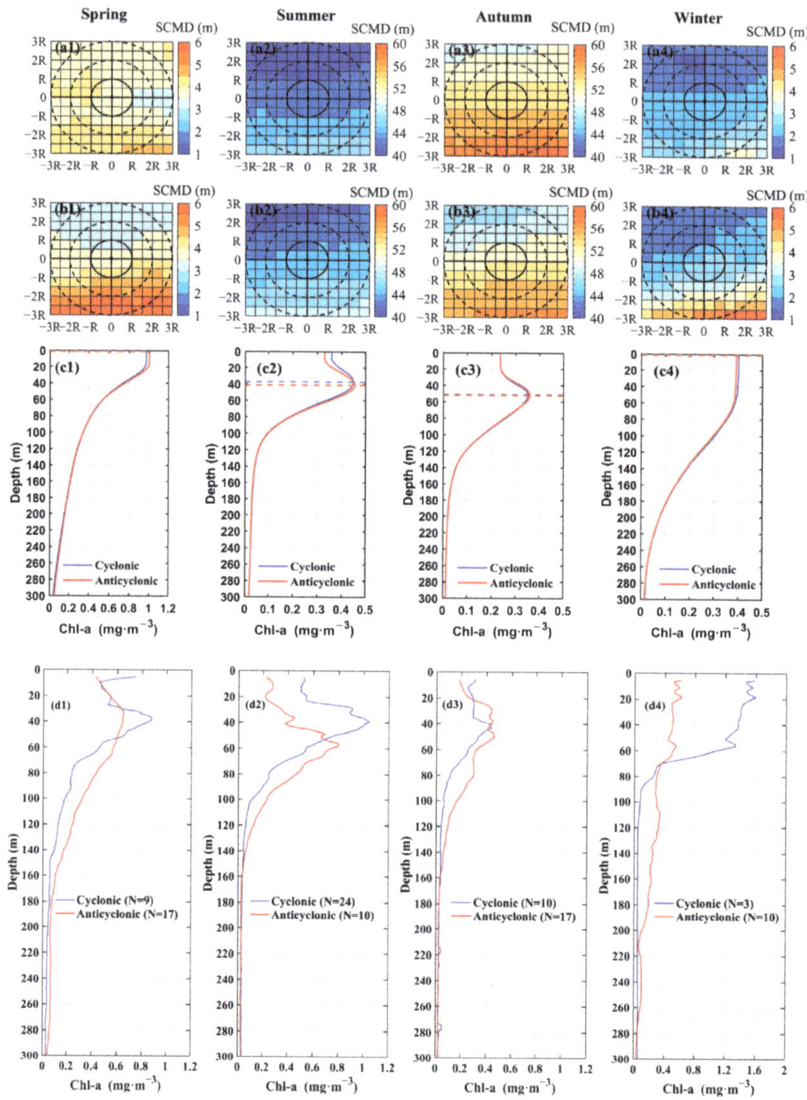

Figure 5. Spatial distributions of the SCMD in CEs and ACEs in different seasons. (**a1–a4**) show the horizontal distribution of the SCMD of the CEs, and (**b1–b4**) show the horizontal distribution of the SCMD of the ACEs. The solid black lines represent the mean eddy radius and the dashed black lines represent twice the eddy radius. (**c1–c4**) show the vertical profile of the average Chl-A concentration within the mean eddy radius. The dashed blue and red lines denote the SCMDs in the CEs and ACEs, respectively. (**d1–d4**) show the observed chlorophyll profiles in the eddies. "N" indicates the number of profiles.

The average profile of Chl-A within 1R shows that it attained its maximum value below the sea surface in the summer and autumn (Figure 5(c2,c3)). The vertical changes in Chl-A were particularly pronounced in the summer, when it initially increased and then decreased with an increase in depth from 20 to 60 m. The SCMD was about 37 m in the CEs, with the Chl-A concentration exceeding 0.45 mg/m^3, while it was deeper in the ACEs (~41 m). There was a significant increase in the Chl-A concentration at 30–80 m in the autumn, with maximum values of 0.35 mg/m^3 at 51 m in the CEs, and 0.36 mg/m^3 at 52 m in the ACEs. The changes in Chl-A concentration were similar to those in the temperature during winter and spring. Its concentration was mixed at a depth of 20 m in spring (>1 mg/m^3), while its distribution was more uniform, and above 60 m in winter (0.4 mg/m^3). Due to the limited number of BGC-Argo observations, the mean Chl-A profile within the eddies shows significant vertical variability (Figure 5(d1–d4))). In summer, the SCMD in CEs is typically around 40 m, with SCM averaging about 1 mg/m^3. In contrast, ACEs have a deeper SCMD of approximately 60 m and a slightly lower SCM of around 0.8 mg/m^3 (Figure 5(d2)). Notably, at depths shallower than 50 m, CEs exhibit significantly higher Chl-A concentrations compared to ACEs. However, at depths greater than 50 m, this trend is reversed, with ACEs showing higher Chl-A concentrations than CEs. In autumn, the SCMD in both CEs and ACEs is generally around 40 m (Figure 5(d3)). Chl-A concentrations shallower than 60 m in CEs exceed 1.2 mg/m^3 in winter, while in ACEs, Chl-A concentrations range from 0.4 to 0.5 mg/m^3 (Figure 5(d4)).

Figures 6 and 7 show that the SCMD was more pronounced in summer and autumn, and exhibit the variations in the SCMD during the evolution of the eddies. The SCMD gradually became shallower from south to north during the evolution of the CEs (Figure 6(a1–a3,c1–c3)), where this is consistent with the average spatial distribution shown in Figure 5. The SCMD of the CEs decreased first and then gradually rose with the evolution of the eddies in summer, and reached the minimum depth in the stable stage (Figure 5(a,b)). It increased first and then decreased in autumn, with the largest value occurring in the stable stage of eddy evolution (Figure 6(c1–c3,d1–d3)). The high values of the SCMD on the southern side of the eddy expanded northward, from the edge of 2R–3R, from the stage of development to that of stability of the CEs in autumn. The radially averaged monopole structures reflected significant spatial disparities in the SCMD within the CEs: in summer, the SCMD was internally shallow and externally deep, with a pronounced gradient of depth that increased from the center to the edge of the eddy (Figure 6(b1–b3)). The three stages of evolution of the CEs in autumn resembled their pattern in the summer, with the SCMD being internally shallow and externally deep. Compared with those in the summer, changes in the SCMD were more pronounced at the center in autumn during the stable period (Figure 6(d2), with the contour lines of the SCMD more densely distributed within the average radius of the eddy). This was different from its pattern of increase from the center to the edge of the eddy, with its maximum value occurring between 1R and 2R. Changes in the SCMD were smaller than 10 m during the evolution of the CE within 3R. The mean SCMD ranged from 42.6 to 43.6 m in summer, and from 51.8 to 53.4 m in autumn.

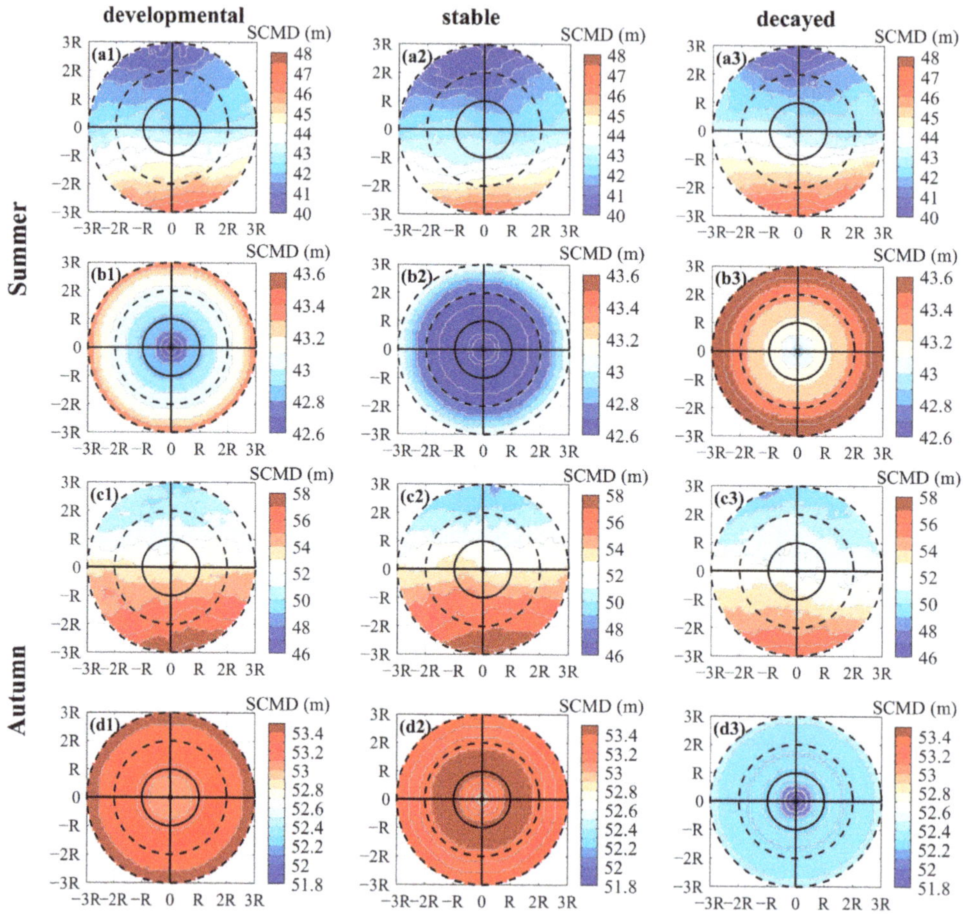

Figure 6. Spatiotemporal distributions of the SCMD of the CEs in summer (**a1**–**a3**,**b1**–**b3**) and autumn (**c1**–**c3**,**d1**–**d3**) in different stages of eddy evolution. (**a1**–**a3**,**c1**–**c3**) show the distributions of the SCMD in the stages of development, stability, and decay of the eddy, while (**b1**–**b3**,**d1**–**d3**) show the radially averaged monopoles of the SCMD.

The SCMD within the ACEs also exhibited a pattern whereby it was deeper in the south and shallower in the north (Figure 7(a1–a3,c1–c3)). The high SCMD gradually expanded from the southern side toward the eddy center from the stage of development to that of stability of the ACEs, although this trend weakened in the stage of decay. The SCMD in the ACEs increased first and then decreased with the evolution of the eddies, reaching its highest values in the stable stage (Figure 7(b1–b3,d1–d3)). The values of SCMD at the eddy center were 44.3 m, 45.2 m, and 44 m in the stages of development, stability, and decay of the eddies, respectively, in summer. It had a distinctly high value that gradually decreased from the inside to the outside of the eddies (Figure 7(b1–b3)). The monopole structures show that the SCMD was deeper near the eddy center and shallower along its edges in the stages of stability and decay (Figure 7(d2–d3)). Compared with the CEs in summer, the monopole structure of the ACEs exhibited the opposite characteristics, whereby the internal SCMD was more uniform. The SCMD in the ACEs was slightly deeper than that in the CEs during their evolution in summer, which is consistent with the structures of the profiles shown in Figure 5.

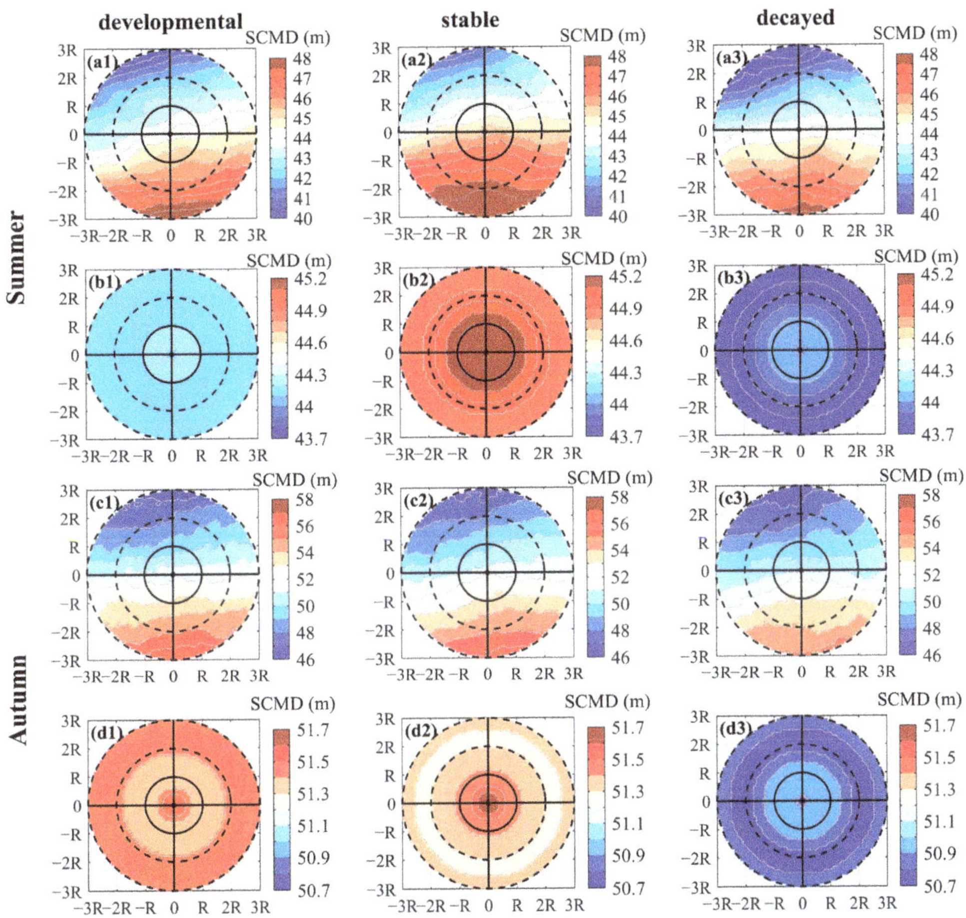

Figure 7. Spatiotemporal distributions of the SCMD of the ACEs in summer (**a1–a3**,**b1–b3**) and autumn (**c1–c3**,**d1–d3**) in different stages of eddy evolution. (**a1–a3**,**c1–c3**) show the distributions of the SCMD in the stages of development, stability, and decay of the eddy, while (**b1–b3**,**d1–d3**) show the radially averaged monopoles of the SCMD.

3.3. SCMD Response to Eddy Evolution

The time series of the SCMD, *APD*, sea surface temperature (SST), and mixing layer depth (MLD) in Figure 8 attest to the influence of the evolution of the eddies on the SCMD. The SCMD, SST, and MLD exhibited significant seasonal variations within the eddies, and the intensity of ACEs also exhibited prominent seasonal differences (Figure 8(b1)). It represents occasional, abrupt changes occurring due to the scarcity of eddies and their short lifetimes in the Kuroshio–Oyashio confluence region (Figure 8(a1)). The values of the SCMD corresponding to both the CEs and ACEs were positively correlated with the *APD* and SST, and were negatively correlated with the MLD. The SCMD was predominantly located in shallower layers of water from February to April of each year, corresponding to a lower SST (<15 °C) and a deeper MLD (>75 m). Conversely, significant ocean warming occurred from July to October, with the SST exceeding 18 °C. The SCMD reached its maximum values of 40–70 m during this period, when the MLD decreased to above 25 m. Meanwhile, the *APD* of the ACEs exceeded 0.03 m as the SST and MLD remained large and stable. The maximum SCMD in the ACEs was about 50 m, slightly smaller than that in the CEs (>60 m).

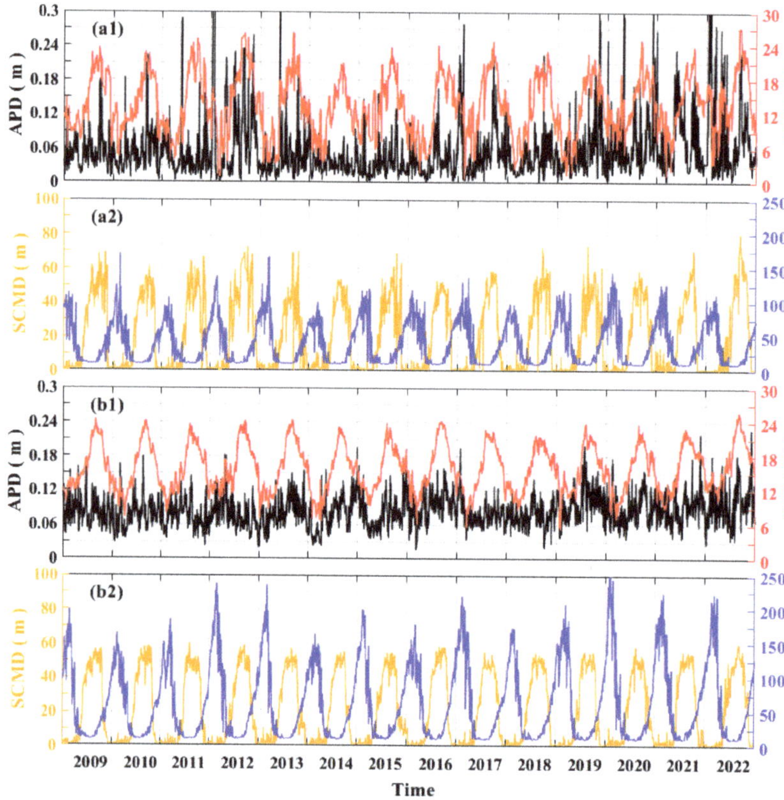

Figure 8. Time series of the SCMD, *APD*, SST, and MLD within the mean radius of the eddies on each day. (**a1**,**a2**) show the sequences of parameters of the CEs, while (**b1**,**b2**) show those of the ACEs.

Table 1 lists the correlation coefficients of the SCMD, *APD*, SST, and MLD. Consistently with the characteristics shown in Figure 8, we noted a strong positive correlation between the SCMD and the SST in the CEs (0.83) and ACEs (0.873). The SCMD was strongly negatively correlated with the MLD, with a value of −0.622 for CEs and −0.687 for ACEs. The correlation coefficient between the *APD* and SST of the CEs was 0.335 and that of the ACEs was 0.312. These results show that the SCMD was primarily controlled by the temperature, with additional effects stemming from the dynamic mixing associated with the eddies.

Table 1. Correlation coefficients of the SCMD, *APD*, SST, and MLD within eddies.

	CEs				ACEs			
	SCMD	*APD*	SST	MLD	SCMD	*APD*	SST	MLD
SCMD	1	0.171	0.83	−0.622	1	0.144	0.873	−0.687
APD	0.171	1	0.335	−0.06	0.144	1	0.312	−0.056
SST	0.83	0.335	1	−0.508	0.873	0.312	1	−0.632
MLD	−0.622	−0.06	−0.508	1	−0.687	−0.056	−0.632	1

To preserve the integrity of the table and facilitate comparative analysis, non-significant correlation coefficients are retained.

In light of seasonal variations in the SCMD, Figure 9 shows the curves of variations in the SCMD, *APD*, SST, and MLD within the average radius of the eddies in the summer and autumn (June to November), to explore the response of the SCMD to eddy evolution. Figure 9 shows the positive correlation of the SCMD with the *APD*/SST, as well as its negative correlation with the MLD. These correlations were more pronounced in the CEs in the summer (Figure 9(a1); Table 2) and the ACEs in the autumn (Figure 9(b1); Table 3). ACEs with a long lifetime and a large radius were the dominant feature of the Kuroshio–Oyashio confluence region. They had a higher SST, and their values of the *APD* in the summer and autumn were greater than those of the CEs, while they had a deeper MLD, as well.

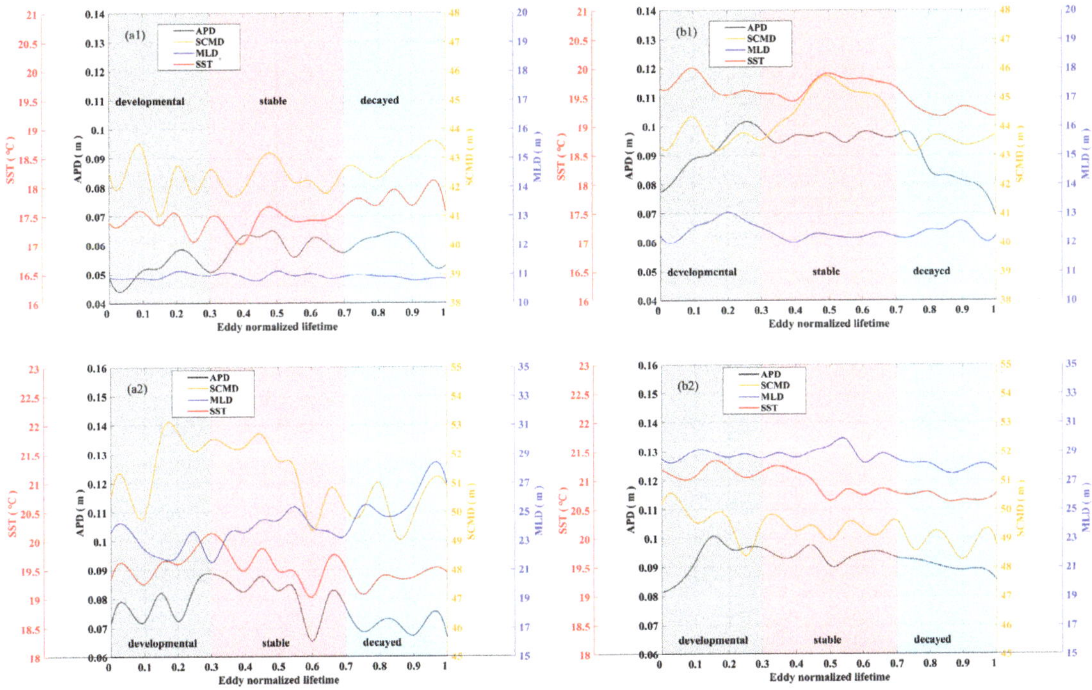

Figure 9. Changes in the SCMD, *APD*, SST, and MLD within the mean radius of the eddies during their evolution in summer (**a1**,**b1**) and autumn (**a2**,**b2**). (**a**) shows the CEs and (**b**) shows the ACEs.

Table 2. Correlation coefficients of the SCMD, *APD*, SST and MLD during the evolution of eddies in summer.

	CEs				ACEs			
	SCMD	*APD*	SST	MLD	SCMD	*APD*	SST	MLD
SCMD	1.000	0.035	0.723	−0.252	1.000	0.464	0.626	−0.360
APD	0.035	1.000	0.157	0.166	0.464	1.000	0.489	0.134
SST	0.723	0.157	1.000	−0.293	0.626	0.489	1.000	−0.062
MLD	−0.252	0.166	−0.293	1.000	−0.360	0.134	−0.062	1.000

To preserve the integrity of the table and facilitate comparative analysis, non-significant correlation coefficients are retained.

The SCMD first decreased and then increased in the ACEs in summer (Figure 9(b1)) and the CEs in autumn (Figure 9(a1)). The maximum SCMD of the ACEs in summer was slightly deeper (>45.5 m), owing to a higher intensity of the eddies and a higher SST (Figure 9(b1)). A similar variation was observed in the CEs in autumn, but with a higher maximum SCMD (>52.5 m). This shows that the impact of the type and intensity of the eddies on the SCMD exhibited seasonal differences. Changes in the SCMD and SST were highly consistent from 20% to 80% of the lifetime of the CEs. The SST and SCMD exhibited significant fluctuations throughout the lifetime of the CEs in summer, where this was influenced by the instability and energy of the eddies (Figure 9(a1)). As the CEs evolved in autumn (Figure 9(a2)), the MLD gradually deepened (enhanced mixing), and exhibited a similar trend in variation to those of the *APD* and SST. Two distinct increases in the value of the SST, at about 30% and 65% of the lifetime of the CEs, corresponded to a deepening SCMD and a rise in the MLD. This phenomenon highlights the positive effect of the SST on the SCMD, moderated by the eddies, as well as the negative effect of seawater mixing on it. We also observed instances of sea surface cooling and an increase in the MLD at 60% of the lifetime of the eddies while the SCMD decreased. That is, the SST had a more significant influence on the SCMD than the MLD. The MLD was relatively shallower from 30% to 70% of the lifetime of the ACEs in summer (Figure 9(b1)) than in autumn, while both the SST and SCMD first increased and then decreased. The SCMD varied within a range of 2 m, with a maximum value of 45.8 m. By contrast, the environmental parameters of the ACEs remained relatively stable during autumn (Figure 9(b2)), with the SCMD ranging from 48 to 50 m. Compared with that in the CEs, stronger mixing within the ACEs led to a slightly lower SCMD.

Table 3. Correlation coefficients of the SCMD, *APD*, SST, and MLD during the evolution of eddies in autumn.

	CEs				ACEs			
	SCMD	***APD***	**SST**	**MLD**	**SCMD**	***APD***	**SST**	**MLD**
SCMD	1.000	0.763	0.748	−0.372	1.000	−0.134	0.582	−0.687
APD	0.763	1.000	0.842	−0.401	−0.134	1.000	0.401	−0.056
SST	0.748	0.842	1.000	−0.394	0.582	0.401	1.000	−0.632
MLD	−0.372	−0.401	−0.394	1.000	0.308	0.395	0.381	1

To preserve the integrity of the table and facilitate comparative analysis, non-significant correlation coefficients are retained.

4. Discussion

4.1. Impact of Eddy Evolution on SCM

The subsurface Chl-A maximum (SCM) is an important indicator of the primary productivity of the ocean. Figure 10 shows the composite distribution of the SCM in different evolutionary stages of the eddies. Our temporal analysis revealed a downward trend in the SCM for both CEs and ACEs throughout their evolution. The SCM exhibited a pattern whereby it was high in the north and low in the south of the study area during the stages of development and stability of the eddies (Figure 10(a1,b1,a2,b2)). Combining the distributions of the SCMD (Figure 6(a1–a3,c1–c3) shows that a high Chl-A concentration in the northern region was located in shallow waters, and peaked at 0.62 mg/m^3, while the southern SCM was approximately 0.06 mg/m^3 lower, and was located in deep waters. Moreover, the horizontal gradient of the SCM within the mean radius of the eddy was relatively low in its stages of development and stability. There were significant variations in Chl-A in the CEs and ACEs during the decay stage, and might have been related to the differences in eddy energy (Figure 10(a3,b3)).

By contrast, the SCM in the CEs was lower than that in the ACEs during their evolution. It had a high value (>0.6 mg/m^3) inside the CEs during their development. Once the CEs had stabilized, the corresponding SCM decreased. A low-value center (~0.49 mg/m^3) was positioned about 2R east of the eddy center in the stage of decay of the

CEs. Figure 10(b1–b3) show that the SCM outside the ACEs (from 1R to 3R) was higher than that at the eddy center during the stages of development and stability. When the ACEs decayed, the SCM exhibited a notable horizontal disparity that was characterized by two low-value centers in the southwest and northeast.

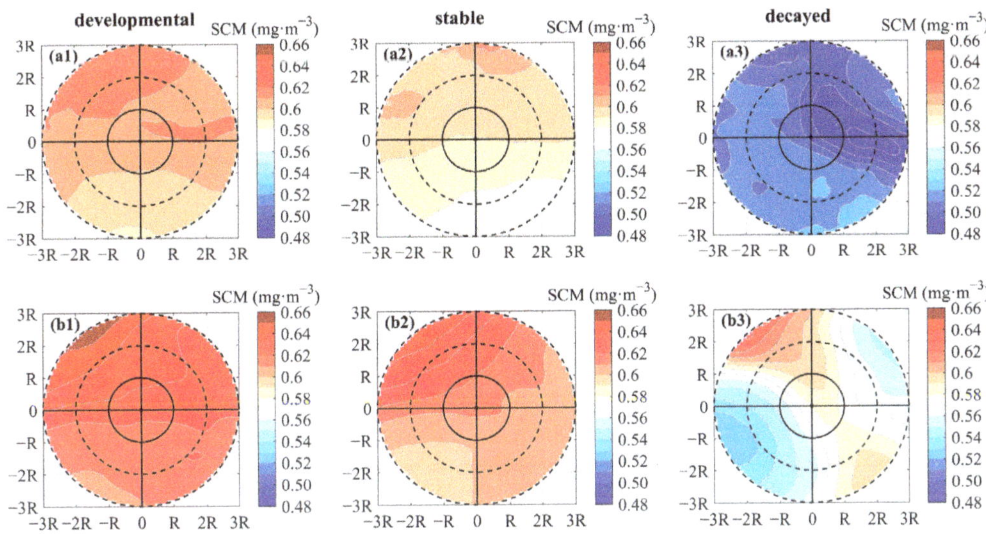

Figure 10. Spatial distributions of the SCM in the stages of development, stability, and decay of eddy evolution. (**a1–a3**) CEs. (**b1–b3**) ACEs.

Although upwelling and downwelling in the stages of formation of the CEs and ACEs likely lead to an increase and a decrease in the primary productivity of the ocean surface, respectively [5,13,30], the intrinsic properties of water masses carried by eddies are also important for the concentration of deep chlorophyll [7]. CEs are mostly generated in the northern regions of the study area, while many ACEs originate from the Kuroshio Extension. However, some are formed in the Sea of Okhotsk, the Bering Sea, and the Alaskan Stream [41–43]. Due to differences in their mechanisms of formation, there are significant variations within the ACEs in this region. Studies of the 3D structure of eddies have verified the presence of ACEs with a cold and saline core. Moreover, some ACEs with a distinct, warm core in the upper layer have a cold and fresh core in the lower layer [44]. This suggests the need to comprehensively explore the mechanism whereby eddy evolution influences the SCM, by integrating the 3D temperature and salinity of the eddy with the transport of the water mass [11,20,45].

4.2. Transport of SCM Due to Eddies

Hundreds of eddies exhibit distinct differences in regulating the transport of Chl-A due to their generation, movement, and lifetime in the Kuroshio–Oyashio confluence region [46]. The structure of eddies in the region has been investigated, and the results have shown a general trend of westward strengthening in them [45,47,48]. Figure 11 shows the spatiotemporal transport of the SCM by the CEs and ACEs based on the mean transport distances of Chl-A along 16 cardinal directions. Overall, the eddies in this region generally prompted the westward transport of the SCM. Long-distance SCM transport was predominantly observed in the range of 225–315°. The ACEs had a more pronounced effect, and achieved a maximum transport distance of 260 km. However, the volume of transported water varied with the type and stage of development of the eddy. For CEs (Figure 11a), the transport distance of high SCM (0.9–1.8 mg/m^3) ranged from 40 to 120 km in the stage of development, and increased to 60–180 km in the stable stage of their evolution. In the stage of eddy decay, the distance to which the SCM was transported by the CEs decreased along all directions and became more uniform. The ACEs had the greatest influence on SCM transport in their developmental stage (Figure 11b), and this led to the transport of a high SCM (0.9–1.2 mg/m^3) over 220–260 km along the 270–315° direction. Compared with those in the stages of their development and decay (Figure 11(b1,b3)), the ACEs exhibited a distinct characteristic of westward transport while they were stable, and the transport of the SCM was approximately equal in all directions within a distance of 100 km (Figure 11(b2)). The westward transport of the SCM intensified during the decay of the eddy, with a high SCM (>0.9 mg/m^3) concentrated in the range of 80–120 km within a range of angles of 67.5–203°.

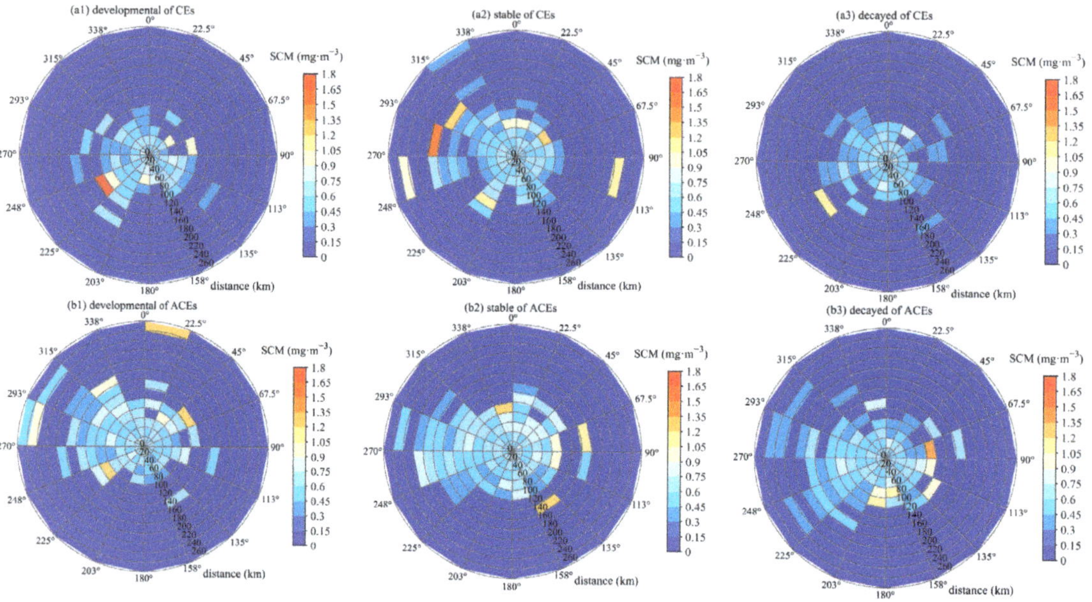

Figure 11. Spatiotemporal transport of the SCM by the CEs (**a**) and ACEs (**b**) during their evolution. (**a1–a3**) represent SCM transport during the stages of development, stability, and decay of the CEs, while (**b1–b3**) show the corresponding results for the ACEs. The direction of the polar coordinates was divided into 16 azimuthal angles, and the radius was assumed to be the relative distance from the eddy center.

4.3. Comparison of Observed Chlorophyll Profiles

Reanalysis datasets do not always perfectly replicate the observed features that rapidly change. Therefore, in situ data are needed to validate the results in general. Many researchers have explored the relationship between eddies and Chl-A concentrations by using BGC–Argo [28,29]. However, recent studies have focused on either vast oceanic areas or continuous observations from specific buoys [29,35]. Owing to the limited number and distribution of data on BGC–Argo profiles, we were unable to present the 3D structure of Chl-A during the evolution of the eddies based solely on observational data. Following the approach reported by Wang et al. (2023) [49], we supplemented the average vertical structure of Chl-A profiles inside the eddies in different seasons, as shown in Figure 5(d1–d3). The observed and analyzed data showed similar seasonal patterns: the Chl-A concentrations in the subsurface layer reached significant maximum values in the summer and autumn, with the peak typically occurring in the range of depths of 30 to

60 m (Figure 5(d2,d3)). In winter, the Chl-A concentrations were higher in the water layers above 60 m in both the CEs and ACEs. Owing to the limited number of available profiles, Chl-A concentrations shallower than 60 m were significantly higher in the CEs than in the ACEs during winter. We noted differences in the magnitudes of Chl-A concentrations between the observed and analyzed data.

This study includes a total of 725 CEs and 1115 ACEs for analysis, with each eddy lasting anywhere from fourteen days to over a hundred days. Given the large number of samples, the SCMD differences across different stages of eddy evolution are reduced, due to the averaging effect and the background gradient distribution within the study area. As a result, these differences are less pronounced compared to those observed in in situ profiles (e.g., BGC-Argo). Additionally, although the Chl-A data used in this study are daily-averaged, they are reanalysis data derived from a biological model. The filtering process removes some real-time information, and the multi-year composite averaging further diminishes the differences in SCMD between CEs and ACEs.

5. Conclusions

In this study, we considered 725 CEs and 1115 ACEs in the Kuroshio–Oyashio confluence region with lifetimes exceeding 14 days from 2009 to 2022, and used them in conjunction with the composite and normalization analyses of high-resolution reanalysis data to investigate seasonal changes in the SCMD and its responses to the evolution of eddies. Mesoscale eddies can cause spatiotemporal changes in the vertical distribution of Chl-A in each stage of the evolution of eddies. The SCMD primarily occurred in the summer (at a depth of 30–50 m) and autumn (40–70 m), due to the mixing of the upper layer of seawater. The SCMD within the eddies was deeper in the south and shallower in the north. BGC–Argo-based observational data showed similar seasonal patterns, with only a difference in the magnitude of the Chl-A concentration.

Within the same season, the SCMD difference between CEs and ACEs is relatively small. It first decreased and then increased as the CEs evolved, and recorded its maximum depth in the stable stage of their evolution in autumn. The deeper SCMD horizontally extended from the edge of the CEs (2R–3R) toward their center. The monopoles of the SCMD in the CEs were internally shallow and externally deep. Higher SCMD values in the ACEs expanded from the south to the center during their stages of development and stability, and weakened as they decayed. The monopoles showed that the SCMD in the ACEs was deeper near the eddy center and shallower along its edges. Compared with that in the CEs, the spatial distribution of the SCMD within 1R of the ACEs was relatively uniform. The SCMD exhibited a positive correlation with the SST (0.83 for CEs and 0.873 for ACEs), and a negative correlation with the MLD (−0.622 for CEs and −0.687 for ACEs). This was influenced by seasonal variations over the long term. Time-series analyses showed that the ACEs had higher values of the *APD*, SST, and MLD. Variations in the SST and SCMD within the CEs were more pronounced, due to their stability and energy. The maximum SCMD within the CEs (>60 m) was marginally deeper than that in the ACEs (~50 m). This verifies that the intensity of the CEs enhanced the SCMD. The positive correlation between the SCMD and *APD* was more pronounced in the ACEs in summer (0.464) and the CEs in autumn (0.763), as the eddies evolved. The SST had more a significant effect on the SCMD than the dynamic eddies. A higher intensity of the ACE, as well as a higher SST in the summer, led to a deeper SCMD, while stronger mixing resulted in a slightly shallower SCMD. This relationship needs to be analyzed in greater detail by considering the intensity of photosynthesis and the biomass of phytoplankton in the SCM layer [29,35,50,51].

Finally, we examined the distribution of the SCM within the eddy and its spatiotemporal transport due to eddy movement. The results showed that in the stages of development and stability during the evolution of the eddy, the northern SCM was larger and concentrated in the upper, shallower layers, while the southern SCM exhibited the opposite characteristics. It decreased, and had a prominent horizontal gradient in the stage of eddy decay. The transport of the SCM followed a pattern of westward intensification, with a majority of its long-distance transport concentrated in the range of 225° and 315°. The ACEs transported the SCM over a longer distance than the CEs, but the mean volume of transport of the SCM in CEs was higher than that in the ACEs. The seasonal changes in the SCMD and its responses to mesoscale eddies were then discussed, based on the composite average. Future research in the area needs to examine the mechanism of action of different types of eddies on the SCM in relation to their 3D structure and material transport. Further investigation is also needed to explore the real-time responses of Chl-A, SCMD, and local environmental parameters to the movement of eddies. This requires a large number of high-resolution observations from within these eddies.

Author Contributions: Conceptualization, Z.C. and C.Z.; methodology, Z.C.; validation, Z.C., J.F. and H.Y.; formal analysis, C.Z.; investigation, J.F. and H.Y.; data curation, Z.C. and H.Y.; writing—original draft preparation, Z.C. and C.Z.; writing—review and editing, Z.C and J.F.; visualization, Z.C. and C.Z.; project administration, Z.C. All authors have read and agreed to the published version of the manuscript.

Funding: This research was funded by the National Natural Science Foundation of China under contract no. 42106090, the Program on the Survey, Monitoring and Assessment of Global Fishery Resources (comprehensive scientific survey of fisheries resources at the high seas) sponsored by the Ministry of Agriculture and Rural Affairs under Grant D-8021-22-0129.

Institutional Review Board Statement: Not applicable.

Informed Consent Statement: Not applicable.

Data Availability Statement: Data are contained within the article.

Acknowledgments: We would like to express our sincere gratitude to our supervisor for the valuable guidance on the eddy detection program, and to the Copernicus Marine Service for providing the data used in this study.

Conflicts of Interest: The authors declare no conflicts of interest.

References

1. Kida, S.; Mitsudera, H.; Aoki, S.; Guo, X.; Ito, S.; Kobashi, F.; Komori, N.; Kubokawa, A.; Miyama, T.; Morie, R. Oceanic fronts and jets around Japan: A review. In *"Hot Spots" in the Climate System: New Developments in the Extratropical Ocean-Atmosphere Interaction Research*; Springer Nature: Berlin/Heidelberg, Germany, 2016; pp. 1–30.
2. Qiu, B. Kuroshio and Oyashio currents. *Ocean. Curr. A Deriv. Encycl. Ocean. Sci.* **2001**, *2*, 61–72.
3. Chen, X.; Chen, Y.; Tian, S.; Liu, B.; Qian, W. An assessment of the west winter–spring cohort of neon flying squid (*Ommastrephes bartramii*) in the Northwest Pacific Ocean. *Fish. Res.* **2008**, *92*, 221–230. [CrossRef]
4. Xu, W.; Wang, G.; Cheng, X.; Xing, X.; Qin, J.; Zhou, G.; Jiang, L.; Chen, B. Mesoscale eddy modulation of subsurface chlorophyll maximum layers in the South China Sea. *J. Geophys. Res. Biogeosciences* **2023**, *128*, e2023JG007648. [CrossRef]
5. Siegel, D.A.; Peterson, P.; McGillicuddy, D., Jr.; Maritorena, S.; Nelson, N. Bio-optical footprints created by mesoscale eddies in the Sargasso Sea. *Geophys. Res. Lett.* **2011**, *38*, L13608. [CrossRef]
6. Chelton, D.B.; Schlax, M.G.; Samelson, R.M. Global observations of nonlinear mesoscale eddies. *Prog. Oceanogr.* **2011**, *91*, 167–216. [CrossRef]
7. Ueno, H.; Bracco, A.; Barth, J.A.; Budyansky, M.V.; Hasegawa, D.; Itoh, S.; Kim, S.Y.; Ladd, C.; Lin, X.; Park, Y. Review of oceanic mesoscale processes in the North Pacific: Physical and biogeochemical impacts. *Prog. Oceanogr.* **2023**, *212*, 102955. [CrossRef]
8. Maximenko, N.A.; Koshlyakov, M.N.; Ivanov, Y.A.; Yaremchuk, M.I.; Panteleev, G.G. Hydrophysical experiment "Megapolygon-87" in the northwestern Pacific subarctic frontal zone. *J. Geophys. Res. Ocean.* **2001**, *106*, 14143–14163. [CrossRef]
9. Ferrari, R.; Wunsch, C. Ocean circulation kinetic energy: Reservoirs, sources, and sinks. *Annu. Rev. Fluid Mech.* **2009**, *41*, 253–282. [CrossRef]

10. Zhang, Z.; Qiu, B. Surface chlorophyll enhancement in mesoscale eddies by submesoscale spiral bands. *Geophys. Res. Lett.* **2020**, *47*, e2020GL088820. [CrossRef]
11. Zhao, D.; Xu, Y.; Zhang, X.; Huang, C. Global chlorophyll distribution induced by mesoscale eddies. *Remote Sens. Environ.* **2021**, *254*, 112245. [CrossRef]
12. Chelton, D.B.; Gaube, P.; Schlax, M.G.; Early, J.J.; Samelson, R.M. The influence of nonlinear mesoscale eddies on near-surface oceanic chlorophyll. *Science* **2011**, *334*, 328–332. [CrossRef] [PubMed]
13. Gaube, P.; McGillicuddy, D.J., Jr.; Chelton, D.B.; Behrenfeld, M.J.; Strutton, P.G. Regional variations in the influence of mesoscale eddies on near-surface chlorophyll. *J. Geophys. Res. Ocean.* **2014**, *119*, 8195–8220. [CrossRef]
14. Falkowski, P.G.; Ziemann, D.; Kolber, Z.; Bienfang, P.K. Role of eddy pumping in enhancing primary production in the ocean. *Nature* **1991**, *352*, 55–58. [CrossRef]
15. Oschlies, A.; Garçon, V. Eddy-induced enhancement of primary production in a model of the North Atlantic Ocean. *Nature* **1998**, *394*, 266–269. [CrossRef]
16. Siegel, D.A.; McGillicuddy, D.J., Jr.; Fields, E.A. Mesoscale eddies, satellite altimetry, and new production in the Sargasso Sea. *J. Geophys. Res. Ocean.* **1999**, *104*, 13359–13379. [CrossRef]
17. McGillicuddy, D.J., Jr.; Robinson, A.; Siegel, D.; Jannasch, H.; Johnson, R.; Dickey, T.; McNeil, J.; Michaels, A.; Knap, A. Influence of mesoscale eddies on new production in the Sargasso Sea. *Nature* **1998**, *394*, 263–266. [CrossRef]
18. Xu, G.; Dong, C.; Liu, Y.; Gaube, P.; Yang, J. Chlorophyll rings around ocean eddies in the North Pacific. *Sci. Rep.* **2019**, *9*, 2056. [CrossRef] [PubMed]
19. He, Q.; Zhan, H.; Cai, S.; Li, Z. Eddy effects on surface chlorophyll in the northern South China Sea: Mechanism investigation and temporal variability analysis. *Deep. Sea Res. Part I Oceanogr. Res. Pap.* **2016**, *112*, 25–36. [CrossRef]
20. Zhang, Z.; Wang, W.; Qiu, B. Oceanic mass transport by mesoscale eddies. *Science* **2014**, *345*, 322–324. [CrossRef] [PubMed]
21. Baldry, K.; Strutton, P.G.; Hill, N.A.; Boyd, P.W. Subsurface chlorophyll-a maxima in the Southern Ocean. *Front. Mar. Sci.* **2020**, *7*, 671. [CrossRef]
22. Yasunaka, S.; Ono, T.; Sasaoka, K.; Sato, K. Global distribution and variability of subsurface chlorophyll a concentration. *Ocean. Sci. Discuss.* **2021**, *2021*, 1–22.
23. Matthes, L.; Bélanger, S.; Raulier, B.; Babin, M. Impact of subsurface chlorophyll maxima on satellite-based Arctic spring primary production estimates. *Remote Sens. Environ.* **2023**, *298*, 113795. [CrossRef]
24. Cullen, J.J. Subsurface chlorophyll maximum layers: Enduring enigma or mystery solved? *Annu. Rev. Mar. Sci.* **2015**, *7*, 207–239. [CrossRef]
25. He, Q.; Zhan, H.; Xu, J.; Cai, S.; Zhan, W.; Zhou, L.; Zha, G. Eddy-induced chlorophyll anomalies in the western South China Sea. *J. Geophys. Res. Ocean.* **2019**, *124*, 9487–9506. [CrossRef]
26. Barone, B.; Church, M.J.; Dugenne, M.; Hawco, N.J.; Jahn, O.; White, A.E.; John, S.G.; Follows, M.J.; DeLong, E.F.; Karl, D.M. Biogeochemical dynamics in adjacent mesoscale eddies of opposite polarity. *Glob. Biogeochem. Cycles* **2022**, *36*, e2021GB007115. [CrossRef]
27. Huang, J.; Xu, F. Observational evidence of subsurface chlorophyll response to mesoscale eddies in the North Pacific. *Geophys. Res. Lett.* **2018**, *45*, 8462–8470. [CrossRef]
28. Su, J.; Strutton, P.G.; Schallenberg, C. The subsurface biological structure of Southern Ocean eddies revealed by BGC-Argo floats. *J. Mar. Syst.* **2021**, *220*, 103569. [CrossRef]
29. Cornec, M.; Laxenaire, R.; Speich, S.; Claustre, H. Impact of mesoscale eddies on deep chlorophyll maxima. *Geophys. Res. Lett.* **2021**, *48*, e2021GL093470. [CrossRef]
30. Guo, M.; Xiu, P.; Li, S.; Chai, F.; Xue, H.; Zhou, K.; Dai, M. Seasonal variability and mechanisms regulating chlorophyll distribution in mesoscale eddies in the South China Sea. *J. Geophys. Res. Ocean.* **2017**, *122*, 5329–5347. [CrossRef]
31. Damien, P.; Pasqueron de Fommervault, O.; Sheinbaum, J.; Jouanno, J.; Camacho-Ibar, V.F.; Duteil, O. Partitioning of the open waters of the Gulf of Mexico based on the seasonal and interannual variability of chlorophyll concentration. *J. Geophys. Res. Ocean.* **2018**, *123*, 2592–2614. [CrossRef]
32. Damien, P.; Sheinbaum, J.; Pasqueron de Fommervault, O.; Jouanno, J.; Linacre, L.; Duteil, O. Do Loop Current eddies stimulate productivity in the Gulf of Mexico? *Biogeosciences* **2021**, *18*, 4281–4303. [CrossRef]
33. Aumont, O.; Éthé, C.; Tagliabue, A.; Bopp, L.; Gehlen, M. PISCES-v2: An ocean biogeochemical model for carbon and ecosystem studies. *Geosci. Model Dev. Discuss.* **2015**, *8*, 1375–1509. [CrossRef]
34. de Boyer Montégut, C.; Madec, G.; Fischer, A.S.; Lazar, A.; Iudicone, D. Mixed layer depth over the global ocean: An examination of profile data and a profile-based climatology. *J. Geophys. Res. Ocean.* **2004**, *109*, C12003. [CrossRef]
35. Xing, X.; Boss, E.; Chen, S.; Chai, F. Seasonal and daily-scale photoacclimation modulating the phytoplankton chlorophyll-carbon coupling relationship in the mid-latitude Northwest Pacific. *J. Geophys. Res. Ocean.* **2021**, *126*, e2021JC017717. [CrossRef]

36. Nencioli, F.; Dong, C.; Dickey, T.; Washburn, L.; McWilliams, J.C. A vector geometry–based eddy detection algorithm and its application to a high-resolution numerical model product and high-frequency radar surface velocities in the Southern California Bight. *J. Atmos. Ocean. Technol.* **2010**, *27*, 564–579. [CrossRef]
37. Liu, Y.; Dong, C.; Guan, Y.; Chen, D.; McWilliams, J.; Nencioli, F. Eddy analysis in the subtropical zonal band of the North Pacific Ocean. *Deep. Sea Res. Part I Oceanogr. Res. Pap.* **2012**, *68*, 54–67. [CrossRef]
38. Chuang, Z.; Su, H.; Zhang, C.; Shao, Y.; Hu, S. An eddy evolution based on the three-dimensional structure of chlorophyll A in the Kuroshio-Oyashio Confluence Region. *Mar. Sci. Bull.* **2024**, *43*, 437–451.
39. Guo, M.; Chai, F.; Xiu, P.; Li, S.; Rao, S. Impacts of mesoscale eddies in the South China Sea on biogeochemical cycles. *Ocean. Dyn.* **2015**, *65*, 1335–1352. [CrossRef]
40. Chen, J.; Gong, X.; Guo, X.; Xing, X.; Lu, K.; Gao, H.; Gong, X. Improved perceptron of subsurface chlorophyll maxima by a deep neural network: A case study with BGC-Argo float data in the northwestern Pacific Ocean. *Remote Sens.* **2022**, *14*, 632. [CrossRef]
41. Rogachev, K.A. Recent variability in the Pacific western subarctic boundary currents and Sea of Okhotsk. *Prog. Oceanogr.* **2000**, *47*, 299–336. [CrossRef]
42. Yasuda, I.; Ito, S.; Shimizu, Y.; Ichikawa, K.; Ueda, K.-I.; Honma, T.; Uchiyama, M.; Watanabe, K.; Sunou, N.; Tanaka, K. Cold-core anticyclonic eddies south of the Bussol'Strait in the northwestern subarctic Pacific. *J. Phys. Oceanogr.* **2000**, *30*, 1137–1157. [CrossRef]
43. Rogachev, K.; Shlyk, N.; Carmack, E. The shedding of mesoscale anticyclonic eddies from the Alaskan Stream and westward transport of warm water. *Deep. Sea Res. Part II Top. Stud. Oceanogr.* **2007**, *54*, 2643–2656. [CrossRef]
44. Itoh, S.; Yasuda, I. Water mass structure of warm and cold anticyclonic eddies in the western boundary region of the subarctic North Pacific. *J. Phys. Oceanogr.* **2010**, *40*, 2624–2642. [CrossRef]
45. Yao, H.; Ma, C.; Jing, Z.; Zhang, Z. On the Vertical Structure of Mesoscale Eddies in the Kuroshio-Oyashio Extension. *Geophys. Res. Lett.* **2023**, *50*, e2023GL105642. [CrossRef]
46. Cui, W.; Yang, J.; Zhou, C. Mesoscale Eddy Properties in Four Major Western Boundary Current Regions. *Remote Sens.* **2024**, *16*, 4470. [CrossRef]
47. Dong, D.; Brandt, P.; Chang, P.; Schütte, F.; Yang, X.; Yan, J.; Zeng, J. Mesoscale eddies in the northwestern Pacific Ocean: Three-dimensional eddy structures and heat/salt transports. *J. Geophys. Res. Ocean.* **2017**, *122*, 9795–9813. [CrossRef]
48. Sun, W.; Dong, C.; Wang, R.; Liu, Y.; Yu, K. Vertical structure anomalies of oceanic eddies in the Kuroshio Extension region. *J. Geophys. Res. Ocean.* **2017**, *122*, 1476–1496. [CrossRef]
49. Wang, T.; Zhang, S.; Chen, F.; Xiao, L. The Seasonality of Eddy-Induced Chlorophyll-a Anomalies in the Kuroshio Extension System. *Remote Sens.* **2023**, *15*, 3865. [CrossRef]
50. Itoh, S.; Yasuda, I.; Saito, H.; Tsuda, A.; Komatsu, K. Mixed layer depth and chlorophyll a: Profiling float observations in the Kuroshio–Oyashio Extension region. *J. Mar. Syst.* **2015**, *151*, 1–14. [CrossRef]
51. Wang, Y.; Tang, R.; Yu, Y.; Ji, F. Variability in the sea surface temperature gradient and its impacts on chlorophyll-a concentration in the Kuroshio extension. *Remote Sens.* **2021**, *13*, 888. [CrossRef]

Journal of
*Marine Science
and Engineering*

MDPI

Article

Seasonal Variability and Underlying Dynamical Processes of Sea Surface Temperature Fronts in Zhoushan and Its Adjacent Seas

Hui Chen [1], Qiyan Ji [1,*], Qiong Wu [1], Tengteng Peng [2], Yuting Wang [1] and Ziyin Meng [1]

[1] Marine Science and Technology College, Zhejiang Ocean University, Zhoushan 316022, China; chenhui@zjou.edu.cn (H.C.); qiongwu_edu@yeah.net (Q.W.); wangyuting@zjou.edu.cn (Y.W.); mengziyin@zjou.edu.cn (Z.M.)

[2] Polar Research Institute of China, Shanghai 201209, China; pengtengteng@pric.org.cn

* Correspondence: jiqiyan@zjou.edu.cn

Abstract: The oceanic fronts play an important role in marine ecosystems and fisheries. This study investigates the seasonal variability of sea surface temperature (SST) fronts in Zhoushan and its adjacent seas for the period 1982–2021. The influences of various underlying dynamic processes on the fronts are also discussed. The horizontal gradient of SST is calculated as frontal intensity, and a threshold value of 0.03 °C/km is set to count the frontal frequency. The fronts in Zhoushan and its adjacent seas show significant seasonal variability, with high (0.1 °C/km and 60–90%) and low (0.03 °C/km and 30–60%) frontal activity in winter and summer, respectively. In summer, the fronts along Jiangsu and the north of the Changjiang River Estuary show higher frontal intensity and frequency, which is mainly influenced by the Changjiang diluted water and southerly wind, and fronts around Zhoushan Island are highly related with Zhoushan upwelling. In winter, the fronts strengthen into regular bands offshore and parallel to the coast, which are mainly influenced by coastal currents. Frontal intensity and frequency show a more significant long-term increasing trend in winter than in summer.

Keywords: sea surface temperature fronts; seasonal variability; dynamical processes; frontal detection; Zhoushan

1. Introduction

Ocean fronts typically occur in a narrow strip with a high horizontal gradient in physically, chemically, or biologically relevant properties and often form between two water masses with significantly different characteristics, such as temperature, salinity, and density [1–3]. Frontal waters typically contain abundant nutrients [4], which are beneficial in promoting phytoplankton growth [5], increasing marine primary productivity, providing abundant bait for fish and other species, and further affecting marine fisheries and marine ecosystems [6–8]. Fronts also play an important role in marine pollution. In the frontal zone, the convergence of seawater is commonly enhanced, which can concentrate substances, such as oil pollution, microplastics, and heavy metals, posing a threat to the ecosystem [9–11]. The strong non-linear properties of mesoscale and submesoscale fronts in the ocean further increase the heat flux coefficient, enhancing heat exchange at the air–sea interface and tuning air–sea interactions [12–15]. Therefore, understanding the variability and the dynamic mechanisms of the main fronts in offshore and nearshore waters is of considerable importance.

In recent decades, high-resolution remote sensing data have been widely used in the study of variability in regional ocean fronts on multiple spatial-temporal scales [16–18]. Various studies have shown that frontogenesis involves many different dynamic processes [19,20]. Strong western boundary currents, such as the Kuroshio and

Gulf Stream, intersect with low-temperature and low-salinity continental shelf waters to form major ocean fronts [21]. Coastal upwelling is another important driver of ocean fronts, which brings low-temperature bottom water to the sea surface [22,23]. Topography is also an important factor in front generation, and sloping terrain can easily induce front formation [24–26]. Some studies have found a significant correlation between sea surface temperature (SST) fronts and wind, suggesting that an increase in the wind-driven upwelling increases the frequency of fronts [22,27]. Meanwhile, global climate change would have a profound effect on large-scale oceanic fronts, resulting in several degrees of latitudinal shift [7,28].

Numerous previous studies focus on the detection methods, characteristics, and mechanisms of ocean fronts near the East China Sea (ECS) and pay attention to the influence of ocean fronts on the biochemical environment. The types of ocean fronts in ECS have been systematically summarized based on hydrological observations from 1934 to 1988, along with the seasonal variations in the Kuroshio front and the coastal front in Zhejiang [29,30]. He et al. (1995) [31] identified five types of fronts, namely shelf-break front, thermohaline front, estuarine front, tidal front, and upwelling front, using satellite SST and altimeter data and presented the locations of eight main frontal zones in eastern China Seas. Ning et al. (1998) [32] pointed out that SST data are more suitable for distinguishing and identifying water masses and fronts in ECS during winter, while water color data are more effective in detecting the fronts during summer and fall. Hickox et al. (2000) [33] and Park et al. (2006) [34] calculated the front frequency to approximate the distribution of the main fronts in the Yellow Sea, ECS, and northern South China Sea and analyzed the seasonal variation in the frontal intensity.

However, the majority of the above-mentioned studies focus on large- to meso-scale fronts near the China seas and are based on relatively short-term satellite observation data. Zhoushan and its adjacent seas, as a special part of ECS, are located in the coastal area of Zhejiang, China. Major surface features surrounding Zhoushan Island can be found in the previous studies, including those in the ECS. The SST isotherms are mostly parallel to the isobaths, with a southwest–northeast orientation [35–37]; they are mainly dominated by a monsoon with a northerly wind during winter and southerly wind during summer, and the alongshore component of wind stress can drive coastal upwelling during the summer [35]. Along the 20 m isobaths, from the north to the east and south of Zhoushan Island, two major southward coastal currents, namely, the Yellow Sea coastal current and ECS coastal current, persist along the Jiangsu and Zhejiang coasts [38]. These currents can be enhanced by the discharge of Changjiang River and weakened by the southerly summer wind. The inshore branch of the Taiwan Warm current, a northward coastal current, flows near the 50 m isobaths off the southeast of the Zhejiang coast. Changjiang River carries a substantial amount of freshwater and terrestrial materials passing through Zhoushan Island [33,39,40]. The ocean fronts in this area show great diversity and are influenced by numerous physical processes of ocean dynamics. However, few direct studies have been performed on the frontal variability around the Zhoushan Island, and the underlying mechanisms of frontogenesis remain to be extensively discussed.

This study aims to investigate the variability in the fronts in Zhoushan and its adjacent seas based on 40 years of high-resolution SST data. The influences of oceanic dynamics on the front are discussed in different seasons. This study will comprehensively describe the characteristics of seasonal evolution of fronts and their underlying mechanisms in this area. The remainder of this paper is organized as follows: Section 2 describes the data and methodology used in this study, Section 3 analyzes the results, Section 4 provides a comprehensive discussion, and Section 5 presents the conclusions.

2. Materials and Methods

2.1. Study Area

Zhoushan and its adjacent seas are located in the east of Zhejiang Province, China, bounded by the Changjiang River Estuary and Hangzhou Bay to the north, ECS to the east, and Fujian offshore to the south (Figure 1). The coastal bathymetry is primarily northeast–southwest, shallower than 100 m, and has relatively rapid changes in the depth of the water in the east of the Zhoushan Island, as shown by the isobaths in Figure 1. The study area had abundant fishing resources. However, the occurrence of marine heatwaves poses a threat to the marine ecosystem, resulting in alterations in the productivity of marine fish catching and the structure of the community.

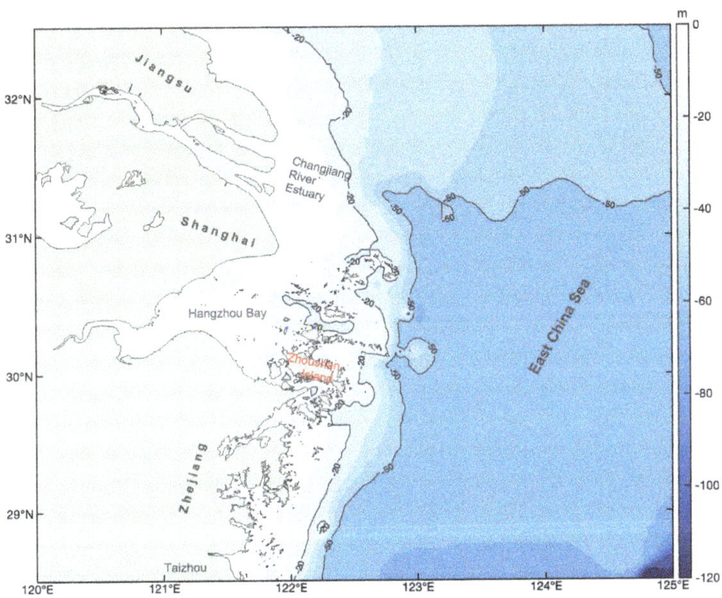

Figure 1. Bathymetry of Zhoushan Island and its adjacent seas. The topography, bathymetry, and shoreline data come from 2-minute Gridded Global Relief Data (ETOPO2) v2, National Geophysical Data Center, NOAA. https://doi.org/10.7289/V5J1012Q (accessed on 1 October 2024).

2.2. Data

2.2.1. SST Data

The global Operational SST and Sea Ice Analysis (OSTIA) reprocessed analysis product used in this study was obtained from the Copernicus Marine Environment Monitoring Service (CMEMS) [41]. This product provides daily gap-free maps of foundation SST and ice concentration (referred to as an L4 product). This product is a satellite and in situ foundation SST analysis created using the OSTIA system using reprocessed ESA SST CCI, C3S EUMETSAT, and REMSS satellite data and in situ data from the HadIOD dataset. The product is available from 1 October 1981 on a global regular grid at 0.05° × 0.05° resolution. The time range of the OSTIA SST data used in this study is from 1 January 1982 to 31 December 2021. Data can be found at DOI (product): https://doi.org/10.48670/moi-00168 (accessed on 4 June 2023).

Since the OSTIA SST is a product that has already been well quality controlled and provided by CMEMS, these data have been pre-processed for analysis of seasonal and spatial variability based on the calculation of climatological monthly mean SST and SST fronts. The climatological monthly mean SST is calculated from the daily SST data. For example, the monthly mean SST of January each year is calculated first, and then the mean SST of January each year from 1982 to 2021 is used to calculate the monthly mean SST of the climate state. The same method is used for the other months. Then, the horizontal gradient of the climatological monthly mean SST is calculated to analyze the spatial distribution characteristics of the SST fronts.

2.2.2. Current Data

The current data were obtained from the GLORYS12V1 reanalysis product, which is also designed and implemented in the CMEMS framework. GLORYS12 is a global ocean eddy-resolving oceanic numerical model output on a standard regular grid at 1/12° (approximately 8 km) and at 50 standard levels from 1993 to the present. The model is largely based on the current real-time global forecasting CMEMS system. The model component is the NEMO platform driven at the surface by ECMWF ERA-Interim, followed by ERA5 reanalysis for recent years. Observations are assimilated by means of a reduced-order Kalman filter. Along-track altimeter data (sea level anomaly), satellite SST, sea ice concentration, and in situ temperature and salinity vertical profiles are jointly assimilated. This product includes daily and monthly mean files for temperature, salinity, currents, sea level, mixed layer depth, and ice parameters from the top to the bottom. The monthly climatology is downloaded directly and used in this study. This data can be accessed from DOI (product): https://doi.org/10.48670/moi-00021 (accessed on 28 December 2023).

2.2.3. Wind Data

The wind data were obtained from ERA5 monthly averaged data on single levels from 1940 to the present. The ERA5 is the fifth-generation ECMWF reanalysis for the global climate and weather. The reanalysis involves model data with observations from across the world into a globally complete and consistent dataset using the laws of physics. This study used monthly data with a spatial resolution of 0.25°, and a period from January 1982 to December 2021 were used. The climatological monthly average wind was obtained by averaging the corresponding monthly data from 1982 to 2021.This data can be accessed at: https://doi.org/10.24381/cds.f17050d7 (accessed on 10 October 2024).

2.3. *Methods*

2.3.1. Front Detection

The fronts in Zhoushan and its adjacent seas were detected by calculating the horizontal gradient of the SST. The gradient magnitudes are treated as frontal intensity and are calculated as follows:

$$grad_{(x,y)} = \sqrt{grad_x{}^2 + grad_y{}^2}, \tag{1}$$

where $grad_{(x,y)}$ is the horizontal gradient at the SST data grid point (x, y), and $grad_x$ and $grad_y$ are the gradient components in the x and y direction as follows:

$$grad_x = \frac{1}{4}\frac{\overline{T} \cdot Gx}{dx}, \tag{2}$$

$$grad_y = \frac{1}{4}\frac{\overline{T} \cdot Gy}{dy}, \tag{3}$$

in which

$$\overline{T} = \begin{bmatrix} T_{(x-1,y+1)} & T_{(x,y+1)} & T_{(x+1,y+1)} \\ T_{(x-1,y)} & T_{(x,y)} & T_{(x+1,y)} \\ T_{(x-1,y-1)} & T_{(x,y-1)} & T_{(x+1,y-1)} \end{bmatrix}, \tag{4}$$

$$Gx = \begin{bmatrix} -1 & 0 & 1 \\ -2 & 0 & 2 \\ -1 & 0 & 1 \end{bmatrix}; Gy = \begin{bmatrix} -1 & -2 & 1 \\ 0 & 0 & 0 \\ 1 & 2 & 1 \end{bmatrix}. \tag{5}$$

In Formula (5), the Sobel operators Gx and Gy, consisting of two 3 × 3 convolution kernels, are used to calculate the gradients $grad_{(x,y)}$ in the x and y directions, respectively. And Gy is Gx rotated by 90° counter-clockwise. The Sobel operator is known as a simple and effective way of enhancing the visibility of edges in digital images [42,43] and is widely used in a variety of applications. Meanwhile, this operator is easy to use and not only produces good detection results, but also has a smooth suppression effect on noise. $\overline{T}$ is equal to the SST data T in a 3 × 3 convolution kernels composed of the Sobel operator (as shown in Formula (4)). In Formulas (2) and (3), dx is the distance in the x direction between points $(x-1, y)$ and $(x+1, y)$, and dy is the distance in the y direction between points $(x, y-1)$ and $(x, y+1)$. The magnitude of the horizontal SST gradient computed by Formulas (1)–(5) is described as the frontal intensity in this study.

2.3.2. Frontal Frequency

After detecting the daily SST gradients based on the daily SST data, the frontal frequency (Fp) can be calculated in each daily gradient map as a ratio of the number of fronts (N_{front}) to the number of valid data (N_{valid}) during a given time period [44,45]. The Fp is calculated as follows:

$$Fp = \frac{N_{front}}{N_{valid}}. \tag{6}$$

A gradient threshold (0.03 °C/km) for the occurrence of coastal fronts is determined based on the results of reprocessing analyses of daily SSTs from OSTIA over a 40-year period, from 1982 to 2021, rather than a pre-defined threshold. Thus, a gradient greater than 0.03 °C/km is considered to be a place or event where an SST front occurred in this study. This threshold is of similar magnitude to the thresholds (0.028 °C/km) used in previous studies [38].

3. Results

3.1. Seasonal Variability

Monthly climatological SSTs are calculated for the period 1982–2021, and the spatial pattern of monthly SST variability is shown in Figure 2. The SST around Zhoushan Island and its adjacent seas show significant monthly variations. In spring (from March to May), the SST isotherms trend from the southwest to the northeast. In March and April, the north is dominated by cold water, which mainly comes from the remnants of winter. In May, the 18 °C isotherm began to converge around Zhoushan Island, indicating that a cold central area caused by upwelling began to form. In summer (from June to August), the average SST in the study area reaches its maximum. However, an evident cold-water center is observed in this study area due to the influence of upwelling. The temperature difference between the center and the periphery of the cold water mass is less than 0.5 °C in June, but exceeds 0.5 °C in July and August. In autumn (from September to November), when the summer upwelling enters the stage of subsidence, the cold-water mass gradually disappears. In September, weak cold-water masses still remained in the center of the study area. Meanwhile, in October and November, the temperature of near-shore water in Zhejiang gradually dropped, and the isothermal distribution returned to the southwest–northeast direction. In winter (from December to January), the isotherms of the mean SST are parallel to the coast of Zhejiang province and the densest in four seasons. The SST gradually increases from the west to the east surrounding the sea waters of Zhoushan Island due to the mixing of cold coastal waters from the north (offshore of Jiangsu) and warm currents from the south (offshore of Zhejiang). The difference in SSTs between the east and west of the study area reaches its maximum throughout the year, indicating that mixed water is conducive to the growth of temperature fronts.

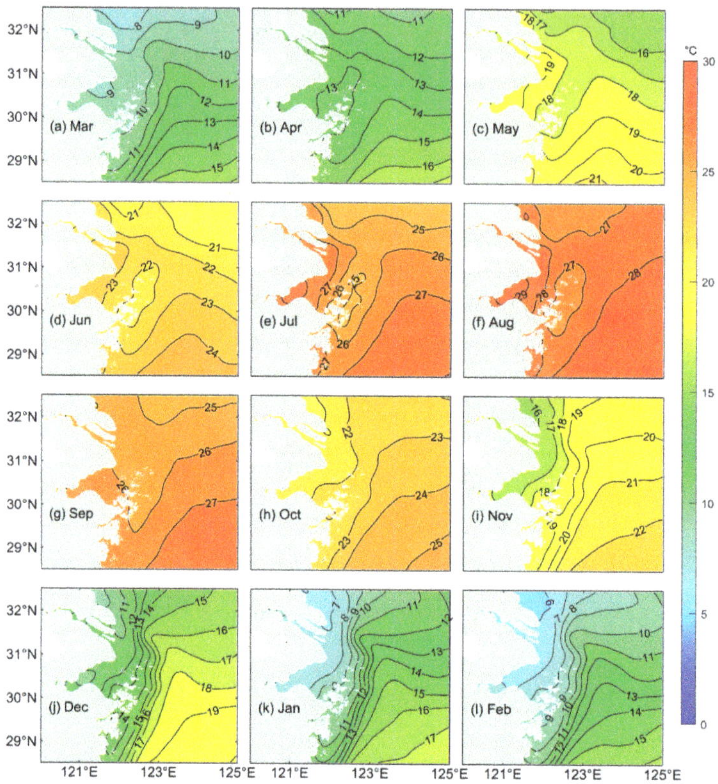

Figure 2. Monthly climatology of SST (1982–2021). (**a**) March; (**b**) April; (**c**) May; (**d**) June; (**e**) July; (**f**) August; (**g**) September; (**h**) October; (**i**) November; (**j**) December; (**k**) January; (**l**) February.

The SST fronts in Zhoushan Island and its adjacent seas also show significant seasonal variability from the spatial distribution of the fronts in each climatological month (Figure 3). In spring, strong temperature fronts are observed along the coast of Zhejiang and the northern part of the Changjiang River Estuary, with an intensity of 0.04 °C/km. However, the intensity of the temperature fronts along the coast of Zhejiang gradually decreased from March to May. Meanwhile, the intensity of temperature fronts in the northern portion of the Changjiang River Estuary shows an increasing trend. In summer, strong temperatures persist on the northern side of the Changjiang River Estuary, and the intensity of the temperature front along the coast of Zhejiang gradually decreases. However, the upwelling front near Zhoushan Island gradually intensifies, and the front intensity reaches more than 0.04 °C/km in July and August. In autumn, the temperature front along the coast of Zhejiang and the northern portion of the Changjiang River Estuary gradually intensifies, while the upwelling front in Zhoushan gradually weakens and disappears. In November, the study area is dominated by the Zhejiang-Fujian coastal front, which forms a coastal front area with an intensity of more than 0.04 °C/km. In winter, the coastal front of Zhejiang-Fujian is the strongest among the four seasons, with a front intensity of more than 0.1 °C/km. This coastal front strengthens from December to January and gradually weakens from February.

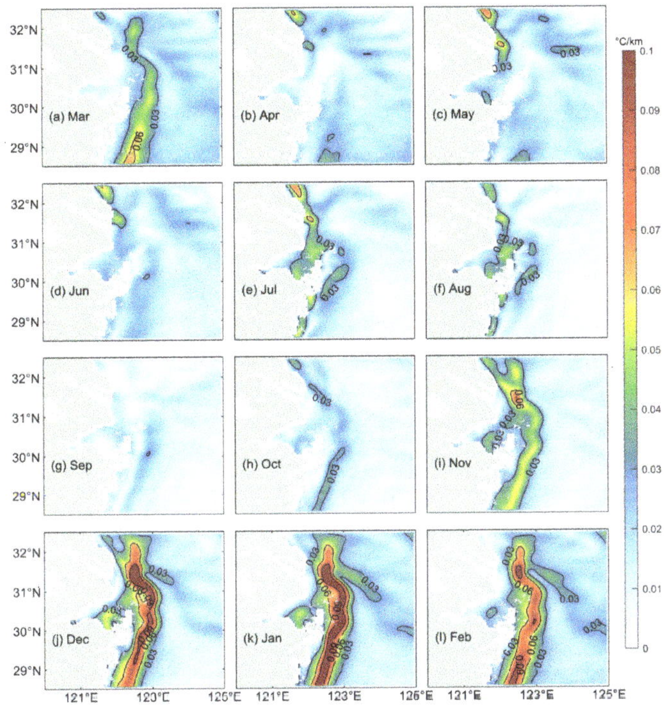

Figure 3. Monthly climatology of the SST frontal intensity (1982–2021). (**a**) March; (**b**) April; (**c**) May; (**d**) June; (**e**) July; (**f**) August; (**g**) September; (**h**) October; (**i**) November; (**j**) December; (**k**) January; (**l**) February.

The high frequency of the front indicates that frontal events are frequently developed, and the sea-water mixing activity is strong. This study used 0.03 °C/km as the frontal threshold. The SST frontal frequency for the climatological month around the Zhoushan Island and adjacent seas is shown in Figure 4. The study area consistently exhibits a high frontal frequency in each month. A significant seasonal variability is also observed. The frontal frequency is highest along the coastal seas in winter (above 90%), and the spatial distribution is similar to the frontal distribution. The frontal activity began to weaken in spring. Two areas with strong frontal activity are separated by the Zhoushan Island in April, May, and June. In the northwest portion of the Zhoushan Island, near the Changjiang River Estuary, the frontal frequency is greater than 60%, with a maximum reaching over 90%. In the southeast of the Zhoushan Island, the total frontal frequency is approximately 60%. The frontal frequency around the Zhoushan Island began to increase in July and August, but it decreased in September and October. The frontal frequency in September and October also reaches a minimum during the year and began to strengthen in November and reached its maximum in winter. The spatial distribution characteristics of frontal activity are also diverse. The frontal frequency is high in the coastal area around the Zhejiang and Changjiang estuaries and low in the offshore area of ECS. In the north of the Changjiang River Estuary, the frontal frequency in spring and summer is higher than in autumn and winter. In the seas around Zhoushan Island, the frontal frequency in summer and winter is higher than that in spring and autumn. The underlying reasons of this phenomenon will be discussed in next section. In the coastal area of Zhejiang-Taizhou, the frontal frequency also shows prominent seasonal variability and is always associated with high frontal activity. The spatial distribution of the frontal frequency and the corresponding frontal intensity is similar, confirming that Zhoushan and its adjacent seas have high frontal activities.

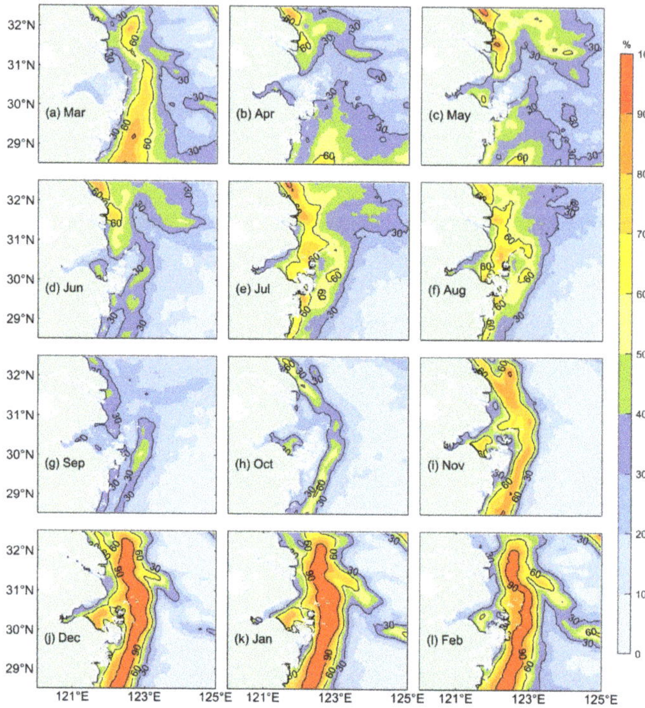

Figure 4. Monthly climatology of the SST frontal frequency (1982–2021). (**a**) March; (**b**) April; (**c**) May; (**d**) June; (**e**) July; (**f**) August; (**g**) September; (**h**) October; (**i**) November; (**j**) December; (**k**) January; (**l**) February.

The analysis of the spatial distribution of the SST frontal intensity and frequency indicated that the SST frontal intensity and frequency in the study area have significant seasonal characteristics. The differences in the seasonal characteristics of the front are influenced by different underlying dynamic processes, especially in the summer and in the winter. However, spring and autumn belong to the transitional season, and the frontal characteristics are the result of alternating and integrating different dynamic processes in summer and winter. Taking summer and winter as the representative seasons, this study analyzes the characteristics of seasonal variability in the fronts around Zhoushan and its adjacent seas and discusses the underlying influencing mechanisms. The climatology of frontal intensity and frequency in summer and winter shows significantly different characteristics (Figure 5). The strong frontal zone in the summer is distributed in different areas of the sea with a distinct regional diversity. Meanwhile, the strong frontal zones in the winter show as a regular band. The frontal zone is roughly divided into three subareas (indicated by the dotted lines in Figure 5) based on the intensity and frequency of the frontal zone in summer and winter. Area A1 refers to the north of the Changjiang River Estuary and the coastal waters of Jiangsu, area A2 indicates the southeast of Hangzhou Bay and the seas around the Zhoushan Island, and area A3 denotes the coastal area of Taizhou, which is also part of the coastal waters of Zhejiang. The different potential dynamic processes are not consistent. Consequently, the frontal characteristics in the three defined areas exhibit significant differences, especially in summer. In area A1, the frontal intensity (more than 0.06 °C/km) and the frontal frequency (more than 60%) are concentrated along the coastal areas of Jiangsu in summer. In winter, the maximum frontal intensity (more than 1 °C/km) and frontal frequency (more than 90%) are observed around the Changjiang River Estuary. In area A2, the summer frontal intensity is greater than 0.03 °C/km, with a summer frontal frequency of about 30–70%. Meanwhile, the frontal frequency in the northeast portion of A2 reaches over 60%. In area A3, a strong frontal zone is located near the coast of Taizhou, with a frontal intensity of about 0.03 °C/km and a frontal frequency near 60% in summer. In winter, the frontal intensity is more than 0.09 °C/km, with a frontal frequency of about 90%.

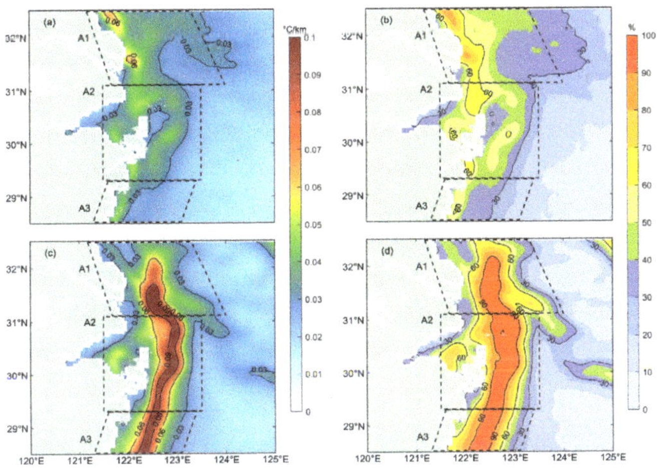

Figure 5. Spatial distribution of the climatological (**a**) summer frontal intensity, (**b**) summer frontal frequency, (**c**) winter frontal intensity, and (**d**) winter frontal frequency from 1982–2021.

3.2. Trends in Summer and Winter Fronts

This study aims to further investigate the long-term trends in frontal intensity and frontal frequency in the three subareas in both winter and summer. The annual mean anomalies (annual mean minus multiyear mean) are calculated for each area, and the long-term linear trend from 1982 to 2021 is analyzed (Figure 6). In area A1 (Figure 6a,b), the frontal intensity and frontal frequency show an increasing trend both in summer and winter, but they are more significant in winter. For example, the linear increasing rates of frontal intensity in summer and winter are 2.5×10^{-6} °C/km/year and 1.6×10^{-4} °C/km/year, respectively, but the latter is significantly higher than the former. The linear increasing rates in frontal frequency for summer and winter are 0.05%/year and 0.23%/year, respectively, while the latter is only more than four times that of the former. In area A2, the long-term trends in frontal intensity and frontal frequency in summer and winter are similar to those in region A1, with some differences in specific values. However, the frontal intensity and the frontal frequency in area A3 show a decreasing trend in summer, while they show an increasing trend in winter. In the same region, the long-term trends show significant differences across the different seasons. The trend in winter is larger than that in summer, and the difference in frontal intensity is much larger than the difference in frontal frequency. The regional differences are also evident for the same season. The frontal intensity and frequency in summer show an increasing trend for A1 and A2, but the latter is larger, while A3 shows a decreasing trend. However, the frontal intensity and frontal frequency in winter show an increasing trend in all three regions, but the differences between the specific values are not as significant as those in summer.

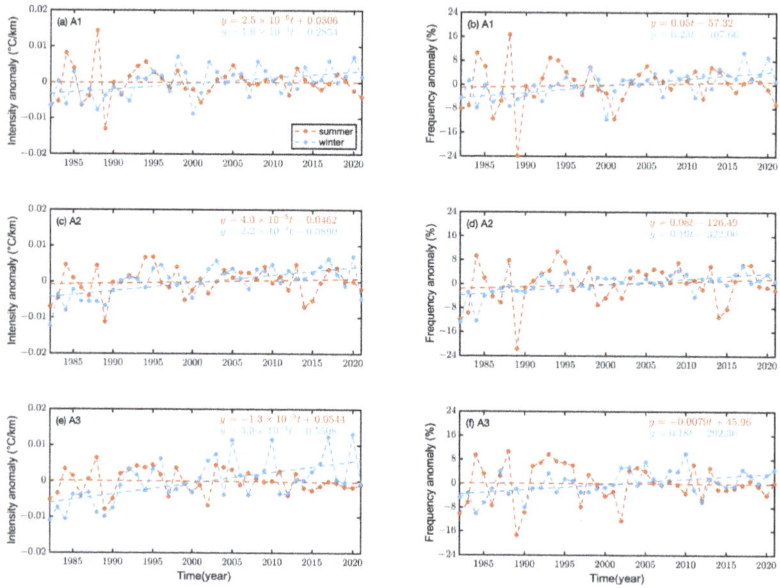

Figure 6. Long-term trends in summer (orange dashed–dotted line) and winter (blue dashed–dotted line) frontal intensity in (**a**) A1, (**c**) A2, and (**e**) A3 and frontal frequency in (**b**) A1, (**d**) A2, and (**f**) A3 from 1982 to 2021.

Figure 7 shows the spatial distribution of the long-term trends in frontal intensity and frequency in summer and winter from 1982 to 2021, where the black dots represent the grid points that passed the 95% significance test. Whether in summer (Figure 7a,b) or winter (Figure 7c,d), the spatial distribution of the black dots that passed the 95% confidence test for frontal intensity and frequency is highly consistent, indicating a local increase in frontal intensity, which leads to a corresponding increase in frontal frequency. The frontal intensity and frontal frequency show an increasing trend in the three sub-areas; however, their specific values are higher in the winter than in the summer, with significant differences in different areas. In Figure 7a, the trend in summer frontal intensity throughout the year is statistically significant around Zhoushan Island, where the cold boundary of the summer also rises. The trend values of the summer frontal intensity vary between -4×10^{-4} °C/km/year and 6×10^{-4} °C/km/year, with the highest trend along Zhoushan Island being greater than 6×10^{-4} °C/km/year. The trend values for the winter frontal intensity ranged from 5×10^{-4} °C/km/year to 1×10^{-3} °C/km/year, with the highest trend in the Changjiang River Estuary greater than 1×10^{-3} °C/km/year. However, the frontal intensity in some areas decreased during the period 1982–2021 (Figure 7c).

The spatial distribution of the long-term frontal frequency in A1, A2, and A3 is similar to the frontal intensity (Figure 7c,d) because the appearance of frontal activity determines the magnitude of frontal frequency and then influences the frontal intensity. In A1, an increasing trend in summer is mainly located in the Changjiang River Estuary; however, in winter, it occurs primarily in the coastal areas of Jiangsu. The increased rate in winter is higher than that in summer. In A2, the areas with a significant increase in frontal intensity and frequency tendencies are located mainly in the area surrounding the Zhoushan Island in both summer and winter. In A3, there is a stronger increasing trend observed in winter than in summer, with a double band of increasing trends in the near-shore and offshore areas.

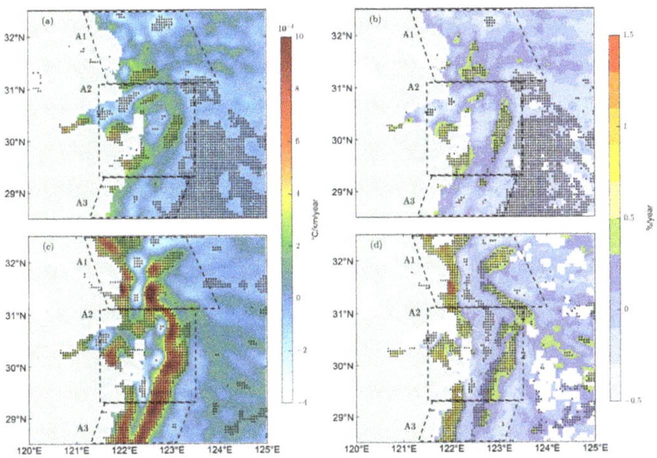

Figure 7. Spatial distribution of the long-term trend of the (**a**) summer frontal intensity, (**b**) summer frontal frequency, (**c**) winter frontal intensity, and (**d**) winter frontal frequency from 1982 to 2021. Black dots represent grids that passed the 95% significance test.

The analysis above indicates that the SST fronts around Zhoushan and its adjacent seas have significant seasonal characteristics and spatial differences. Figure 8 shows the spatial distribution of the summer SST frontal intensity from 1982 to 2021. SST frontal intensity in the summer of the three subareas also significantly varies from year to year. For example, the frontal intensity of A1 in 1984, 1988, 1994, and 1998 are significantly stronger than that in other years, with a frontal intensity of more than 0.05 °C/km. In 1986 and 1989, only a few regions demonstrated a positive intensity of 0.03 °C/km. The frontal intensity in area A2 reached 0.03 °C/km in most years and even 0.05 °C/km in 1995, 2006, 2013, and 2018, while in area A3, it was lower in most years.

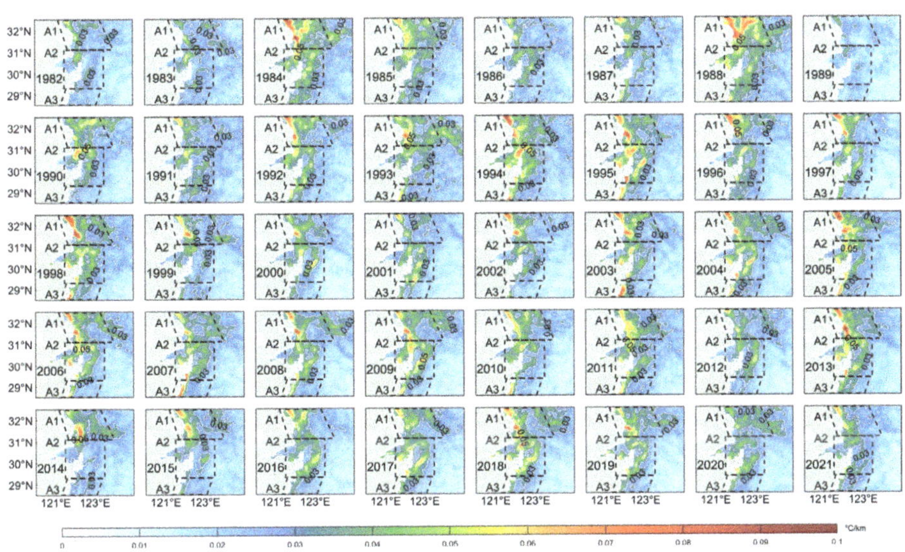

Figure 8. Spatial distribution of the summer SST frontal intensity from 1982 to 2021.

The spatial distribution of the SST frontal frequency in summer from 1982 to 2021 in the three regions also has different interannual characteristics (Figure 9). In area A1, the frontal frequency of more than 60% occupied most of the area in 1984, 1988, 1998, 2011, and 2014, even with a high frontal frequency of more than 90% even in 1998. Area A2 had a higher frontal frequency in 1984, 1988, 1994, 1995, 2009, and 2018, with the lowest frontal frequency in 1988. The frontal frequency in area A3 is approximately between 30% and 60% during the period 1982–2021. Although the spatial distribution of summer SST frontal intensity and frequency shows large variability, the spatial distribution of winter SST frontal intensity and frequency in each year from 1982 to 2021 is very similar to the climatological spatial distribution of winter SST frontal intensity (Figure 5c) and frequency (Figure 5d).

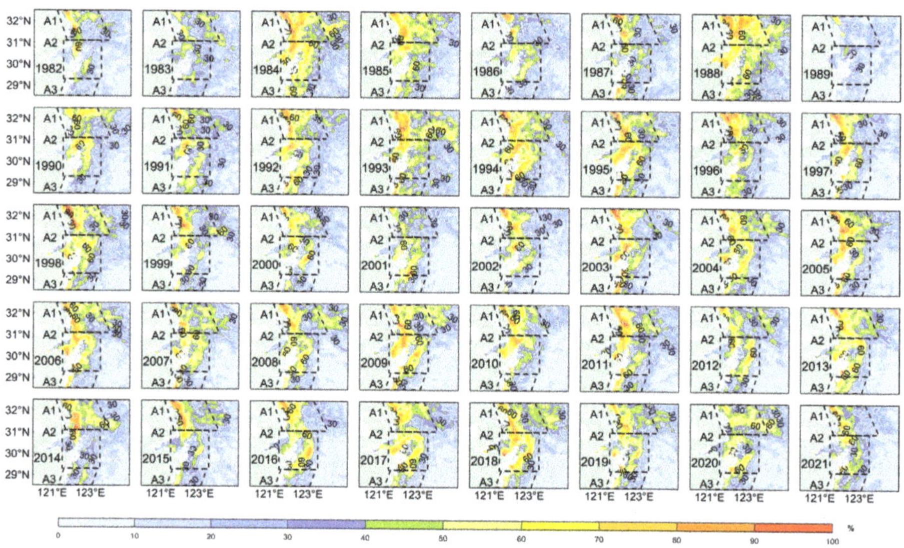

Figure 9. Spatial distribution of the summer SST frontal frequency from 1982 to 2021.

Long-term trends in SST frontal intensity and frequency can be closely related to climate change. SST in ECS is closely related to El Niño/Southern Oscillation (ENSO). ENSO also modulates precipitation and monsoon over ECS, which further affects the discharge of the Changjiang River, coastal upwelling, and coastal currents. However, the trends in different subareas are not the same. This is due to the frontal activities in each subareas are dominated by different physical processes in different seasons. For example, in summer, area A1 is mainly influenced by the Changjiang River, area A2 by the Zhoushan upwelling, and area A3 by the Zhejiang coastal currents. In winter, all three subareas are mainly influenced by coastal currents. These influences are discussed in the next section. For frontal activities in winter are more frequent and stronger than those in summer, the increase trend in SST frontal intensity and frequency in winter is significantly larger than that in summer. This implies that climate change may have a greater influence on the physical processes controlling winter fronts.

4. Discussion

This study analyzes the seasonal characteristics of SST fronts around Zhoushan Island in ECS during 1982–2021. Similar to most previous studies, the SST fronts in Zhoushan and its adjacent seas have significant seasonal variation, with a spatial distribution of frontal frequency similar to that of Cao et al. (2021) [38]. However, the results in this study show higher frontal frequency, especially in summer. The major SST front distributions agree well with Hickox et al. (2000) [33], and the SST gradient aligns with the findings of Cao et al. (2021) [38]. Given that the threshold (0.03 °C/km) of the SST fronts in this study is different from previous studies, the magnitude of the frontal frequency is also different.

The thresholds of 0.1 °C/km and 0.028 °C/km are considered as the threshold in He et al. (2016) [37] and Cao et al. (2021) [38], respectively.

The formation and evolution of oceanic fronts are typically considered to be the results of the interactions between multiple dynamic processes in the ocean and atmosphere. The SST fronts in Zhoushan and its adjacent seas are subjected to the influences of multiple dynamic processes, such as monsoons, coastal upwelling, Changjiang River discharge, and coastal currents. Therefore, the underlying physical mechanisms of the summer and winter fronts in the study area are briefly discussed below, focusing on the influence of wind, summer upwelling, Changjiang River discharge, and coastal currents.

4.1. Influence of Wind on Fronts

Numerous studies suggest that the wind can induce frontal activities through multiple dynamics, and wind-induced coastal upwelling is one of the most well-known factors driving frontogenesis [3,46].

The summer and winter climatological wind patterns are shown in Figure 10 and present a clear characteristic of the monsoon, with southerly wind dominating in summer and northerly wind dominating in winter. Wind-driven upwelling is dominant in the fronts around Zhoushan Island (A2) in summer [38], where the south winds are conducive to Ekman pumping, resulting in significant upwelling and arousing the evolution of the fronts. Ma et al. (2004) [47] suggested that the coastal ocean fronts along Jiangsu (A1) are highly correlated with the alongshore wind. In winter, the north wind could accelerate high-latitude cold coastal water from the north flow south, improving the mixing and interactions with warm water on the shelf in ECS, which is favorable for frontal generation.

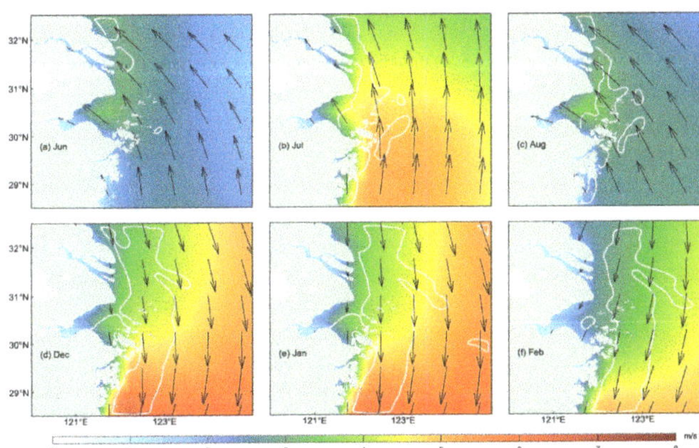

Figure 10. Monthly climatology of the wind (1982–2021). (**a**) June; (**b**) July; (**c**) August; (**d**) December; (**e**) January; (**f**) February. The white contour lines indicates the threshold (0.03 °C/km) of the SST gradient.

The spatial distribution of the frontal intensity and frequency in winter (Figure 5) shows a much higher intensity and frequency than in other seasons. This phenomenon can also be partly explained by the strong north wind in winter. Given that the wind speed in winter is higher than that in other seasons, it can also enhance surface cooling and water mixing in shallow water. However, the research of Cao et al. (2021) [38] demonstrated that the fronts on the shelf in the east of Zhejiang are not completely determined by the wind pattern. He et al. (2016) [48] revealed that there is no obvious relationship between wind stress and front position shift, while the wind can increase the frontal intensity on the nearshore front (A3).

4.2. Influence of Upwelling on Fronts

Upwelling brings cold seawater from the bottom to the upper ocean layer, typically forming a cold central area at the sea surface. When cold seawater collides with the surrounding warm seawater, it always generates upwelling fronts. The frontal intensity and frontal frequency of the upwelling-induced front are directly related to the intensity of the upwelling. The Zhoushan upwelling is an important and famous coastal upwelling in ECS. A number of studies have focused on the Zhoushan upwelling due to the importance of the Zhoushan fishing grounds. Research on Zhoushan summer upwelling was first published in the early 1960s [49]. To date, numerous studies have verified the existence of Zhoushan upwelling using in situ observation data [50] and satellite remote sensing data, and have also analyzed the characteristics of upwelling [51,52].

However, the mechanisms behind the generation of upwelling in Zhoushan are viewed from different perspectives. The remnants northward of the Kuroshio current and the Taiwan Warm Current play an important role in the upwelling of the Zhejiang coastal seas. The non-linear effects of tides and topography can also cause upwelling [53–55]. Early studies have shown that upwelling occurs along the Zhejiang coast in winter [50,56], but no significant low-temperature zone is observed in winter, based on the distribution characteristics of the SST. In this study, the temperature front in area A2 in summer is mainly caused by upwelling, with the intensity and frequency of the front being closely related to the upwelling. In winter, the upwelling current on the eastern side of area A2 is the most closely related to the coastal current.

4.3. Influence of Changjiang River on Fronts

River discharge is an important dynamic factor in estuarine frontogenesis. The Changjiang River contributes approximately 79.7% of the total freshwater input to ECS [32], with a significant seasonal variability in discharge, ranging from approximately 10,000 m^3/s in winter to 500,000 m^3/s in summer [57]. Shelf mixing water plays a key role in the distribution and variation of the fronts due to a significant differences in temperature and salinity between rivers and seas [30]. When rivers discharge large amounts of freshwater into the ocean, strong mixing occurs between river water and seawater, and these interactions can easily generate temperature fronts and salinity fronts in the estuary.

In the northern hemisphere, when rivers flow into the sea, the freshwater is deflected to the right of the flow direction due to the Coriolis force. Accordingly, diluted water from the Changjiang River should flow south to the seas of Hangzhou Bay and Zhoushan Island. However, in summer, Changjiang diluted water mainly influences the south wind, causing large volumes of freshwater to flow to the north and northeast, reaching the Jiangsu coastal areas. This phenomenon explains why the summer frontal intensity and frequency are high in the north of the Changjiang River Estuary and the coastal waters of Jiangsu (area A1). In winter, Changjiang diluted the flow of water south to the seas of Hangzhou Bay and Zhoushan Island, influenced by the Coriolis force and the north wind in the study area. This phenomenon is an important reason for the weak front in the coastal region of Jiangsu in winter.

4.4. Influence of Coastal Currents on Fronts

The Taiwan Strait Current flows from the southwest to the northeast throughout the year in the research area, characterized by high temperature and salinity, mainly outside the 50 m isobaths [39,58].

In summer, the coastal currents in the study area are mainly the warm currents from the Taiwan Strait that move northward. These currents pass through Zhoushan Island and intersect with the cold water caused by the Zhoushan upwelling, which is conducive to the formation of a front around the sea area of the Zhoushan Island in summer (Figure 11a–c). In the area A3, the intensity and frequency of the SST fronts reach 0.03 °C/km and 60%. The frontal activities are mainly influenced by cold coastal waters from the northwest and warm costal currents from the south.

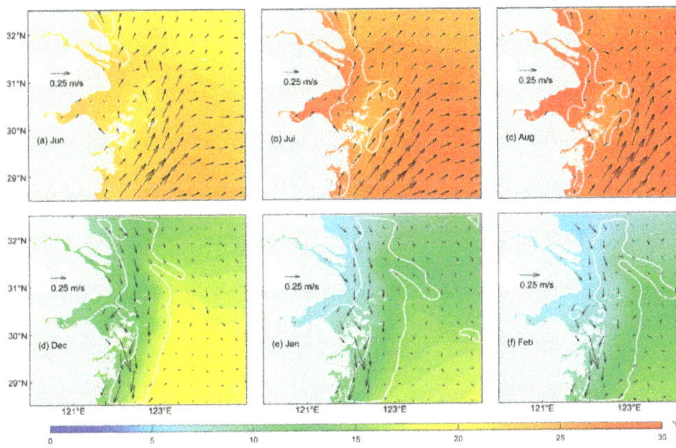

Figure 11. Monthly climatology of the coastal currents and SST (1982–2021). (**a**) June; (**b**) July; (**c**) August; (**d**) December; (**e**) January; (**f**) February. The arrows indicate the coastal currents, filled color represents SST, and the white contour lines are the same as in Figure 10.

In winter, the coastal currents in the study area mainly flow to the south from the Jiangsu coast and invade ECS [59]. The coastal currents pass through the Changjiang River Estuary and mix with the freshwater from the Changjiang River, resulting in characteristics of low temperature and low salinity (Figure 11d–f). Low-temperature coastal currents form a significant winter SST front with water from the shelf in ECS. Low-temperature northerly winds can also strengthen low-temperature coastal currents and reduce surface temperature, and also intensify the front [38]. Therefore, the winter SST fronts are mainly influenced by coastal currents.

5. Conclusions

This study presents the seasonal variability in the SST fronts in Zhoushan and its adjacent seas. The spatial and temporal variabilities in the monthly frontal intensity and frequency display significant seasonal fluctuations. The fronts in the study area are further divided into three sub-areas according to the spatial diversity of the fronts in summer and winter: A1 is the area along the Jiangsu coast and north of the Changjiang River Estuary, A2 is the area around Zhoushan Island and parts of Hangzhou Bay, and A3 is the area along the coast of Zhejiang-Taizhou. The frontal intensity and frequency and the underlying dynamic process of these areas can be summarized as follows.

Frontal intensity and frequency, along with the main underlying process, are especially different in summer. In area A1, the frontal intensity reaches approximately 0.06 °C/km and has more than 60% of the frontal frequency. In areas A2 and A3, the frontal intensity and frequency reach up to 0.03 °C/km and 60%, respectively. The frontal activities in three subareas are mainly controlled by different dynamic processes. The diluted Changjiang water and the south wind are the primary drivers of the abundant frontal activity in area A1. The Zhoushan upwelling is responsible for the frontal activities in the area A2, which are also highly influenced by wind, unique topography, tide, and the northward remnants of the Kuroshio and the Taiwan Warm Current. In area A3, the frontal activities are mainly influenced by cold coastal waters and warm costal currents. The summer frontal intensity and frequency show a weak spatial and temporal variation trend during 1982–2021.

In winter, the SST fronts in A1, A2, and A3 merge into a large-scale strong frontal zone, which appears as a red band located in the offshore areas and almost parallel to the coast. The frontal intensity and frequency of this strong frontal zone reach up to 0.1 °C/km and more than 90%, respectively. The wind and cold coastal currents of the Jiangsu coast are the main factors that influence the frontal activities in winter. The north wind strengthens the cold coastal currents from the north and carries the diluted Changjiang water to the south. The mixing of cold coastal currents and warm water in ECS generates this strong front zone. The frontal intensity and frequency in A1, A2, and A3 in winter exhibit an increasing trend during the period 1982–2021, and the linear rate of increase is greater in winter than in summer.

Oceanic fronts play an important role in marine pollution, air–sea interactions, marine ecosystems, and marine fisheries. The analysis of the seasonal variability in the fronts and the underlying dynamic process in Zhoushan and its adjacent seas is crucial for the management of fisheries and coastal pollution. Due to the limited resolution of the SST data, there are still some missing values in the coastal areas, which is not conducive to the analysis of finer-scale frontal features in the coastal regions. It is recommended that high-resolution and high-frequency field observations in the region be used in future studies. Studies on the detailed mechanisms of frontogenesis and its vertical characteristics are still lacking. The influence of other factors, such as topography, surface heat flux, and tides, on the evolution of SST fronts in this region will be investigated in the future. Therefore, numerical model experiments will be used in the future to verify the detailed mechanisms of frontal formation.

Author Contributions: Conceptualization, Q.J. and H.C.; methodology, Q.W. and T.P.; software, Q.J. and H.C.; investigation, Y.W. and Z.M.; resources, Q.J.; data curation, T.P. and Y.W.; writing—original draft preparation, Q.J. and H.C.; writing—review and editing, Q.W.; visualization, Q.J. and H.C.; project administration, Q.J.; funding acquisition, Q.J. All authors have read and agreed to the published version of the manuscript.

Funding: This study was supported by the National Key Research and Development Program of China (No. 2023YFD2401904) and the Basic Public Welfare Research Project of Zhejiang Province (No. LGF22D060001) and the National Natural Science Foundation of China (No. 41806004).

Institutional Review Board Statement: Not applicable.

Informed Consent Statement: Not applicable.

Data Availability Statement: The SST data can be found at: https://doi.org/10.48670/moi-00168 (accessed on 4 June 2023). The GLORYS12V1 reanalysis product can be accessed from DOI (product): https://doi.org/10.48670/moi-00021 (accessed on 28 December 2023). The ERA5 data can be accessed at: https://doi.org/10.24381/cds.f17050d7 (accessed on 10 October 2024). The topography, bathymetry, and shoreline data come from 2-minute Gridded Global Relief Data (ETOPO2) v2, National Geophysical Data Center, NOAA: https://doi.org/10.7289/V5J1012Q, https://www.ngdc.noaa.gov/mgg/global/relief/ETOPO2 (accessed 1 October 2024).

Conflicts of Interest: The authors declare no conflicts of interest.

References

1. Belkin, I.M.; O'Reilly, J.E. An algorithm for oceanic front detection in chlorophyll and SST satellite imagery. *J. Mar. Syst.* **2009**, *78*, 319–326. [CrossRef]
2. Belkin, I.M.; Cornillon, P.C.; Sherman, K. Fronts in large marine ecosystems. *Prog. Oceanogr.* **2009**, *81*, 223–236. [CrossRef]
3. Wang, Y.; Yu, Y.; Zhang, Y.; Zhang, H.R.; Chai, F. Distribution and variability of sea surface temperature fronts in the south China sea. *Estuar. Coast. Shelf Sci.* **2020**, *240*, 106793. [CrossRef]
4. Greer, A.T.; Cowen, R.K.; Guigand, C.M.; Hare, J.A. Fine-scale planktonic habitat partitioning at a shelf-slope front revealed by a high-resolution imaging system. *J. Mar. Syst.* **2015**, *142*, 111–125. [CrossRef]
5. Mangolte, I.; Lévy, M.; Haëck, C.; Ohman, M.D. Sub-frontal niches of plankton communities driven by transport and trophic interactions at ocean fronts. *Biogeosciences* **2023**, *20*, 3273–3299. [CrossRef]
6. Brandini, F.P.; Tura, P.M.; Santos, P.P. Ecosystem responses to biogeochemical fronts in the South Brazil Bight. *Prog. Oceanogr.* **2018**, *164*, 52–62. [CrossRef]
7. Chapman, C.C.; Lea, M.A.; Meyer, A.; Sallée, J.B.; Hindell, M. Defining Southern Ocean fronts and their influence on biological and physical processes in a changing climate. *Nat. Clim. Chang.* **2020**, *10*, 209–219. [CrossRef]
8. Xing, Q.; Yu, H.; Wang, H. Global mapping and evolution of persistent fronts in Large Marine Ecosystems over the past 40 years. *Nat. Commun.* **2024**, *15*, 4090. [CrossRef]
9. Franks, P.J. Sink or swim: Accumulation of biomass at fronts. *Mar. Ecol. Prog. Ser. Oldendorf* **1992**, *82*, 1–12. [CrossRef]
10. Acha, E.M.; Mianzan, H.W.; Iribarne, O.; Gagliardini, D.A.; Lasta, C.; Daleo, P. The role of the Rıo de la Plata bottom salinity front in accumulating debris. *Mar. Pollut. Bull.* **2003**, *46*, 197–202. [CrossRef]

11. Barnes, D.K.; Galgani, F.; Thompson, R.C.; Barlaz, M. Accumulation and fragmentation of plastic debris in global environments. *Philos. Trans. R. Soc. B: Biol. Sci.* **2009**, *364*, 1985–1998. [CrossRef] [PubMed]
12. O'Neill, L.W.; Chelton, D.B.; Esbensen, S.K. Observations of SST-induced perturbations of the wind stress field over the Southern Ocean on seasonal timescales. *J. Clim.* **2003**, *16*, 2340–2354. [CrossRef]
13. Chelton, D.B.; Schlax, M.G.; Samelson, R.M. Summertime coupling between sea surface temperature and wind stress in the California Current System. *J. Phys. Oceanogr.* **2007**, *37*, 495–517. [CrossRef]
14. Iizuka, S.; Shiota, M.; Kawamura, R.; Hatsushika, H. Influence of the monsoon variability and sea surface temperature front on the explosive cyclone activity in the vicinity of Japan during northern winter. *SOLA* **2013**, *9*, 1–4. [CrossRef]
15. Yang, H.; Chen, Z.; Sun, S.; Li, M.; Cai, W.; Wu, L.; Cai, J.; Sun, B.; Ma, K.; Ma, X.; et al. Observations reveal intense air-sea exchanges over submesoscale ocean front. *Geophys. Res. Lett.* **2024**, *51*, e2023GL106840. [CrossRef]
16. Legeckis, R. A survey of worldwide sea surface temperature fronts detected by environmental satellites. *J. Geophys. Res. Ocean.* **1978**, *83*, 4501–4522. [CrossRef]
17. Nieto, K.; Demarcq, H.; McClatchie, S. Mesoscale frontal structures in the Canary Upwelling System: New front and filament detection algorithms applied to spatial and temporal patterns. *Remote Sens. Environ.* **2012**, *123*, 339–346. [CrossRef]
18. Wang, Y.; Liu, J.; Liu, H.; Lin, P.; Yuan, Y.; Chai, F. Seasonal and interannual variability in the sea surface temperature front in the eastern Pacific Ocean. *J. Geophys. Res. Ocean.* **2021**, *126*, e2020JC016356. [CrossRef]
19. McWilliams, J.C. Oceanic frontogenesis. *Annu. Rev. Mar. Sci.* **2021**, *13*, 227–253. [CrossRef]
20. Du, Y.; Zhang, J.; Wei, Z.; Yin, W.; Wu, H.; Yuan, Y.; Wang, Y.P. Spatio-Temporal Variability of Suspended Sediment Fronts (SSFs) on the Inner Shelf of the East China Sea: The Contribution of Multiple Factors. *J. Geophys. Res. Ocean.* **2022**, *127*, e2021JC018392. [CrossRef]
21. Bai, H.; Hu, H.; Ren, X.; Yang, X.Q.; Zhang, Y.; Mao, K.; Zhao, Y. The impacts of East China Sea Kuroshio front on winter heavy precipitation events in Southern China. *J. Geophys. Res. Atmos.* **2023**, *128*, e2022JD037341. [CrossRef]
22. Castelao, R.M.; Wang, Y. Wind-driven variability in sea surface temperature front distribution in the California Current System. *J. Geophys. Res. Ocean.* **2014**, *119*, 1861–1875. [CrossRef]
23. Oerder, V.; Bento, J.P.; Morales, C.E.; Hormazabal, S.; Pizarro, O. Coastal upwelling front detection off central Chile (36.5–37 S) and spatio-temporal variability of frontal characteristics. *Remote Sens.* **2018**, *10*, 690. [CrossRef]
24. Chen, D.; Liu, W.T.; Tang, W.; Wang, Z. Air-sea interaction at an oceanic front: Implications for frontogenesis and primary production. *Geophys. Res. Lett.* **2003**, *30*, 1745. [CrossRef]
25. Castelao, R.M.; Barth, J.A. Coastal ocean response to summer upwelling favorable winds in a region of alongshore bottom topography variations off Oregon. *J. Geophys. Res. Ocean.* **2005**, *110*, C10S04. [CrossRef]
26. Gan, J.; Cheung, A.; Guo, X.; Li, L. Intensified upwelling over a widened shelf in the northeastern South China Sea. *J. Geophys. Res. Ocean.* **2009**, *114*, C9. [CrossRef]
27. Wang, Y.; Castelao, R.M.; Yuan, Y. Seasonal variability of alongshore winds and sea surface temperature fronts in E astern B oundary C urrent S ystems. *J. Geophys. Res. Ocean.* **2015**, *120*, 2385–2400. [CrossRef]
28. Nishikawa, H.; Nishikawa, S.; Ishizaki, H.; Wakamatsu, T.; Ishikawa, Y. Detection of the Oyashio and Kuroshio fronts under the projected climate change in the 21st century. *Prog. Earth Planet. Sci.* **2020**, *7*, 29. [CrossRef]
29. Tang, Y. Preliminary study on classification of oceanic fronts in East China Sea. *J. Oceanogr. Yellow Bohai Seas* **1995**, *13*, 16–22. (In Chinese)
30. Tang, Y. Distributional features and seasonal variations of temperature fronts in the East China Sea. *Oceanol. Limnol. Sin.* **1996**, *27*, 436–444. (In Chinese)
31. He, M.X.; Chen, G.; Sugimori, Y. Investigation of mesoscale fronts, eddies and upwelling in the China Seas with satellite data. *Glob. Atmos. Ocean. Syst.* **1995**, *3*, 273–288.
32. Ning, X.R.; Liu, Z.L.; Cai, Y.M.; Fang, M.; Chai, F. Physicobiological oceanographic remote sensing of the East China Sea: Satellite and in situ observations. *J. Geophys. Res. Ocean.* **1998**, *103*, 21623–21635. [CrossRef]
33. Hickox, R.; Belkin, I.; Cornillon, P.; Shan, Z. Climatology and seasonal variability of ocean fronts in the East China, Yellow and Bohai Seas from satellite SST data. *Geophys. Res. Lett.* **2000**, *27*, 2945–2948. [CrossRef]
34. Park, S.; Chu, P.C. Thermal and haline fronts in the Yellow/East China Seas: Surface and subsurface seasonality comparison. *J. Oceanogr.* **2006**, *62*, 617–638. [CrossRef]
35. Tseng, C.; Lin, C.; Chen, S.; Shyu, C. Temporal and spatial variations of sea surface temperature in the East China Sea. *Cont. Shelf Res.* **2000**, *20*, 373–387. [CrossRef]
36. Wang, F.; Meng, Q.; Tang, X.; Hu, D. The long-term variability of sea surface temperature in the seas east of China in the past 40 a. *Acta Oceanol. Sin.* **2013**, *32*, 48–53. [CrossRef]
37. He, S.; Huang, D.; Zeng, D. Double SST fronts observed from MODIS data in the East China Sea off the Zhejiang–Fujian coast, China. *J. Mar. Syst* **2016**, *154*, 93–102. [CrossRef]
38. Cao, L.; Tang, R.; Huang, W.; Wang, Y. Seasonal variability and dynamics of coastal sea surface temperature fronts in the East China Sea. *Ocean Dynam.* **2021**, *71*, 237–249. [CrossRef]
39. Beardsley, R.; Limeburner, R.; Yu, H.; Cannon, G. Discharge of the Changjiang (Yangtze river) into the East China sea. *Cont. Shelf Res.* **1985**, *4*, 57–76. [CrossRef]

40. Xuan, J.L.; Huang, D.; Zhou, F.; Zhu, X.H.; Fan, X. The role of wind on the detachment of low salinity water in the Changjiang Estuary in summer. *J. Geophys. Res. Ocean.* **2012**, *117*, C10. [CrossRef]
41. Good, S.; Fiedler, E.; Mao, C.; Martin, M.J.; Maycock, A.; Reid, R.; Roberts-Jones, J.; Searle, T.; Waters, J.; While, J.; et al. The current configuration of the OSTIA system for operational production of foundation sea surface temperature and ice concentration analyses. *Remote Sens.* **2020**, *12*, 720. [CrossRef]
42. Kanopoulos, N.; Vasanthavada, N.; Baker, R. Design of an image edge detection filter using the Sobel operator. *IEEE J. Solid-State Circuits* **1988**, *23*, 358–367. [CrossRef]
43. Gao, W.; Zhang, X.; Yang, L.; Liu, H. An improved Sobel edge detection. In Proceedings of the 2010 3rd International Conference on Computer Science and Information Technology, Chengdu, China, 9–11 July 2010; Volume 5, pp. 67–71. [CrossRef]
44. Kahru, M.; Kudela, R.M.; Manzano-Sarabia, M.; Mitchell, B.G. Trends in the surface chlorophyll of the California Current: Merging data from multiple ocean color satellites. *Deep Sea Res. II Top. Stud. Oceanogr.* **2012**, *77*, 89–98. [CrossRef]
45. Kahru, M.; Jacox, M.G.; Ohman, M.D. CCE1: Decrease in the frequency of oceanic fronts and surface chlorophyll concentration in the California Current System during the 2014–2016 northeast Pacific warm anomalies. *Deep Sea Res. Part I Oceanogr. Res. Pap.* **2018**, *140*, 4–13. [CrossRef]
46. García-Reyes, M.; Largier, J. Observations of increased wind-driven coastal upwelling off central California. *J. Geophys. Res. Ocean.* **2010**, *115*, C4. [CrossRef]
47. Ma, J.; Qiao, F.; Xia, C.; Yang, Y. Tidal effects on temperature iront in the Yellow Sea. *Chin. J. Ocean. Limnol.* **2004**, *22*, 314–321. (In Chinese)
48. He, Q.; Zhang, C.; Gao, G.; Wei, Y.; An, B. Study on the temporal and spatial characteristics of Zhoushan coastal upwelling and relationship with wind field in Summer period. *J. Shanghai Ocean Univ.* **2016**, *25*, 142–151.
49. Mao, H. A preliminary investigation on the application of using TS diagram for a quantitative analysis of the water masses in the shallow water area. *Oceanol. Limnol. Sin.* **1964**, *6*, 1–22. (In Chinese)
50. Xu, J.; Cao, X.; Pan, Y. Evidence for the coastal upwelling off Zhejiang. *Trans. Oceanol. Limnol.* **1983**, *4*, 17–25. (In Chinese)
51. Cao, G.; Song, J.; Fan, W. Mechanism of upwelling evolvement in the Yangtze River Estuary adjacent waters in summer, 2007. *Mar. Sci.* **2013**, *37*, 102–112. (In Chinese)
52. Yin, W.; Ma, Y.; Wang, D.; He, S.; Huang, D. Surface Upwelling off the Zhoushan Islands, East China Sea, from Himawari-8 AHI Data. *Remote Sens.* **2022**, *14*, 3261. [CrossRef]
53. Chen, C.; Beardsley, R.C.; Limeburner, R. A numerical study of stratified tidal rectification over finite-amplitude banks. Part II: Georges Bank. *J. Phys. Oceanogr.* **1995**, *25*, 2111–2128. [CrossRef]
54. Luo, Y. Numerical modelling of upwelling in coastal areas of the East China Sea. *Trans. Oceanol. Limnol.* **1998**, *50*, 555–563. (In Chinese)
55. Lü, X.; Qiao, F.; Xia, C.; Yuan, Y. Tidally induced upwelling off Yangtze River estuary and in Zhejiang coastal waters in summer. *Sci. China Ser. D Earth Sci.* **2007**, *50*, 462–473. [CrossRef]
56. Pan, Y.; Sha, W. Numerical study on the coastal upwelling off Fujian and Zhejiang. *Mar. Forecasts* **2004**, *21*, 86–95. (In Chinese)
57. Pan, Y.; Liang, X.; Huang, S. The evolution of the East China Sea dense water circulation and its influence on the mixing water diffusing off Changjiang mouth. *Dohai Mar. Sci.* **1997**, *15*, 15–24. (In Chinese)
58. Su, J. A review of circulation dynamics of the coastal oceans near China. *Acta Oceanol. Sin.* **2001**, *23*, 1–16. (In Chinese)
59. Qiao, F.; Yang, Y.; Lü, X.; Xia, C.; Chen, X.; Wang, B.; Yuan, Y. Coastal upwelling in the East China Sea in winter. *J. Geophys. Res. Ocean.* **2006**, *111*, C11. [CrossRef]

Journal of
Marine Science and Engineering

MDPI

Article

Typhoon Intensity Change in the Vicinity of the Semi-Enclosed Sea of Japan

Soo-Min Choi [1] and Hyo Choi [2,*]

1 Department of Computer Engineering, Konkuk University, 268 Chungwon-daero, Chungju 27478, Republic of Korea; fuledoc@daum.net
2 Atmospheric & Oceanic Disaster Research Institute, Dalim Apartment 209 ho, Songjungdong 940-23, Gangneung 25563, Republic of Korea
* Correspondence: du8392@hanmail.net; Tel.: +82-10-7240-0357

Abstract: The intensity change of Typhoon Songda (TY-0418) in the vicinity of the semi-enclosed Sea of Japan (SJ) was numerically investigated using 3D-WRF and UM-KMA models and GOES-IR satellite images on 4 to 8 September 2004. After the typhoon originated in the Western Pacific Ocean in August, it moved to the East China Sea. Following the north-eastward Kuroshio Warm Current, it developed with horizontal and vertical asymmetrical wind and moisture patterns until 5 September. On 7 September, closing to the Kyushu Island, it was divided into three wind fields near the surface due to the increased friction from the surrounding lands and shallower sea depth close to the land, but it still maintained its circular shape over 1 km in height. As it passed by the Korea Strait and entered the SJ, it became a smaller, deformed typhoon due to the SJ's surrounding mountains, located between the East Korea and Tsushima Warm Currents inside the SJ. Its center matched a high equivalent potential temperature area, releasing significant latent heat through the condensation of water particles over warm currents. The latent heat converted to kinetic energy could be supplied into the typhoon circulation, causing its development. Moist flux and streamline at 1.5 km in height clearly showed the moisture transportation via the mutual interaction of the cyclonic circulation of the typhoon and the anti-cyclonic circulation of the North Pacific High Pressure from the typhoon's tail toward both the center of the SJ and the Russian Sakhalin Island in the north of Japan, directly causing large clouds in its right quadrant. Simultaneously, the central pressure decrease with time could converge both transported moist air by the typhoon itself and water particles evaporated from the sea, causing them to rise and resulting in the formation of large clouds and the rapid development of the typhoon circulation. The strong downslope winds from the surrounding mountains of the SJ to its center also produced a cyclonic vortex due to the Coriolis force to the right, enhancing the typhoon's circulation.

Keywords: typhoon intensity change; WRF and UK-KMA models; GOES-IR satellite images; equivalent potential temperature; warm current; downslope wind

1. Introduction

In summer, typhoons, otherwise known as tropical cyclones or hurricanes, can be responsible for severe impacts including loss of life, economic hardship, destruction of dwellings, coastal erosion, inundation by floods, strong winds, and storm surges [1,2]. The classification of these systems varies across national institutions. The classification by the Hong Kong Observatory (HKO) [3] consists of six categories, comprising super typhoons, severe typhoons, typhoons, severe tropical storms, tropical storms, and tropical depression in terms of maximum sustained wind speeds averaged over 10 min near the center of the system, as suggested by the World Meteorological Organization (WMO) [4]. The Korean Meteorological Administration (KMA) [5] simplified the classifications into four groups consisting of typhoons, severe tropical storms, tropical storms, and tropical depression, which use different surface atmospheric pressure and wind speed thresholds near the center of the system.

Monaldo et al. [6], Cione and Uhlhorn [7], and Knauss [8] explained that surface divergence of sea surface water due to cyclonic surface winds in a hurricane results in upwelling of bottom colder water to the surface. Gilbes et al. [9] and Babin et al. [10] showed that hurricane-induced phytoplankton blooms were supplied by nutrient-laden waters upwelling colder bottom waters toward the sea surface. Willoughby et al. [11] presented a hurricane structure and evolution simulated by an axisymmetric, non-hydrostatic numerical weather prediction model. They explained that the spiral band structure in Hurricane Gert in 1981 was produced by convergence and divergence on the east and west sides of the eye wall by the movement of the environmental flow through the vortex core at the lower levels.

Using observational data, Marks et al. [12] analyzed an asymmetric wind pattern at several vertical levels for Hurricane Norbert, and similarly, Franklin et al. [13] analyzed these patterns for Hurricane Gloria. Including the research by Bender [14], their results indicated that the relative environmental flow in the direction of the storm motion was most important in Hurricane Norbert, while the cross-track component dominated in Hurricane Gloria as the hurricane moved northwest. Elsner [15] and Jian and Wu [16] explained that when a tropical cyclone approaches complex terrain during its landfall period, a more complicated asymmetric wind structure causes the deflection of the typhoon track. Choi and Lee [17] described the outbreak of cold sea surface temperatures near Cheju Island in response to strong cyclonic winds and a positive geopotential tendency at sea level following the track of a typhoon center based on the weather research and forecasting model simulation.

Shen et al. [18] simulated the prediction of Tropical Cyclone Nargis using a global mesoscale model. Xu at el. [19] investigated the impact of Tropical Storm Bopha on the intensity change of Super Typhoon Saomai in the 2006 typhoon season. They demonstrated that the existence of Typhoon Bopha and its increasing intensity would weaken Typhoon Saomai at its intensifying stage in 2006, while intensifying Saomai at its weakening stage was explained by the Fujiwhara effect [20]. Choi and Choi [21] numerically simulated the multiple interactions of not only three typhoons but also a mid-latitude cloud band-associated trough in the Northwest Pacific, showing their intensification and decay.

Yamasaki [22] examined the development of weak vortices and tropical cyclones by varying the ice microphysics in numerical experiments, and Bu et al. [23] showed the influences of boundary layer mixing and cloud-radiative forcing. Recently, significantly improved spaceborne microwave sensors can penetrate clouds and provide detailed microphysical information regarding clouds and precipitation, greatly improving the detection capability of fine cloud and rain structure inside tropical cyclones. Therefore, Wu et al. [24] explained that the ice water content might be an important indicator for the estimation of rapid intesification of tropical cyclones and the spatial structure of cloud and rain particles. Another paper by Wu et al. [25] described in detail the interaction of cloud dynamics and microphysics during the rapid intensification of Super-Typhoon Namadol (2022) based on multi-satellite observations. Su et al. [26] and Tapiador et al. [27] applied satellite observations, microwave data, and neural network techniques for predicting the rapid intensification of tropical cyclones.

Bruneau et al. [28] explained the impact of ocean mesoscale temperature anomalies on tropical cyclone size and AMS [29]. Tamamadin, et al. [30] showed that the area of high equivalent potential temperature containing significant moisture corresponds to the typhoon area. This is because latent heat released from the cloud cluster through the condensation process via cooling moisture can be converted into significant kinetic energy to be supplied into the typhoon system, resulting in typhoon intensification.

When most typhoons landfall and pass by the Korean Peninsula or Japan Island and moved into the SJ, their circular structures are quickly destroyed and deformed into a low pressure due to not only the increased friction by surrounding lands and the shallower sea depth close to the land, but also because there is an insufficient supply of moisture over the land for maintaining its structure [5,21]. The development and decay of typhoons are affected by many kinds of driving mechanisms based on micro-, meso-, and synoptic scales, which are directly and indirectly associated with each other. Namely, it is not appropriate to say which of them is the most important. However, our research focused on the intensification of a smaller reformed typhoon in the SJ that has not been previously explained its meso- and synoptic-scale horizontal and vertical structures dynamically and thermodynamically in detail.

Thus, the intensity change of Typhoon Songda was investigated using numerical simulations using both the WRF-3.6.1 model and UM-KMA model and evaluating wind velocity, relative humidity, moist flux, streamline, atmospheric pressure tendency, equivalent potential temperature, and sea surface temperature during the passage of Songda from 5–8 September 2004. The typhoon track and its horizontal and vertical structures were also analyzed using Geostationary Operational Environmental Satellite (GOES)-Infrared cloud images with no index that could be only obtained at that time (2004), differently from high-resolution images supplied by the Korean COMS satellite launched in recent and surface weather charts supplied by the KMA. We compared these images with the numerical simulation results of horizontal and vertical moist (relative humidity), wind and equivalent potential temperature distribution, and so on.

2. Study Area

Figure 1a,b show the Northwest Pacific Ocean, including the South China Sea, the East China Sea, the Yellow Sea, and the semi-enclosed Sea of Japan, as well as topographical features surrounding the Korean peninsula, Russia, China, and Japan. More than 10 typhoons originated in the Western Tropical Pacific Ocean near Guam Island and passed through these areas from early summer (June) until late fall (November) 2004 in Figure 1c. They made landfalls on the Chinese coast or penetrated the Korean peninsula or Japanese islands, resulting in huge economic losses from flooding and destructive winds.

The sea depth from Okinawa Islands close to the East China Sea and South Sea of Korea is much shallower than 200 m, different from the approximately 2000 m depth in the West Pacific Ocean. Figure 1d shows the ocean currents that flow along the coasts around the Korean Peninsula, China, and Japan, such as the Yellow Sea Warm Current, the East Korea Warm Current, the North Korea Cold Current, the Liman Cold Current, the Tsushima Warm Current, the Tsugaru Warm Current, the Soya Warm Current, etc. All typhoons traveling to Northeast Asia develop or decay partially under the influence of the ocean currents mentioned above.

The track of Typhoon Songda (TY 0418) formed in the Western Pacific Ocean is shown in Figure 1c from 27 August to 10 September 2004. Figure 1d displays the ocean currents flowing around the Korean peninsula, such as the Yellow Sea Warm Current, the East Korea Warm Current, the North Korea Cold Current, the Liman Cold Current, the Tsushima Warm Current, the Tsugaru Warm Current, and the Soya Warm Current, respectively.

Typhoon Songda passed by the East China Sea and the korea Strait and moved into the Sj, where both the East Korea Warm Current (EK) and the Tsushima Warm Current (TS) originated from the Kuroshio Current pass by the Korea Strait, consisting of a shallow depth of about 100 m, to the deepest depth of the SJ, extending down to about 3800 m. The SJ covers a three times larger area than the Yellow Sea, with a maximum depth of 74 m and a mean depth of about 40 m. It is enclosed by the mountains of Korea (west), Hokkaido, and Honshu islands of Japan (east) and Russia (north).

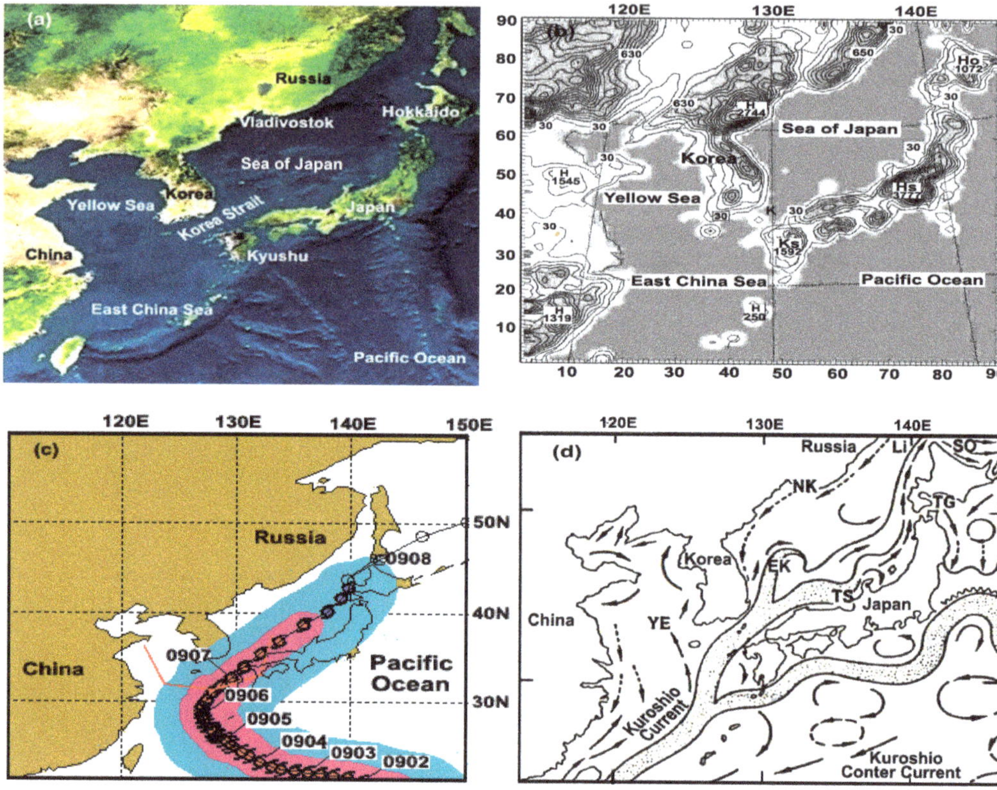

Figure 1. (**a**) Map of Northeast Asia with seas and ocean, (**b**) topography surrounding the Korean peninsula, China, and Japan, (**c**) track of TY Songda from 2 to 9 September, and (**d**) ocean currents surrounding Korea (modified from Lee [31]). In (**b**), Ho, Hs, and Ks denote Hokkaido, Honshu, and Kyushu Islands, and in (**d**), YE, EK, NK, Li, TS, TG, and SO denote the Yellow Sea Warm Current, the East Korea Warm Current, the North Korea Cold Current, the Liman Cold Current, the Tsushima Warm Current, the Tsugaru Warm Current, and the Soya Warm Current.

3. Numerical Method and Input Data

The three-dimensional Weather Research and Forecasting (WRF)-Version 3.6.1 model released on 18 September 2014 [32] was adapted for the numerical simulation of wind speed and direction, precipitation, atmospheric pressure tendency, potential temperature, potential vorticity, etc., in Northeast Asia. In the numerical simulation, one-way triple nesting from a coarse-mesh domain to a fine-mesh domain was performed using a horizontal grid spacing of 27 km, covering a 91 × 91 grid square in the first coarse mesh domain. The second (third) domain consists of a 9 km (3 km) horizontal grid interval of the same number grid as the first domain.

As meteorological input data to the model, NCEP/NCAR reanalysis–final analyses (FNL) $1.0^0 \times 1.0^0$ resolution data were used by interpolating and extrapolating the data at each grid point onto the first model domain with vertical interpolation between the surface and the upper boundary level of 100 hPa onto 36 levels, being more closely stacked in the lower levels of the atmosphere [33]. The meteorological elements evaluated in the first model domain were set up at each grid point of the second domain, and these values were calculated again with a 30-second topography resolution for each grid point.

In the model, the WSM 6 scheme was used for heat and moisture budgets and microphysical processes in the atmospheric boundary layer, and the YSU PBL scheme was adopted for the planetary boundary layer. The Kain–Fritsch scheme (new Eta) was adopted for cumulus parameterization, and the fifth thermal diffusion model was employed for land surfaces. The RRTM long-wave radiation scheme and Dudhia short-wave radiation schemes were used [21]. However, all the figures created using the model in this study were shown in the first domain.

Another model, the British Unified Model (UM) [34] adopted and modified by the Korean Meteorological Administration, is called the UM-KMA model (weather forecasting model used by KMA at present), which consists of a software suite written in Fortran-90. It uses initial input data provided by observations from satellites, ground-based weather stations, marine buoys, data from radar, radiosonde weather balloons, wind profilers, commercial aircrafts, and a background meteorological field derived from preceding model runs. The UM-KMA model uses nesting techniques for horizontal grid domains of 1.5km and 4 km for representing areas such as the size of a city, 12 km for regional-scale domains, and 40 km for global-scale domains.

The streamline analyses, moisture flux fields, and equivalent potential temperature distribution at a 1.5 km height were evaluated to investigate how a typhoon can be intensified through moisture and momentum transfer into its system through complex interactions between moisture, wind, and temperature fields, as well as where the typhoon center exists.

4. Results

4.1. Wind Field before the Surface Split of TY-Songda

The Joint Typhoon Warning Center, USA (JTWC) reported a new area of convection at 11:00 UTC (20:00 LST = UTC + 09:00 LST (Korean Local Standard Time)), on 26 August 2004. It continued to develop at 210 nautical miles northeast of Kwajalein and was first warned as a tropical depression 22 W at 12:00 UTC, 27 August, with its center at 270 nautical miles east of Eniwetak Atoll in the Pacific Ocean (Figure 1c). This depression was re-assigned the name of Songda as a tropical storm with a maximum surface wind speed of 35 kts (~18.5 m/s) at 00:00 UTC, 28 August.

When it was located about 17 nautical miles north–northeast of Agrihan Island in the Northern Mariana Islands at 18:00 UTC, it became a typhoon, with a maximum wind speed of 95 kts (~47.5 m/s). After Typhoon Songda formed in the Western Pacific Ocean with a central pressure of 950 hPa and maximum surface wind speed of 40 m/s (10 min average), it maintained its track toward the northwest until 00:00 UTC (09:00 LST) on 5 September, and the typhoon reached the East China Sea at 09:00 LST on 5 September, as shown in the surface weather map including a square covering the WRF course-mesh model domain and GOES-9-IR satellite images in Figure 2a,b.

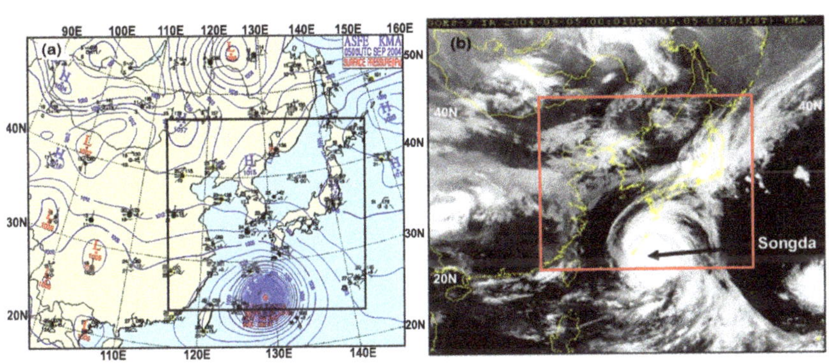

Figure 2. (**a**) Surface weather map (hPa) supplied by the Korean Meteorological Administration (KMA) including a square area covered by the WRF course-mesh model domain at 09:00 LST, 5 September. (**b**) GOES-9 IR satellite image supplied by KMA at 09:01 LST, 2004.

As shown in Figure 3, with the horizontal and vertical structure of the typhoon, the typhoon became weak due to not only the geographical friction of surrounding lands (China; west), Korean peninsula (north), and Kyushu Island (Japan; east)) but also the increase in shallower bottom friction created by its moving over a 200 m shallower sea on its left compared to a depth of about 1000 m on its right. Thus, the mutual interaction among the reduction of surface wind speeds in the left quadrant of the typhoon center due to the shallower bottom friction, the surface friction of surrounding lands, and the northward moving speed of a typhoon can cause an asymmetry, with weaker winds in the left than the right of the typhoon center, horizontally and vertically (Figure 3a–c). The typhoon eye, a small yellow circle with a 150 km diameter, has very weak wind or is calm (x = 43, y = 3).

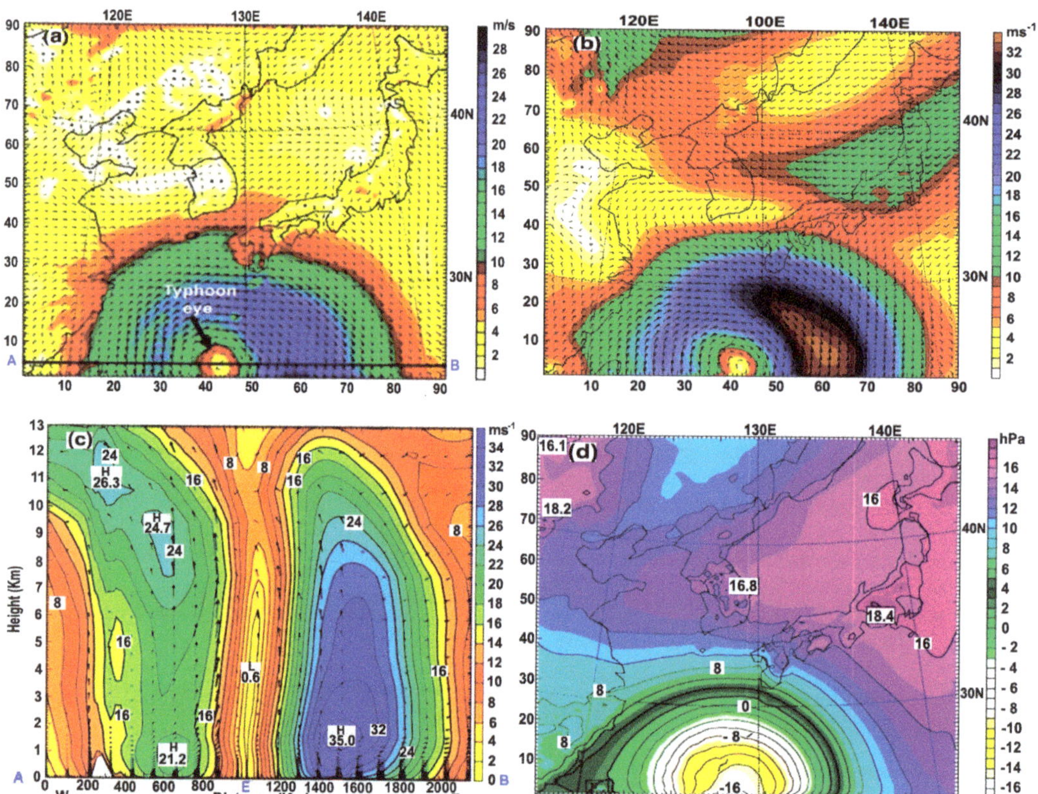

Figure 3. (**a**) Surface wind (m/s) at 10 m height at 09:00 LST, 5 September; (**b**) 3 km height; (**c**) vertical profile of horizontal winds (m/s) on a line A–B (typhoon eye; E) of Figure 3a; (**d**) surface atmospheric pressure tendency (−16h Pa/day) causing typhoon development. As it approaches the lands, there are horizontal and vertical asymmetries of stronger winds to its right and increased sea bottom friction in the East China Sea as well as surface friction from surrounding lands.

In Figure 3d, at 09:00 LST on 5 September, a negative rate of atmospheric pressure change with time ($\partial p/\partial t$) of −16 hPa/day for 12 h at the typhoon center reflected the deepening (falling) atmospheric pressure, strengthening the typhoon's cyclonic wind. Hesselberg [35], Brunt and Douglas [36], Haurwitz [37], Gill [38], Choi [39], and AMS [40] defined the contours of atmospheric pressure change in a horizontal distance with time (atmospheric pressure tendency) as isallobars. The effects of friction on the ground and turbulence produce a flow of air across the isallobars toward falling pressure, proportional to the isallobaric gradient, resulting in the convergence of airflow and resultantly the uprising of air, causing the formation of clouds and severe weather.

Haurwitz [37] and AMS [40] showed mathematically that the isallobaric wind $V_{is} = \mathbf{k} \times \frac{1}{f}\frac{\partial V_g}{\partial t} = -\frac{\alpha}{f^2}\nabla_H \frac{\partial p}{\partial t}$ is defined in terms of the vertical unit vector (k), the Coriolis parameter (f), the local accelerations of geostrophic wind ($\frac{\partial V_g}{\partial t}$), the specific volume ($\alpha$), and the horizontal spatial variation ($\nabla_H \frac{\partial p}{\partial t}$; $\frac{\partial}{\partial x}\left(\frac{\partial p}{\partial t}\right) + \frac{\partial}{\partial y}\left(\frac{\partial p}{\partial t}\right)$) of atmospheric pressure tendency ($\frac{\partial p}{\partial t}$). The resulting surface wind is made up of geostrophic winds blowing parallel to the isobars, with an added component directed into the isallobaric low. Consequently, if the absolute value of negative atmospheric pressure tendency increases ($|\partial p/\partial t|$; $|-16$ hPa/day$|$), the isallobaric wind increases, causing the convergence of airflow to rise up and the rapid development of the typhoon circulation.

TY Songda was on a northerly track by 09:00 LST, 6 September in Figure 4a,b. At this time, the edge of the typhoon became torn by surrounding land masses like China, Korea, and Japan, and its circulation was prohibited owing to both the greater frictions of the adjacent lands in Eastern China, Southern Korea, and Southwestern Japan and the shallower sea depths of the East China Sea to the left of its eye in Figure 5a.

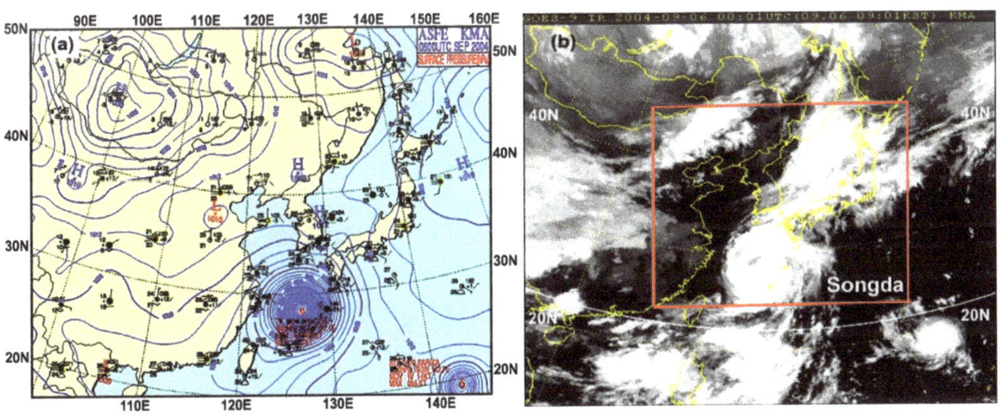

Figure 4. (**a**) Surface weather chart (hPa) at 09:00LST, 6 September, 2004; (**b**) GOES-9 IR satellite image before TY-Songda reached the Korea Strait and the west of Kyushu Island.

This kind of geographical and topographical feature causes the horizontal wind pattern to be asymmetric, and a vertical cross-section of the horizontal wind on a straight cutting line (A-typhoon eye-B) in Figure 5a,b shows that the horizontal wind is asymmetric (E; an ellipse of about 200 km in the x-axis and 300 km in the y-axis). This corresponds to the vertical asymmetry of the horizontal winds of 20 m/s (21 m/s) near the surface in the left of the typhoon eye compared to 26 m/s (37.2 m/s) at a 1.5 km height in its right. This asymmetry is attributed to a weakening of the surface winds from the increased friction created by both the surrounding lands and the shallower sea depth closer to the land.

In Figure 6a,b, the majority of moist fluxes over 80% relative humidity took place at levels of 900~850 hPa (about 1~1.5 km height) in the right quadrant of the typhoon eye, from its tail in the East China Sea toward Honshu Island, Japan under the influence of strong southerly wind in Figure 5a,b. This kind of moist transport is clearly seen in the figure of relative humidity at a 3 km height in Figure 6a, and the vertical transport of moisture over 80% relative humidity extended from the sea surface (about 100 m) to a 9.5 km height in the right side of the typhoon eye in Figure 6b, in contrast to the vertical limitation of moisture to a 2 km height in its left side, showing an asymmetric distribution of the cloud.

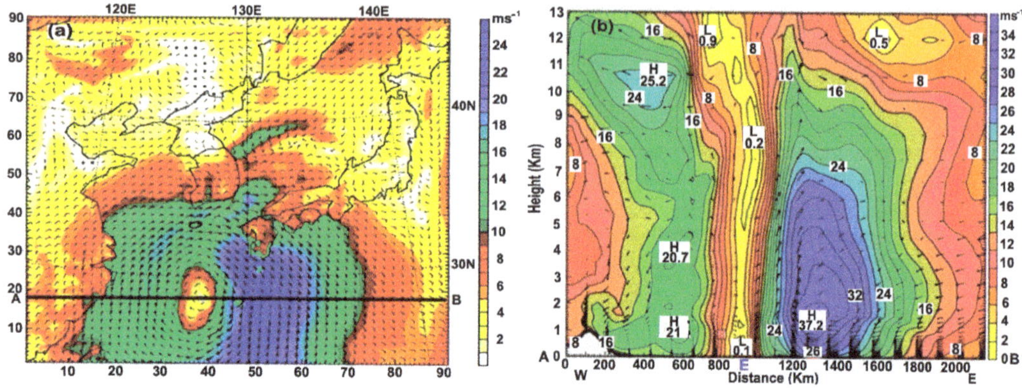

Figure 5. (**a**) Surface wind (m/s) at 10 m height; (**b**) vertical profile of the horizontal wind (m/s) along the line A–B (across the typhoon eye; E) in (**a**) based on WRF model simulation at 09:00 LST, 6 September 2004, showing the asymmetrical deformation of the typhoon circulation.

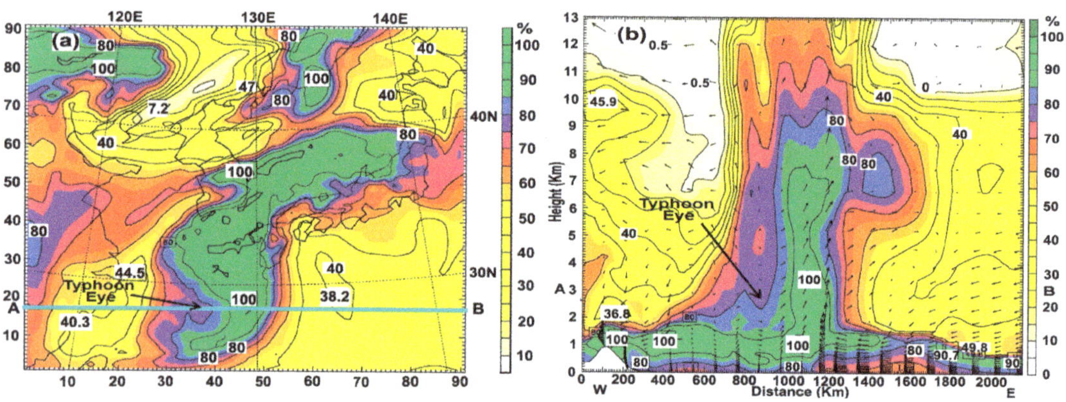

Figure 6. (**a**) Relative humidity (%) with wind speed (m/s) at 3 km height; (**b**) relative humidity (%) with wind (m/s) from 0 m to 13 km height along a line A–B in (**a**); refer to Figure 5b.

In the area containing significant moisture with a large cloud cluster in the vicinity of the typhoon center, it is easily expected that more latent heat released from the cloud cluster forming by cooling the moisture can be converted into more kinetic energy to be supplied into the typhoon system, resulting in the production of stronger winds in the right quadrant of the typhoon circulation, horizontally and vertically. Thus, the transport of momentum (wind) and moisture (relative humidity) in the right-hand side of the typhoon center can drive stronger cyclonic circulation of the typhoon in its right, intensifying TY Songda.

Figure 7a indicates that even if the deepened negative surface pressure tendency of −12.9 hPa/day on 6 September was less than −16 hPa/day on 5 September, the typhoon was still strengthened. Figure 7b shows that the area of high equivalent potential temperature (EPT) (357 K) contains much more moisture where the typhoon is located. It is expected that the latent heat released from the cloud cluster through the condensation process via significant moisture cooling can be converted into more kinetic energy to be transferred into the typhoon system, resulting in typhoon intensification, as referred to by Tamamadin et al. [30] and AMS [29]. Thus, it can be seen that the typhoon moves into the high equivalent potential temperature area.

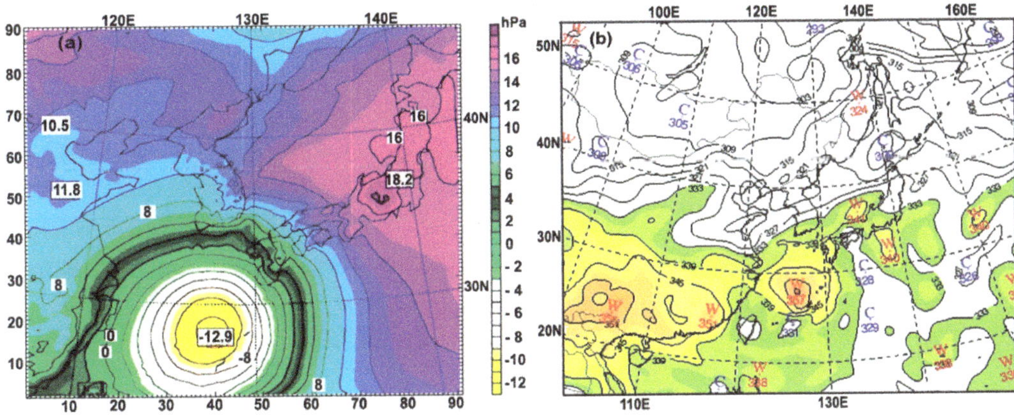

Figure 7. (**a**) Surface pressure tendency (∂p/∂t; hPa/day) for 12 hrs at 10 m height based on WRF model simulation; (**b**) equivalent potential temperature (EPT; K) at 1.5 km height (850 hPa) based on UM-KMA model simulation at 09:00 LST, 6 September 2004. The typhoon center corresponds to −12.9 hPa/day and a 367 K area.

4.2. Intensification of a Smaller Typhoon Split by the Surrounding Mountains of the SJ

At 09:00 LST, 7 September, Typhoon Songda was located between the west of Kyushu Island, Japan and the South Sea of the Korean Peninsula (Figures 8 and 9).

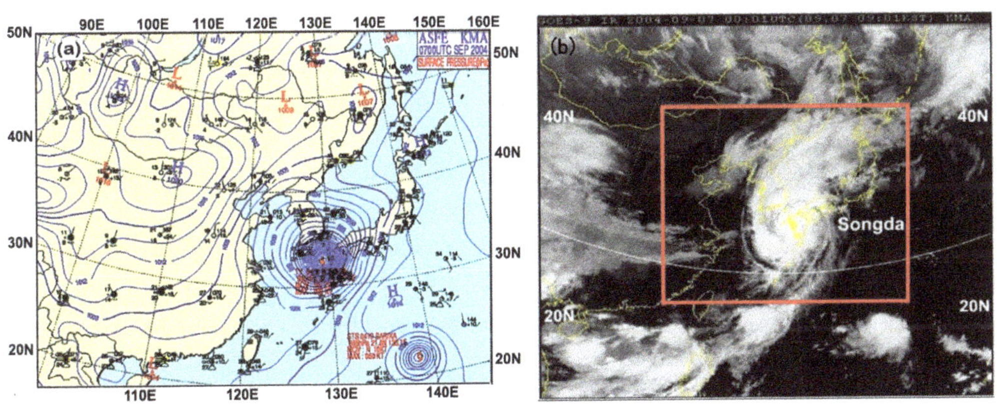

Figure 8. (**a**) Surface weather chart (hPa) at 09:00 LST; (**b**) GOES−9−IR satellite image at 09:01 LST, 7 September 2004.

According to the numerical simulation using the WRF model, as shown in Figure 9a,b, the typhoon was divided into three wind fields near the surface of 10 m over the Yellow Sea (I), the South Sea near Japan (II), and the Japan Sea (III) due to the increased friction from the surrounding lands of the moving typhoon and shallower sea depth approaching the lands. In particular, the vertical distribution of the horizontal winds showed a reinforced asymmetric structure with a maximum wind speed of 44.1 m/s at 1.5 km height at 09:00 LST, 7 September, in contrast to 37.2 m/s on 6 September. The vertical extension of the strong horizontal wind band was in the right of the typhoon eye, much higher than its left, still showing calm and weak wind in its eye (Figure 9b).

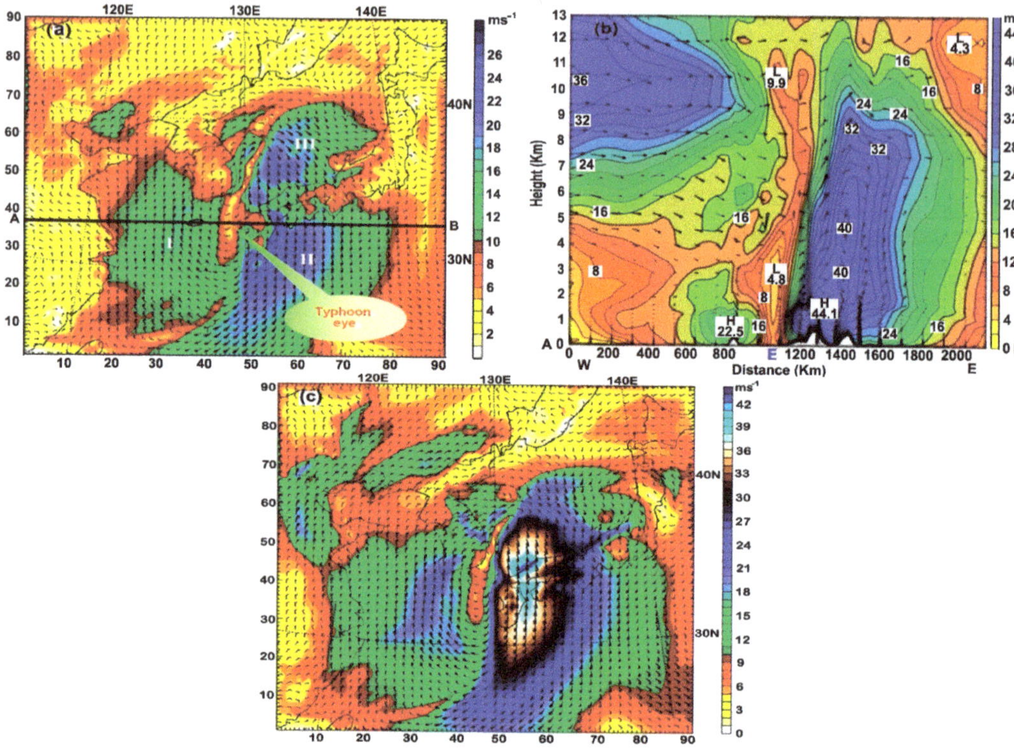

Figure 9. (**a**) Surface wind (m/s) at 10 m height at 09:00 LST, 7 September; (**b**) vertical profile horizontal wind (m/s) along the line A–B (E; the typhoon eye) from the surface to 13 km height in (**a**) based on WRF model simulation; (**c**) wind (m/s) at 1 km height, still maintaining a circular shape.

However, at a 1 km height, in Figure 9c, the typhoon structure still has a circular shape with a maximum wind speed in its right quadrant. This is because the friction effect of the land's topography at an upper altitude was greatly reduced compared to the surface, so the typhoon could have still a circular shape. Thus, it is very important to recognize that although the edges of the circular typhoon are somewhat torn, it still maintains its circular structure at altitudes above 1 km. Namely, the intensification of vertical flow can draw more moisture from the sea surface to the atmosphere, which can be cooled to form large cloud clusters through the condensation process vertically on the right side of the typhoon eye.

Further consideration was given to moist flux and streamline at 850 (hPa; about 1.5 km height) at 09:00 LST, 7 September, in Figure 10a,b. The majority of moisture transportation took place in the right quadrant of the typhoon, from its tail toward Kyushu Island and the center of the SJ, pulling significant moisture via mutual interactions between the cyclonic circulation by TY Songda and the anti-cyclonic circulation by the North Pacific High pressure (H) between 40 N and 20 N.

Thus, the merging area of the moisture corresponds to an area of significant amounts of latent heat being released through the moisture condensation to form clouds. Sequentially, the heat should be converted into sufficient amounts of kinetic energy to be supplied to the typhoon circulation, causing greater intensification of the typhoon.

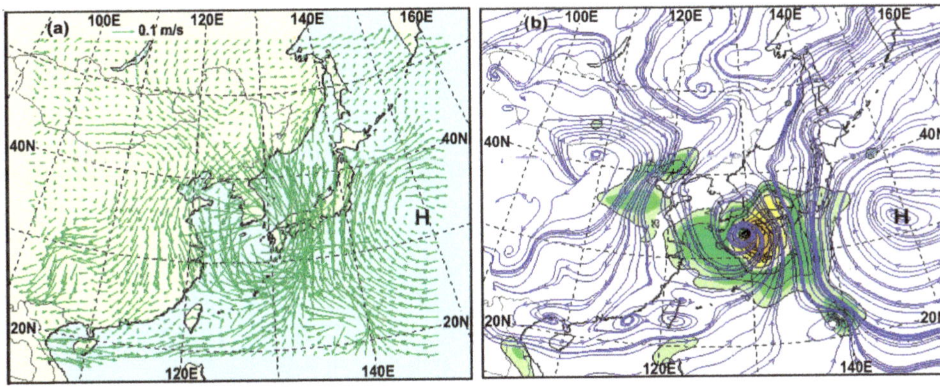

Figure 10. (**a**) Moist flux (0.1 m/s); (**b**) streamline and isotach (wind speed > 25 kt (green color and 50 kt (yellow color) at 850 (hPa; approximately 1.5 km height) based on UK-KMA model simulation at 09:00 LST, 7 September 2015. H denotes the North Pacific High pressure. The majority of moisture flux and streamline occurred in the right quadrant of the typhoon center, from its tail toward the right of Kyushu Island into the SJ (Japan), pulling significant moisture via mutual interactions between the cyclonic typhoon and the anti-cyclonic H from 40 N to 20 N.

In Figure 11a, the surface pressure tendency in its center changed from −12.9 hPa/day at 09:00 LST, 5 September to −6.0 hPa/day in the Korea Strait and −8.1 hPa/day to the left of Kyushu Island at 09:00 LST, 7 September. This may be attributed to the increased friction of the shallower sea depth and surrounding lands, as the typhoon passed by the Korea Strait with less than 100 m depth. Thus, its circulation should be prohibited to be slightly weakened compared to the one at 09:00 LST on 6 September.

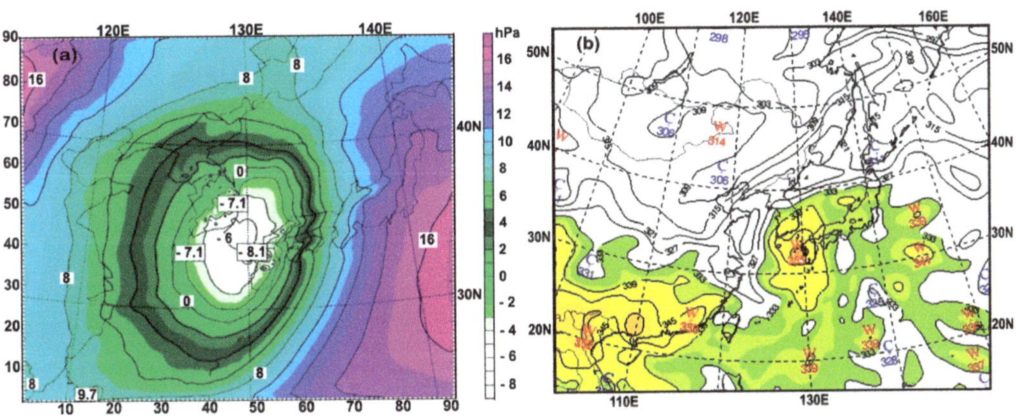

Figure 11. (**a**) Surface pressure tendency (hPa/day) for 12 hours at a 10 m height based on WRF model simulation; (**b**) equivalent potential temperature (EPT; K) at 1.5 km height (850 hPa) based on UM-KMA model simulation at 09:00 LST, 7 September. The typhoon center corresponds to −8.1 hPa/day and a 355 K area.

Figure 11b displays the EPT (K) at a 1.5 km height (850 hPa) at 09:00 LST on 7 September 2004. Even if the EPT in the typhoon center slightly decreased from 357 k to 355 k, the area of high EFT still contains significant moisture, and latent heat released from the cloud condensation process via cooling of the air parcel can be converted into significant kinetic energy to be transferred into the typhoon system, enhancing its circulation. Thus, we can recognize the moving track of the typhoon, where significant energy exists.

As shown in Figure 12a,b, at 21:00 LST, 7 September, when TY-Songda entered the SJ, passing the Korea Strait, the original Songda was reformed into a smaller split typhoon with a circular shape, located in the central part of the SJ between the East Korea Warm Current and Tsushima Warm Current, closer to the west of Honshu Island, Japan, as shown in Figures 1d and 12a.

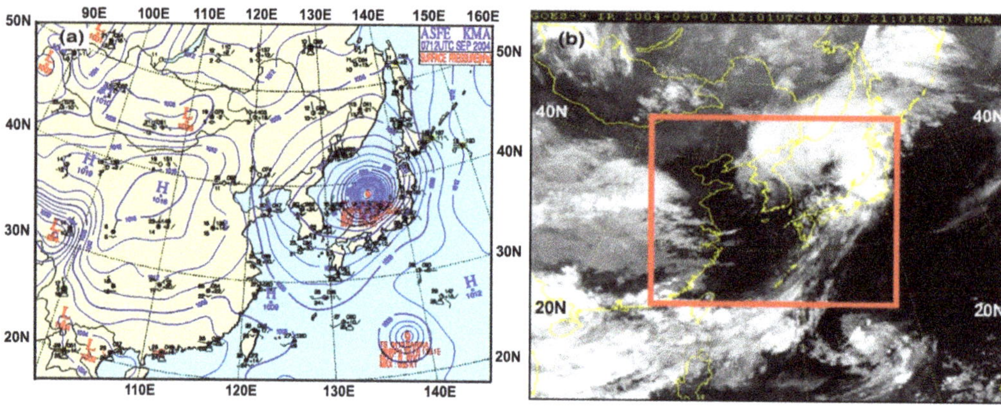

Figure 12. (**a**) Surface weather chart (hPa) at 21:00 LST; (**b**) GOES-9 IR satellite image at 21:00 LST, 7 September 2004. Red frame in (**b**) denotes a coarse domain of the model.

Its intensification was detected at heights of 10 m, 1 km, and 3 km at 21:00 LST, 7 September in the mid part of the SJ (Figures 13a and 14a,b). When TY Songda passed by the Korea Strait and entered the semi-enclosed SJ with a maximum depth of 3800 m near its center, it was blocked like the Japan Sea (III) wind field, and a small split typhoon could be intensified due to the channeling effects of cyclonic rotating winds from a shallow sea depth of the Korea Strait into the SJ's center and the surrounding high terrain of Korea, Japan, and Russia, as well as sharp increasing sea depths of the SJ from the coast to its center.

Another effect of its intensification was the intrusion of a strong downslope wind from the Japan western coastal high mountains of Honshu and Hokkaido islands toward the SJ's center in Figure 1b, which could be changed into cyclonic winds due to the Coriolis force to the downslope wind at 90 degrees toward the right in the northern hemisphere, resulting in the strengthening of the smaller deformed cyclonic circulation like a small split typhoon in Figures 13a and 14a.

Figure 15a,b display horizontal relative humidity (%) at a 3 km height and vertical relative humidity (%) with wind (m/s) from 0 m to 13 km height along a line A–B in Figure 13a at 21:00 LST. The horizontal relative humidity (%) at a 3 km height and its vertical distributions correspond to the wind fields in Figures 13b and 14b.

As shown in Figure 16, major moisture flux and streamline at a 1.5 km height occurred in the right quadrant of the typhoon center from its tail toward the right of Kyushu Island, Japan, pulling significant moisture via mutual interactions between the cyclonic TY Songda and the anti-cyclonic North Pacific High pressure (H) from 40 N to 20 N. Thus, the combination of the moisture transported by the typhoon itself and the moisture evaporated from the sea was cooled to form large clouds to its right, and the significant latent heat released from the clouds was converted into kinetic energy to drive its circulation intensification and the asymmetric structures of wind and moisture.

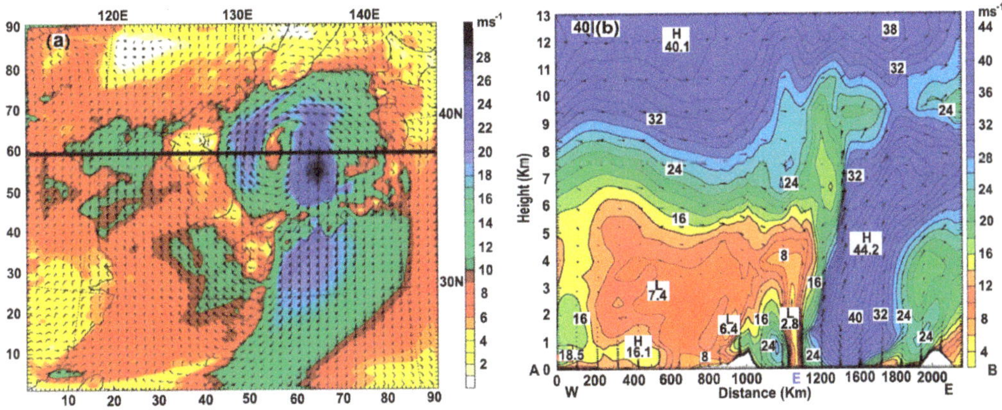

Figure 13. (**a**) Wind velocity (m/s) at 10 m height; (**b**) vertical distribution of horizontal wind (m/s) on a black line in Figure 13a at 21:00 LST, 7 September 2004. When TY-Songda moved to the semi-enclosed SJ, the typhoon reformed as a smaller typhoon, still maintaining a circular shape with asymmetric wind patterns horizontally and vertically.

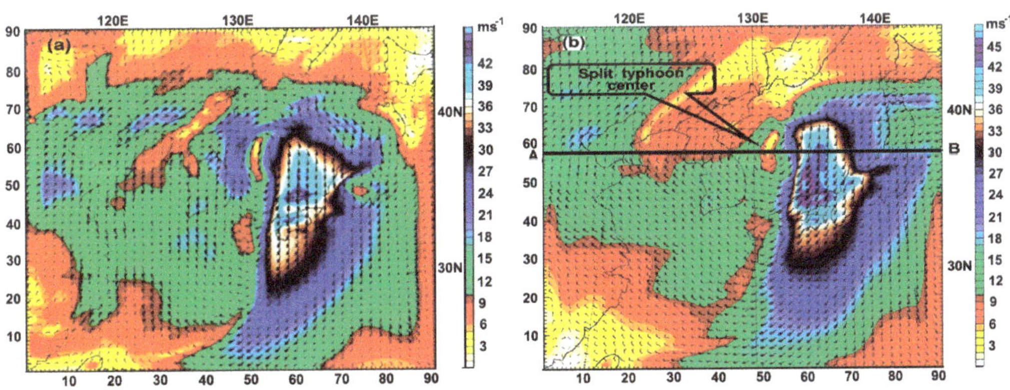

Figure 14. (**a**) Wind velocity (m/s) at 1.5 km; (**b**) 3 km heights at 21:00 LST on 7 September.

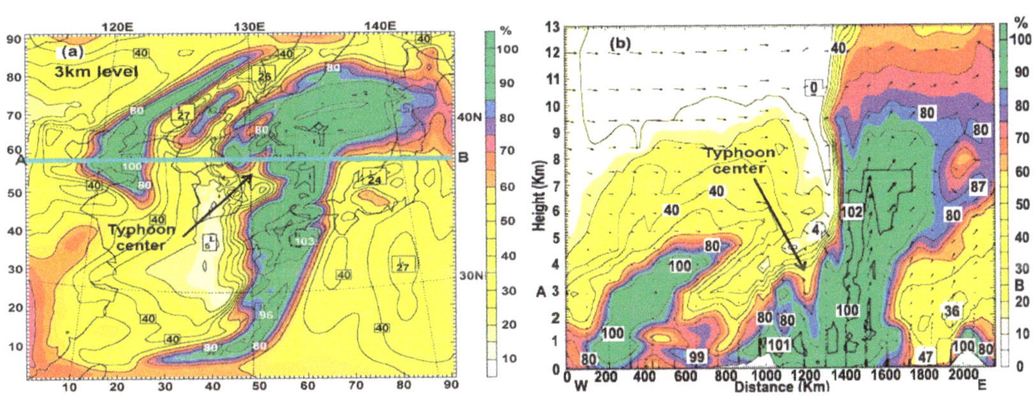

Figure 15. (**a**) Horizontal relative humidity (%) at a 3 km height; (**b**) vertical relative humidity (%) with wind (m/s) from 0 m to 13 km height along a line A–B in (**a**) at 21:00 LST, 7 September, showing the moisture asymmetries corresponding to the wind structures in Figures 13b and 14b.

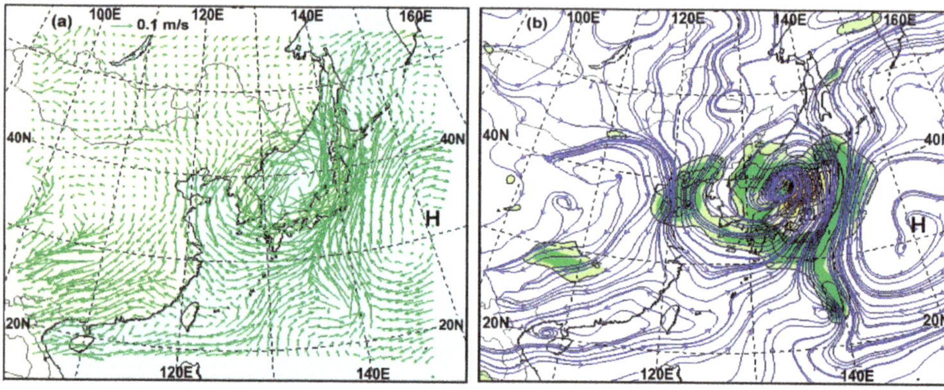

Figure 16. (**a**) Moist flux (0.1 m/s); (**b**) streamline and isotach (wind speed > 25 kt (green color) and 50 kt (yellow color) at 850 (hPa; about 1.5 km height) based on the UK-KMA model at 21:00 LST. Strong moisture flux between the cyclonic typhoon and the anti-cyclonic North Pacific High pressure (H) occurred from south to north.

In Figure 17, the deepened negative change in atmospheric pressure with time (isallobar) had from −8.1 hPa/day to −17.6 hPa/day for 12 h at the typhoon center, enhancing the convergence of moist air rising up and causing the formation of large clouds and rain. As the pressure tendency deepens, severe weather like typhoons can develop.

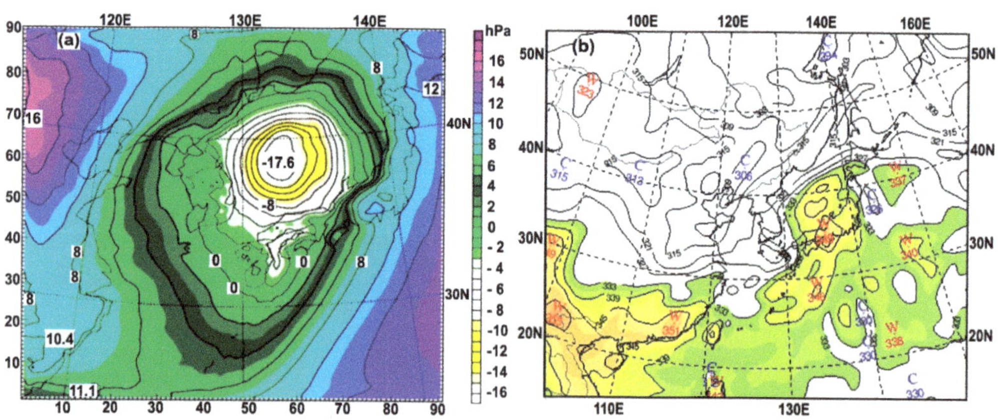

Figure 17. (**a**) Surface pressure tendency (hPa/day) at 10 m height based on WRF model simulation; (**b**) equivalent potential temperature (EPT; K) at a 1.5 km height (850 hPa) based on UM-KMA model simulation at 21:00 LST. The typhoon center corresponds to the area of −17.6 hPa/day and 349 K.

At the same time, the reformed typhoon center was detected at a high equivalent potential temperature (349 K), which contained significant amounts of moisture to be cooled down to form clouds, and the latent heat leased from the cloud could be converted into kinetic energy to induce the intensification of its circulation. The typhoon center in the SJ was closer to the west of Honshu Island, Japan, along which the Tsushima Warm Current had a higher SST than the east of the Korean peninsula, as shown in Figures 1d and 18a–c.

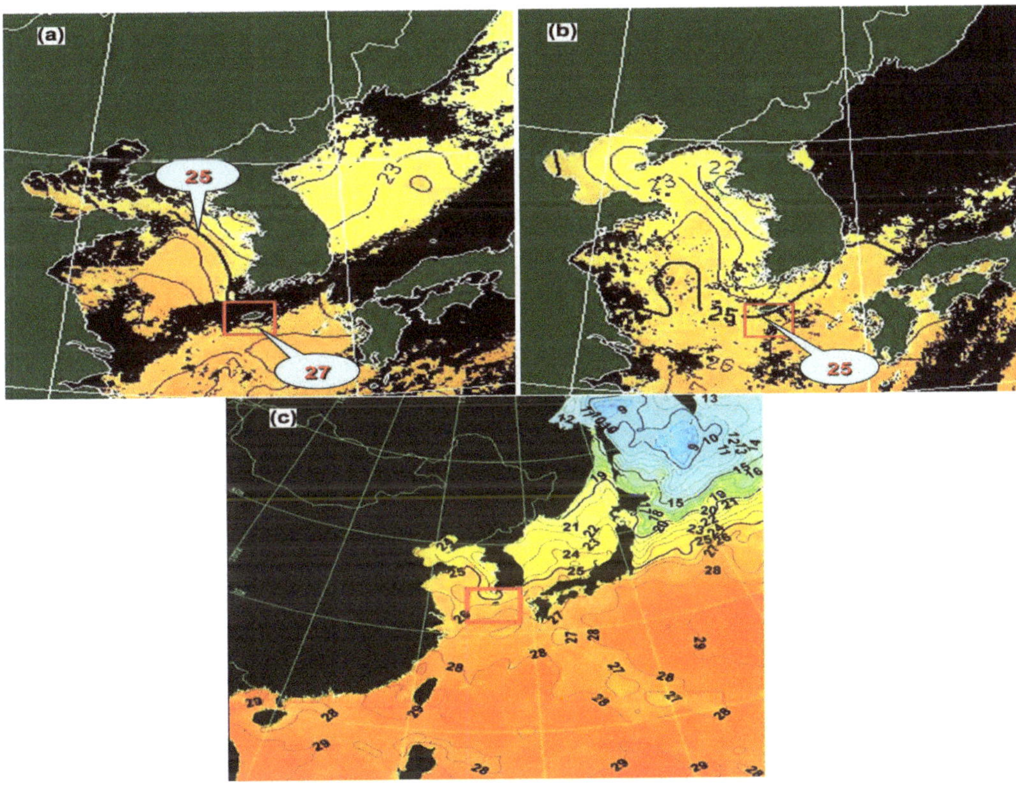

Figure 18. (**a**) Daily mean sea surface temperature (SST; °C) (GOES-9 MCSST) near the Korean peninsula on 4 September; (**b**) 7 September, and (**c**) weekly mean SST (1 to 7 September). The red square includes Jeju Island. The typhoon center corresponds to 24 °C area near the SJ center, closer to the Tsushima Warm Current in Figure 18c. The northeastward Kuroshio Warm Current flows from the eastern seas of Philippine and Taiwan toward the southern and eastern seas of Japan Island (28 °C) in Figure 18c.

Figure 18 shows the daily and weekly mean sea surface temperatures (SST; °C) in the vicinity of the Korean peninsula on 4 and 7 September. The SSTs on 4 September before Typhoon Songda moved into the Sea of Japan (SJ) were higher than the ones on 7 September after it entered. Typhoon Songda, following the Kuroshio Warm Current and passing by the East China Sea, could be continuously developed until 5 September. In particular, on 7 September, the SSTs in the Yellow Sea and the East China Sea were higher than ones in the SJ, but high SSTs existed in the path of the Tsushima Warm Current in the west coast of Honshu Island, shown in Figure 1c (ocean current map).

Thus, the split typhoon located between the East Korea Warm Current along the eastern coast of Korea and the Tsushima Warm Current along the west coast of Honshu Island could be developed through the thermal convection of moist air parcels, resulting in significant amounts of kinetic energy being converted from latent heat through the cloud condensation process with significant moisture. In Figure 18c, the weekly mean SSTs in the Yellow Sea and the East China Sea from Sept 1 to 7 were higher than ones in the SJ, but the SSTs were near the Tsushima Warm Current flowing along the west coast of Honshu Island compared the northeastern sea of the SJ. Thus, the intensification of the split typhoon in the SJ was greatly attributed to thermal convection over the warm currents, as shown in Figure 1c.

5. Conclusions

In summary, the change in intensity of a split typhoon originating from TY-Songda was investigated in the vicinity of the semi-enclosed Sea of Japan (SJ) using 3D-WRF-3.6.1 and UM-KMA numerical models, GOES-IR satellite images, and sea surface temperature distribution. This research provided the following results.

(1) After the typhoon developed in the Western Pacific Ocean in late August, it moved to the East China Sea along the northward Kuroshio Warm Current. It continuously developed, showing horizontally and vertically asymmetrical wind patterns on 5 September.
(2) As it turned to the northeast to the west of Kyushu Island, Japan on 6 September, it began to weaken, owing to the increased friction due to both the shallower seafloor of the East China Sea and the surrounding topography (China, Korea, and Japan), resulting in a tropical depression.
(3) As the typhoon approached Kyushu Island, it was deformed into three divided wind fields over the Yellow Sea (I), the South Sea near Japan (II), and the SJ (III) near the surface, unique from its circular shape at a 1 km altitude.
(4) As the split typhoon (III) from Typhoon Songda entered the SJ and changed into a small circle-shaped typhoon between the East Korea Warm Current and Tsushima Warm Current, it was located in an area of very high equivalent potential temperature of the air parcel, with significant amounts of kinetic energy being converted from the latent heat released from the cloud condensation process of water vapors evaporated from the warm current surface. The moisture was accompanied by the typhoon itself, from its tail to the SJ and the Russian Sakhalin Island.
(5) The majority of moisture flux and streamline occurred in the right quadrant of the typhoon center, from its tail toward the right of Kyushu Island, Japan, pulling significant moisture via mutual interactions between the cyclonic TY Songda and the anti-cyclonic North Pacific High pressure (H) on 40 N to 20 N.
(6) The significantly deepened atmospheric pressure tendency in the typhoon center can cause a convergence of air, which can induce its ascension, forming large clouds and generating severe weather such as the development of a typhoon.
(7) Simultaneously, as the strong downslope winds from the surrounding high mountains of the SJ into its center were deformed by the Coriolis force to cyclonic winds, the circulation of the split Songda inside the SJ could be intensified until 21:00 LST, 7 September.

Author Contributions: Conceptualization, S.-M.C. and H.C.; methodology, S.-M.C. and H.C.; software, S.-M.C. and H.C.; validation, S.-M.C. and H.C.; formal analysis, S.-M.C. and H.C.; investigation, S.-M.C. and H.C.; resources, S.-M.C. and H.C.; data curation, S.-M.C. and H.C.; writing—original draft preparation, S.-M.C. and H.C.; writing—review and editing, S.-M.C. and H.C.; visualization, S.-M.C. and H.C.; supervision, H.C.; project administration, H.C. All authors have read and agreed to the published version of the manuscript.

Funding: This research received no external funding.

Institutional Review Board Statement: Not applicable.

Informed Consent Statement: Not applicable.

Data Availability Statement: The surface weather charts, satellite images, and auxiliary analysis charts were obtained from the Korean Meteorological Administration: http://www.kma.go.kr (accessed on 10 August 2005).

Acknowledgments: The authors would like to thank the Korean Meteorological Administration for access to the surface weather charts, satellite images, and auxiliary analysis charts.

Conflicts of Interest: The authors declare no conflicts of interest.

References

1. Anthes, R.A.; Chang, S.W. Response of the hurricane boundary layer to changes in sea surface temperature in a numerical model. *J. Atmos. Sci.* **1978**, *35*, 1240–1255. [CrossRef]
2. Cheung, T.C.; Chan, P.W. Improving wind and rain simulations for tropical cyclones with the assimilation of Doppler radar data. In Proceeding of the 10th Annual WRF Users' Workshop, Boulder, CO, USA, 23–26 June 2009; pp. 1–835.
3. Hong Kong Meteorological Observatory (HKO). Classification of Tropical Cyclones. 2012. Available online: https://www.hko.gov.hk/tc/index.html (accessed on 20 December 2012).
4. World Meteorological Organization (WMO). Guidelines for Converting between Various Wind Averaging Periods in Tropical Cyclone Conditions. 2010, pp. 1–63. Available online: https://library.wmo.int/index.php?lvl=notice_display&id=135#.ZDPa33ZByUk (accessed on 20 September 2011).
5. Korean Meteorological Administration (KMA). Typhoon. 2003. Available online: https://www.kma.go.kr/neng/typhoon/typhoon-images.do (accessed on 20 September 2005).
6. Monaldo, F.M.; Sikora, T.D.; Babin, S.M.; Sterner, R.E. Satellite imagery of sea surface temperature cooling in the wake of Hurricane Edouard. *Mon. Weather. Rev.* **1997**, *125*, 2716–2721. [CrossRef]
7. Cione, J.J.; Uhlhorn, E.W. Sea surface temperature variablilty in hurricanes: Implications with respect to intensity change. *Mon. Weather. Rev.* **2003**, *131*, 1783–1796. [CrossRef]
8. Knauss, J.A.; Garfield, N. *Introduction to Physical Oceanography*, 2nd ed.; Waveland Press, Inc.: Long Grove, IL, USA, 2005; pp. 1–310.
9. Gilbes, F.; Armstrong, R.A.; Webb, R.M.T.; Muller-Karger, F.E. SeaWiFs helps assess hurricane impact on phytoplankton in Caribbean Sea, Eos, Transactions. *Eos Trans. Am. Geophys. Union* **2001**, *82*, 529–533. [CrossRef]
10. Babin, S.M.; Carton, J.A.; Dickey, T.D.; Wiggert, J.D. Hurricane-induced phytoplankton blooms in the Sargasso Sea. In Proceeding of the 2002 AGU/ASLO Ocean Sciences Meeting, Honolulu, HI, USA, 11–15 February 2002.
11. Willoughby, H.E.; Jin, H.L.; Lord, S.J.; Piotrowicz, J.M. Hurricane structure and evolution as simulated by an axisymmetric, non-hydrostatic numerical model. *J. Atmos. Sci.* **1984**, *41*, 1169–1186. [CrossRef]
12. Marks, F.D.; Houze, R.A., Jr.; Gamache, J.F. Dual-aircraft investigation of the inner core of Hurricane Norbert. Part I: Kinematic Structure. *J. Atmos. Sci.* **1992**, *49*, 919–942. [CrossRef]
13. Franklin, J.L.; Lord, S.J.; Marks, F.D., Jr. The kinematic structure of Hurricane Gloria (1985) determined from nested analyses of dropwindsonde and Doppler radar data. *Mon. Weather. Rev.* **1993**, *121*, 2433–2451. [CrossRef]
14. Bender, M.A. The Effect of Relative Flow on the Asymmetric Structure in the Interior of Hurricanes. *J. Atmos. Sci.* **1997**, *54*, 703–724. [CrossRef]
15. Elsner, J.B. Tracking hurricanes. *Bull. Am. Meteorol. Soc.* **2003**, *84*, 353–356. [CrossRef]
16. Jian, G.J.; Wu, C.C. A numerical study of the track deflection of Super typhoon Haitang (2005) prior to its landfall in Taiwan. *Mon. Weather. Rev.* **2008**, *136*, 598–615. [CrossRef]
17. Choi, H.; Lee, M.S.; Choi, S.M. Cold sea surface temperature near Cheju Island responding to strong cyclonic wind cyclonic wind and positive geopotential tendency behind a typhoon center along its track. *J. Mar. Sci. Tech.* **2012**, *20*, 684–692. [CrossRef]
18. Shen, B.W.; Tao, W.K.; Lau, W.K.; Atlas, R. Predicting tropical cyclogenesis with a global mesoscale model: Hierarchical multiscale interactions during the formation of tropical cyclone Nargis (2008). *J. Geophys. Res. Atmos.* **2010**, *115*, D14102. [CrossRef]
19. Xu, H.; Zhang, X.; Xu, X. Impact of Tropical Storm Bopha on the Intensity Change of Super Typhoon Saomai in the 2006 Typhoon Season. *Adv. Meteorol.* **2013**, *2013*, 487010. [CrossRef]
20. Fujiwhara, S. The natural tendency towards symmetry of motion and its application as a principle in meteorology. *Q. J. R. Meteorol. Soc.* **1921**, *47*, 287–292. [CrossRef]
21. Choi, S.-M.; Choi, H. Influence of multiple interactions of three typhoons and a mid-latitude cloud band associated trough in the North West Pacific upon severe tropical storm Linfa. *Remote Sens.* **2023**, *15*, 2170. [CrossRef]
22. Yamasaki, M. A study on the effect of the ice microphysics on tropical cyclones. *Adv. Meteorol.* **2013**, *2013*, 573786. [CrossRef]
23. Bu, Y.P.; Fovell, R.G.; Corbosiero, K.L. The influences of boundary layer and cloud-radiative forcing on tropical cyclone size. *J. Atmos. Sci.* **2017**, *74*, 1273–1292. [CrossRef]
24. Wu, S.-N.; Soden, B.J.; Alaka Jr, G.J. Ice water content as a precursor to tropical cyclone rapid intensification. *Geophys. Res. Lett.* **2020**, *47*, 32020GL089669. [CrossRef]
25. Wu, Z.; Zhang, Y.; Zhnag, L.; Zheng, H. Interaction of cloud dynamics and microphysics during the rapid intensification of super-typhoon Nanmadol (2022) based on multi-satellite observations. *Geophys. Res. Lett.* **2023**, *50*, e2023GGL104541. [CrossRef]
26. Su, H.; Wu, L.; Jiang, J.H.; Pai, R.; Liu, A.; Zhai, A.J.; Tavallali, P.; DeMaria, M. Applying satellite observations of tropical cyclone internal structure to rapid intensification forecast with machine learning. *Geophys. Res. Lett.* **2020**, *47*, e2020GL089102. [CrossRef]
27. Tapiador, F.J.; Navarro, A.; Martin, R.; Hristova-Veleva, S.; Haddad, Z. Predicting tropical cyclone rapid intensification from satellite microwave data and neural networks. *IEEE Trans. Geosci. Remote Sens.* **2021**, *60*, 4205213. [CrossRef]
28. Bruneau, N.; Wang, S.; Toumi, R. Long memory impact of ocean mesoscale temperature anomalies on tropical cyclone size. *Geophys. Res. Lett.* **2020**, *47*, e2019GL086165. [CrossRef]
29. American Meteorological Society (AMS). Equivalent Potential Temperature. AMS Glossary of Meteorology. 2020. Available online: https://glossary.ametsoc.org/wiki/Equivalent_potential_temperature (accessed on 3 November 2020).

30. Tamamadin, M.; Lee, C.; Kee, S.-H.; Yee, J.-J. Prediction of convective available potential energy and equivalent potential temperature using a coupled WRF and deep learning for typhoon intensification. *IOP Conf. Ser. Earth Environ. Sci.* **2023**, *1245*, 012034. [CrossRef]
31. Lee, S.W. *Introduction to Physical Oceanography*; Jipmoondang Company: Seoul, Republic of Korea, 1983; pp. 1–268.
32. WRF-3.6.1. User's Guide for the Advanced Research WRF (ARW) Modeling System, Version 3.6.1. 2014. Available online: https://www2.mmm.ucar.edu/wrf/users/docs/user_guide_V3/user_guide_V3.6/contents.html (accessed on 1 January 2020).
33. National Centers for Environmental Prediction/National Weather Service/NOAA/U.S. Department of Commerce. 2000, Updated Daily. NCEP FNL Operational Model Global Tropospheric Analyses, Continuing from July 1999. Research Data Archive at the National Center for Atmospheric Research, Computational and Information Systems Laboratory. Available online: https://rda.ucar.edu/datasets/d083002/ (accessed on 16 September 2021).
34. UM-KMA. The British Unified Model (UM) Adopted by Korean Meteorological Administration. 2015. Available online: https://www.kma.go.kr/neng/image/chart/analysis-chart.do (accessed on 20 September 2020).
35. Hesselberg, T. Uber eine Beziehung zwischen Druckgradient, Wind und Gradienanderungen. *Veroeff. Geophys. Inst. Univ. Leipzig* **1915**, *1*, 207–216.
36. Brunt, D.; Douglas, C.K.M. The modification of the strophic balance for changing pressure distribution and its effect on rainfall. *Mem. Roy. Meteorol. Soc.* **1928**, *3*, 29–51.
37. Haurwitz, B. *Dynamic Meteorology*; McGraw Hill Book Company, Inc.: New York, NY, USA, 1941; pp. 155–159.
38. Gill, A.E. *Atmosphere-Ocean Dynamics*; Academic Press: New York, NY, USA, 1982; pp. 1–662.
39. Choi, H. Prediction of marine surface wind and sea surface elevation in the East Sea of Korea. *J. Kor. Meteorol. Soc.* **1996**, *32*, 71–83.
40. American Meteorological Society (AMS). Isallobaric Wind. AMS Glossary of Meteorology. 2012. Available online: https://glossary.ametsoc.org/wiki/Isallobaric_wind (accessed on 5 August 2014).

MDPI AG
Grosspeteranlage 5
4052 Basel
Switzerland
Tel.: +41 61 683 77 34

Journal of Marine Science and Engineering Editorial Office
E-mail: jmse@mdpi.com
www.mdpi.com/journal/jmse

www.ingramcontent.com/pod-product-compliance
Lightning Source LLC
Chambersburg PA
CBHW040829250926
51850CB00007BA/1295
* 9 7 8 3 7 2 5 8 5 7 7 9 1 *